WYOMING WILDLIFE: A NATURAL HISTORY

71 Selected Books by Paul Johnsgard

Handbook of Waterfowl Behavior, 1965
Animal Behavior, 1967, 1972
Waterfowl: Their Biology and Natural History, 1968
Grouse and Quails of North America, 1973
Song of the North Wind: A Story of the Snow Goose, 1974
North American Game Birds of Upland and Shoreline, 1975
Waterfowl of North America, 1975
The Bird Decoy: An American Art Form, 1976
Ducks, Geese, and Swans of the World, 1978
The Birds of Nebraska, 1978, 2018
Birds of the Great Plains: Breeding Species and Their Distribution, 1979
A Guide to North American Waterfowl, 1979
The Plovers, Sandpipers, and Snipes of the World, 1981
Those of the Gray Wind: The Sandhill Cranes, 1981, 2017
Teton Wildlife: Observations by a Naturalist, 1982
Dragons and Unicorns: A Natural History, 1982 (with K. Johnsgard)
The Grouse of the World, 1983
The Hummingbirds of North America, 1983, 1997
Cranes of the World, 1983
The Platte: Channels in Time, 1984, 2008
Prairie Children, Mountain Dreams, 1985
The Pheasants of the World: Biology and Natural History, 1986, 1999
Birds of the Rocky Mountains, 1986, 1992
Diving Birds of North America, 1987
The Waterfowl of North America: The Complete Ducks, Geese, and Swans, 1987
The Quails, Partridges, and Francolins of the World, 1988
North American Owls: Biology and Natural History, 1988, 2002
Hawks, Eagles, and Falcons of North America: Biology and Natural History, 1990
Crane Music: A Natural History of American Cranes, 1991
Bustards, Hemipodes, and Sandgrouse: Birds of Dry Places, 1991
Ducks in the Wild: Conserving Waterfowl and Their Habitats, 1992
Cormorants, Darters, and Pelicans of the World, 1993
Arena Birds: Sexual Selection and Behavior, 1994
This Fragile Land: A Natural History of the Nebraska Sandhills, 1995
Ruddy Ducks and Other Stifftails: Their Behavior and Biology, 1996 (with M. Carbonell)
The Avian Brood Parasites: Deception at the Nest, 1997
Baby Bird Portraits by George Miksch Sutton: Watercolors in the Field Museum, 1998
Earth, Water, and Sky: A Naturalist's Stories and Sketches, 1999

Trogons and Quetzals of the World, 2000
Prairie Birds: Fragile Splendor in the Great Plains, 2001
The Nature of Nebraska: Ecology and Biodiversity, 2001
Grassland Grouse and Their Conservation, 2002
Great Wildlife of the Great Plains, 2003
Lewis and Clark on the Great Plains: A Natural History, 2003
Faces of the Great Plains: Prairie Wildlife, 2003
Prairie Dog Empire: A Saga of the Shortgrass Prairie, 2004
A Nebraska Bird-Finding Guide, [1997], 2005, 2011
The Niobrara: A River Running through Time, 2007
Wind Through the Buffalo Grass: A Lakota Story Cycle, 2008
Sandhill and Whooping Cranes: Ancient Voices over America's Wetlands, 2011
Rocky Mountain Birds: Birds and Birding in the Central and Northern Rockies, 2011
Nebraska's Wetlands: Their Wildlife and Ecology, 2012
Wings over the Great Plains: Bird Migrations in the Central Flyway, 2012
Wetland Birds of the Central Plains: South Dakota, Nebraska, and Kansas, 2012
Birds of the Central Platte River Valley and Adjacent Counties, 2013 (with M. Brown)
Yellowstone Wildlife: Ecology and Natural History of the Greater Yellowstone Ecosystem, 2013
Birds and Birding in Wyoming's Bighorn Mountains Region, 2013 (with J. Canterbury and H. Downing)
Seasons of the Tallgrass Prairie: A Nebraska Year, 2014
Global Warming and Population Responses among Great Plains Birds, 2015
At Home and at Large in the Great Plains: Essays and Memories, 2015
A Chorus of Cranes: The Cranes of North America and the World, 2015
Birding Nebraska's Central Platte Valley and Rainwater Basin, 2015
Swans: Their Biology and Natural History, 2016
The North American Grouse: Their Biology and Behavior, 2016
The North American Geese: Their Biology and Behavior, 2016
The North American Sea Ducks: Their Biology and Behavior, 2016
The North American Perching and Dabbling Ducks, 2017
The North American Whistling-Ducks, Pochards, and Stifftails, 2017
The North American Quails, Partridges, and Pheasants, 2017
A Naturalist's Guide to the Great Plains, 2018
The Ecology of a Tallgrass Treasure: Audubon's Spring Creek Prairie, 2018

Wyoming Wildlife

A Natural History

Paul A. Johnsgard

Black & white photographs
by Thomas D. Mangelsen

School of Biological Sciences
University of Nebraska–Lincoln

Zea Books
Lincoln, Nebraska 2019

ISBN: 978-1-60962-152-0

doi 10.32873/unl.dc.zea.1076

Composed in Myriad Pro types.

Zea Books are published by the University of Nebraska–Lincoln Libraries

Electronic (pdf) edition available online at
https://digitalcommons.unl.edu/zeabook/

Print edition available from
http://www.lulu.com/spotlight/unllib

UNL does not discriminate based upon any protected status.
Please go to http://www.unl.edu/equity/noitce-nondiscrimination

Abstract

This book surveys Wyoming's mammal, bird, reptile, and amphibian faunas. In addition to introducing the state's geography, geology, climate, and major ecosystems, it provides 65 biological profiles of 72 mammal species, 195 profiles of 196 birds, 9 profiles of 12 reptiles, and 6 profiles of 9 amphibians. There are also species lists of Wyoming's 117 mammals, 445 birds, 22 reptiles, and 12 amphibians. Also included are descriptions of nearly 50 national and state properties, including parks, forests, preserves, and other public-access natural areas in Wyoming. The book includes a text of more than 150,000 words, nearly 700 references, a glossary of 115 biological terms, nearly 50 maps and line drawings by the author, and 33 black & white photographs by Thomas D. Mangelsen. Color photographs can be seen online at https://digitalcommons.unl.edu/zeabook/73/

Full color edition online

Nebraska
UNIVERSITY OF
Lincoln

Frontis: Bobcat, adult

Dedicated to the memory of our parents, who bequeathed us far more than they ever imagined, and to the children of Wyoming, in hope that they will be better stewards of its resources than our generation has been.

Contents

Maps

Figures

Photographs

Preface and Acknowledgments

By the time I was a teenager during the 1940s in Wahpeton, North Dakota, I was already an ardent nature lover, and my associated reading had convinced me that I must someday go to Wyoming to see Yellowstone National Park. In the spring of 1946, I was nearly 15 years old, and one afternoon at the dinner table I announced to my parents that I was going to hitchhike there to photograph wildlife with my new camera. I had recently bought an Argus C-3, a 35mm rangefinder camera, although its short 50mm lens was better suited for photographing friendly people than unfriendly bison.

By 1946, post-war autos were finally becoming available to civilians, and Dad had just bought a new family car, our first since about 1940. Furthermore, the end of gas rationing meant that long road trips were again possible. In any case, my parents quickly decided that they didn't want me to risk hitchhiking, and that we would take a family trip to see the American West for the first time. Thus, we undertook our first long post-war road trip. After briefly touring South Dakota's Black Hills, we drove through northern Wyoming, where the Bighorn Mountains gave me an authentic taste of real mountains. Finally, we reached Yellowstone National Park! My most vivid memories of that trip more than 70 years ago involved stalking and trying (unsuccessfully) to photograph a coyote that was hunting for mice along the Lamar River and seeing ospreys nesting on rocky pinnacles in Yellowstone Canyon. I immediately fell in love with Wyoming and its wildlife, and that passionate attachment has only deepened over the seven decades since.

Two summers during the mid-1970s that I spent at the Jackson Lake Biological Station finally gave me an extended exposure to Grand Teton National Park, while on a research program supported by the University of Wyoming and the New York Zoological Society. These experiences were the basis for my 1982 book on the natural history of Grand Teton National Park, *Teton Wildlife: Observations by a Naturalist*. While I was at the biological station, I invited my friend and one-time graduate student Tom Mangelsen to spend a few days with me, which in turn stimulated him to move from Colorado to Wyoming, and to develop his eventually world-famous photographic career.

Since then, Tom and I have collaborated on two books, using my text and his photographs. The first, *Yellowstone Wildlife: Ecology and Natural History of the Greater Yellowstone Ecosystem*, was published in 2013, and the second, *A Chorus of Cranes: The Cranes of North America and the World*, appeared in 2015, both published by the University Press of Colorado. I offer my sincere thanks for his permission to include some of his wonderful photographs in this book. Additionally, Jackie Canterbury, another of my one-time graduate students, collaborated with me in expanding a preliminary checklist of Bighorn Mountain birds by the late Helen Downing into a comprehensive 2013 book, *Birds and Birding in Wyoming's Bighorn Mountains Region*, and we later produced two books on the birds of local areas in northern Wyoming. All these experiences have deepened my emotional ties to Wyoming.

I treasure all my experiences and memories of Wyoming and owe special thanks to the people who have helped me. Of course, they include Tom Mangelsen and Jackie Canterbury, plus many other friends from my unforgettable summers at the Jackson Lake Biological Station. The University of Nebraska–Lincoln Library staff and the UNL School of Biological Sciences have also continued to support my research efforts. People who read and critiqued various parts of my manuscript included Linda Brown, Scott Johnsgard, and several other friends.

In writing this book, I have of necessity also relied on many previous books and other resources on Wyoming's vertebrate fauna. Most of the measurements and weights cited in the mammal species accounts are from Clark and Stromberg (1987), and their book plus that of Armstrong, Fitzgerald, and Meaney (2011) were extensively used by me in writing the mammal species profiles. Avian clutch sizes, incubation periods, and fledging periods are mostly based on Baicich and Harrison (1997). My information on Wyoming's reptiles and amphibians was in large part similarly based on Baxter and Stone (1985) and secondarily relied on the work of Ballinger, Lynch, and Smith (2010). The map showing Wyoming mountain ranges was derived from Clark and Stromberg (1987), with the areas of montane glaciation based on Long (1965). Chapter 7 was partly derived from my 2014 article "The Lives and Deaths of Yellowstone's Grizzlies" from Lincoln's now sadly discontinued progressive newspaper *Prairie Fire*, with the kind permission of its editor, Cris Trautner. The unsigned drawings are my own, but some have been redrawn from various published sources.

I once again owe a continuing debt of gratitude, and my everlasting personal thanks, to the University of Nebraska's indefatigable Digital Commons coordinator, Paul Royster, and to his wonderful editorial assistant, Linnea Fredrickson, who has cheerfully but with limited success tried to correct my writing peculiarities and punctuation innovations for nearly a decade.

Paul A. Johnsgard
Lincoln, Nebraska

Pacific marten, adult

Chapter 1 • Wyoming Geography, Geology, and Ecosystems

Wyoming (WY) is a state of superlatives for wildlife lovers and naturalists because it has two of our most wildlife-rich and beloved national parks, Yellowstone and Grand Teton (Johnsgard, 1982, 2013), and has seemingly endless amounts of open vistas and the lowest human population of any state. Federal acreage in Wyoming composes 42.3 percent of the state's total land area, the highest percentage of public acreage of any of the Rocky Mountain states. Wyoming also has 450,000 acres of wildlife habitat management areas with public access, along with 225 miles of public-access streams and 21,000 lake acres. Wyoming also has 18.5 million acres of Bureau of Land Management (BLM) properties, 1.9 million acres of national grasslands, 2.4 million acres of national forests, 2.3 million acres of National Park Service lands, and ten state parks.

The predominant landforms of the region are associated with Rocky Mountains topography. In Wyoming, the Rockies include three relatively discrete montane regions (Map 1). The northern Rockies extend from Canada south into western Montana and northern Idaho and merge with the central Rockies in extreme northwestern Wyoming. The central Rockies are centered on the Yellowstone Plateau of northwestern Wyoming and extend southward and eastward through the Teton, Absaroka, and Wind River ranges, collectively making up the Greater Yellowstone ecosystem. The southern Rockies extend from the Sierra Madre and Medicine Bow southward through Colorado and New Mexico. These are all extremely high mountain ranges that form the Continental Divide, their maximum elevations reaching 12,850 feet in Montana, 13,785 feet in Wyoming, 14,431 feet in Colorado, and 13,161 feet in New Mexico.

Wyoming and the entire Rocky Mountain region are characterized by having highly variable weather, with great seasonal and daily changes in temperature, and fairly short and cool summers. Directly to the west of the Rockies is the Columbia plateau of Washington, Oregon, and Idaho, and to the east are the Great Plains in the eastern parts of Montana, Wyoming, and Colorado (Map 1). In the northern Rockies, moist winter air from the Pacific Northwest spills inland to produce the lush forests on the region's west-facing slopes. Most of the regional precipitation is mountain-related (orographic) in nature. That is, it is associated with montane topography, with the heaviest precipitation levels typically occurring on the western slopes and reduced precipitation or "rain-shadow" effects showing on the eastern slopes and valleys. Thus, Wyoming's highest precipitation levels of about 60 inches annually occur in the northwestern portion of the state. Wyoming's Bighorn Mountains represent an exception to this general trend; their eastern slopes receive some precipitation from Great Plains sources, whereas their western slopes are in a major rain shadow resulting from the high mountain ranges farther west.

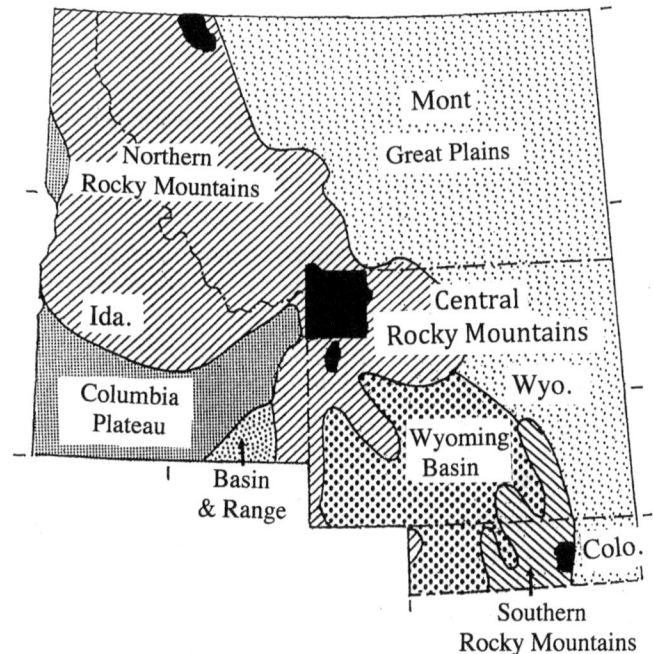

Map 1. Physiography of Wyoming and adjacent Rocky Mountain states. Inked areas indicate national park locations. Modified from Johnsgard (1986).

The geologic forces that shaped the area of the northern and central Rocky Mountains are complex. However, except for the volcanic influences in the Yellowstone region, the mountains are largely the result of folding and thrust-faulting of sedimentary layers, which started roughly 70 million years ago. Lateral pressures on these layers caused folding, buckling, and faulting to occur, with large areas being lifted upward and subsequently eroded away. After these layers had been eroded away, progressively layers of earlier deposits were exposed, until finally layers dating back to more than a half billion years ago have been exposed in some localities, such as on the top of Mount Moran in Grand Teton National Park (Love and Reed, 1971). At similar altitudes in the Snowy Range, fossilized remnants of some of Earth's most ancient plant life, primitive cyanobacteria, occur as stromatolites that may be up to 2.5 billion years old (Knight, 1990). Such fossils represent the oldest evidence of life on earth, with some stromatolites elsewhere in the world believed to be as ancient as 3.5 billion years old.

By roughly 60 million to 70 million years ago, the ancestral mountains in the region were perhaps as rugged as the present-day Rockies but were generally much lower and far more subtropical in climate. At that time a volcanic cone formed in what is northeastern Wyoming, where later erosion gradually exposed an 800-foot basalt pillar known today as Devil's Tower. During various later times from 35 million to 50 million years ago, several periods of volcanic activity resulted in the deposition of great lava plateaus in Idaho's Columbia Plateau and tremendous quantities of windblown ash, which filled mountain valleys and basins mostly toward the eastern plains.

By about 40 million years ago, the first mountains destined to become part of the present-day Rocky Mountain region (the Wind River, Medicine Bow, and Gros Ventre ranges) began to gradually rise (Knight, 1994) until they exceeded their present-day elevations. Other ranges, such as the Bighorns and Black Hills, gradually arose later. About 630,000 years ago, volcanic eruptions of unimaginable magnitude spread deep layers of lava in what is now northwestern Wyoming, forming the present-day Yellowstone Plateau, the third of a series in the region dating back to 2.1 million and 1.4 million years ago. The eruptions also spewed volcanic ash over much of what now makes up the western Great Plains region, exterminating entire populations of animals and plants by burying them in volcanic dust, suffocating them or causing starvation from a lack of food sources (Voorhies, 1981, 1990).

The last major period of mountain building in Wyoming occurred about 10 million years ago with the relatively rapid emergence of the Teton range, the most ruggedly spectacular of all Wyoming's mountains (Love and Reed, 1971). These powerful landscape changes were later supplemented by continued erosion, especially during the past 1.6 million years, when actions of ice, water, and winds removed several thousands of feet from the exposed strata. Montane glaciers also left evidence of their handiwork in the form of U-shaped valleys, cirques, and moraine-formed lakes in most of the higher ranges. Between and below the mountains, rivers and river valleys gradually formed, soils developed, and present-day vegetational patterns began to take shape.

These strong variations of altitude in Wyoming, and their associated microclimate differences, help account for the state's current high species diversity of plants and some animal groups. The elevation range in the five Great Plains states from North Dakota through Oklahoma is less than 7,000 feet, whereas in Wyoming alone it exceeds 9,000 feet. The elevation of Yellowstone Lake is over 7,700 feet, or 500 feet higher than the highest point anywhere in the Great Plains states north of Texas, namely Black Elk Peak, in South Dakota.

Glacial and Post-Glacial Wyoming Climates and Landscapes

The cold climate and several periods of continental glaciation that have occurred during the past 2 million years brought with them a host of northern plant and animal species, which moved variably southward along the Rocky Mountain chain. Subsequently, many of the less mobile species became isolated on high peaks as the climate ameliorated and a period of drying and warming began. This isolation trend, of course, was much truer for plant life than for the more mobile animals. Yet to some degree it can also be observed in such locally variable species

as the rosy-finches and dark-eyed juncos, with their distinctive isolated populations extending from Canada southward. In any case, tundra areas were reduced, fragmented, and isolated during post-glacial times, becoming increasingly restricted to the highest mountains. Likewise, progressively more diverse biological communities evolved throughout the entire Rocky Mountain region, under the influences of variations in altitude, precipitation, slope, soils, and other environmental factors.

The Rocky Mountains now not only form a north-south continental watershed (the Continental Divide) across Wyoming, separating the Great Plains to the east from the Great Basin to the west (Map 2), they also separate several river drainage patterns that have their ultimate ending on three widely separated coastlines. They include the Bighorn and Powder Rivers of northern Wyoming and the North Platte River in eastern Wyoming, all of which feed into the Missouri River and eventually drain via the Mississippi River into the Mississippi delta on the Gulf Coast, nearly 1,500 miles away. In southwestern Wyoming, the Green River flows southward into the Colorado River and ultimately drains into the Gulf of California at the base of Mexico's Baja Peninsula, almost 1,000 miles away. In western Wyoming, the Snake River flows westward to a confluence with the Columbia River, which empties into the Pacific Ocean, about 750 miles away (Map 3). The lowest point in Wyoming is on the Belle Fourche River at 3,099 feet above sea level in extreme northeastern Wyoming at the South Dakota boundary in the Missouri River drainage.

There is a notable gap in Wyoming's high mountains between the Sierra Madre and Wind River ranges, where the Continental Divide drops greatly in elevation. There its splits and crosses two ill-defined areas of low ridges, sand dunes, and buttes at elevations of only about 7,000 to 8,500 feet, enclosing a still-lower region of internal drainage in Sweetwater County, the Great Divide Basin. This driest part of Wyoming covers 9,320 square miles and drops locally in elevation to about 6,600 feet, where annual precipitation may be as little as little as six inches in an area known as the Red Desert (Map 2).

These natural landscapes have been overlain during the past 150 years with the effects of human inhabitation, which became formally recognized as Wyoming statehood in 1890. This event led to the formation of cities and roads, telegraph and telephone networks, the construction of dams and reservoirs along most river systems, and countless other visible hallmarks of civilization (Map 3). The invisible lines of longitude and latitude that originally were applied in defining the geographic limits of Wyoming are still today of critical importance in mapping, navigation, establishing legal boundaries, and the like.

Classifying Wyoming Natural Landscapes and Biological Communities

Today's natural landscapes of Wyoming are marked by broad patterns of native plant and animal life, reflecting past geological effects, climates, biological evolution, migrations, and extinctions. Thus, Long (1965) identified several "faunal areas" of Wyoming in

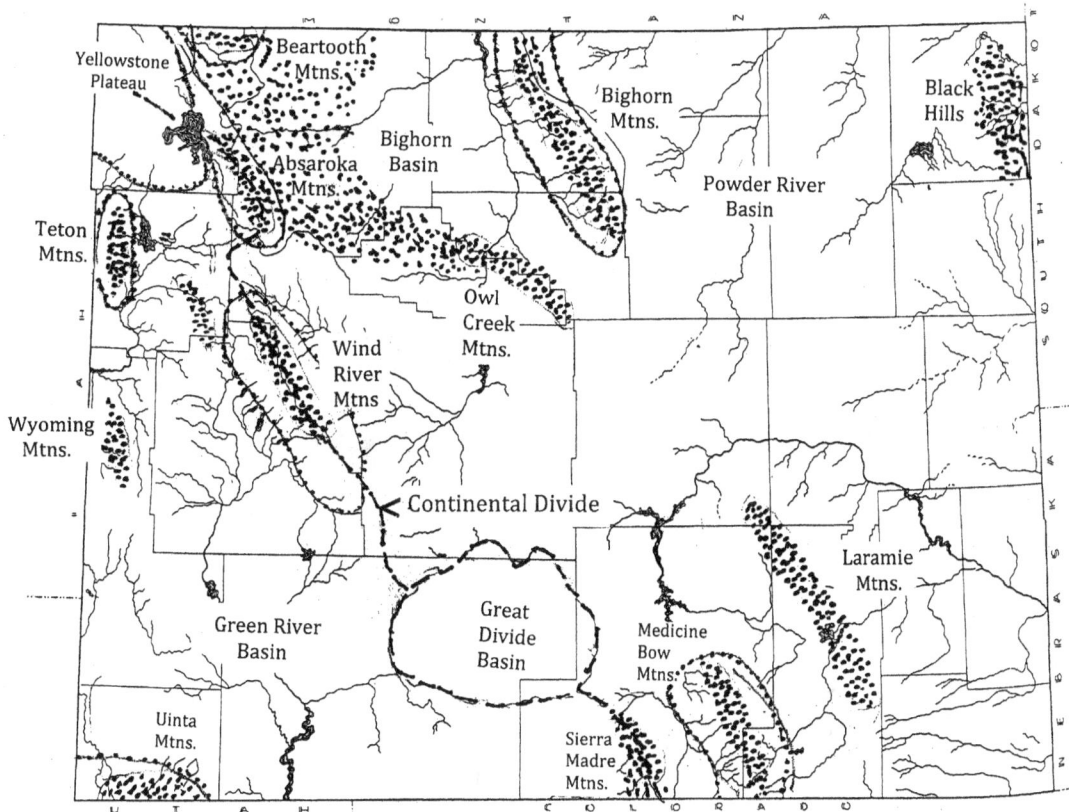

Map 2. County boundaries, mountain ranges, and major basins of Wyoming. The Continental Divide (dashed line) and regions of major montane glaciation (dotted lines) are also indicated. In part after Long (1965), Clark and Stromberg (1987), and Knight (1994).

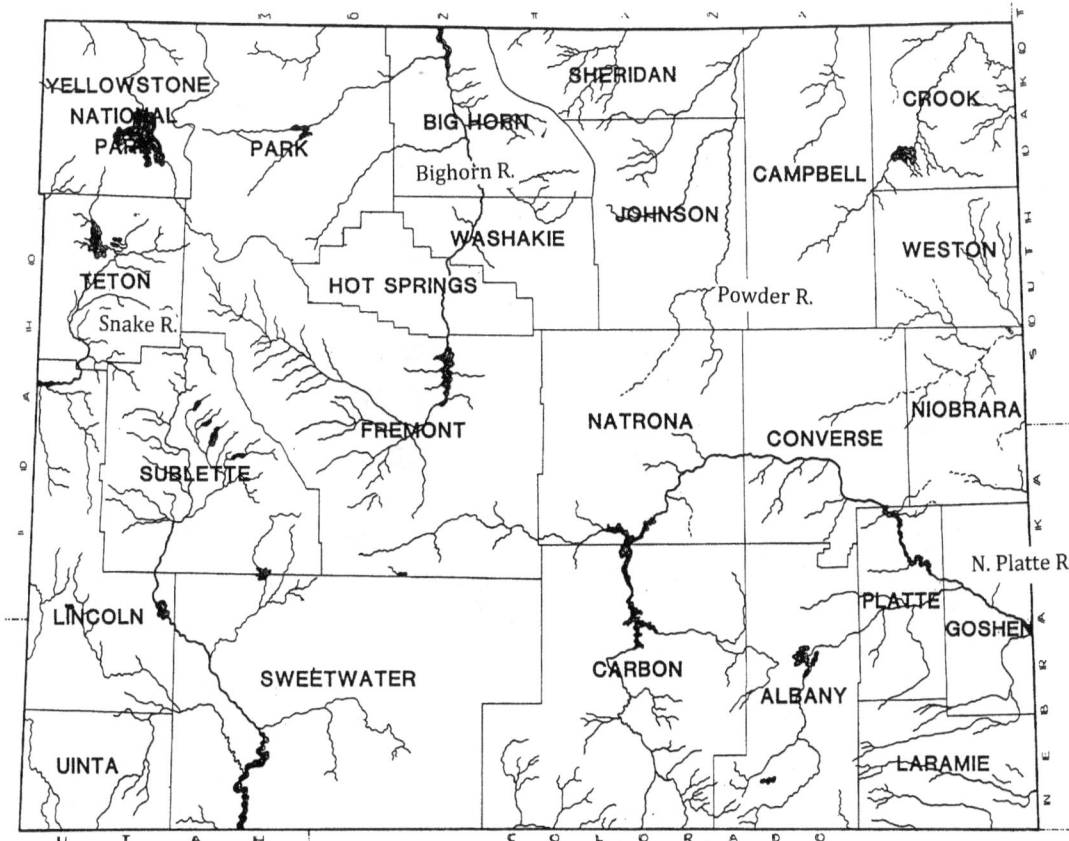

Map 3. County names and river drainage patterns of Wyoming.

relation to their major landscape (physiographic) characteristics and their long-term geologic histories. At much larger national and continental levels, "faunal regions" have long been used to identify broad plant and animal distribution patterns occurring around the world, such as the familiar Nearctic and Neotropical faunal regions of the Western Hemisphere.

At a regional scale, different biological community types ("ecosystems") have often been placed in broad categories variously termed "plant formations" (Daubenmire, 1968), "biomes" (Whittaker, 1975), "biociations" (Kendeigh, 1974), and "ecoregions" (Omernik, 1987). When defining and mapping such large regional and continental ecosystems, the general "life form" of the dominant vegetation is typically used to name them, such as the coniferous forest, grassland, and desert scrub formations (Faber-Langendorn et al., 2012). In Wyoming and adjacent Rocky Mountain states, examples of such "large-grain" vegetational categories include coniferous and deciduous forests, woodlands, sagebrush scrub, and grasslands (Map 4).

Frequently, smaller-scale natural biological communities are identified as "plant associations" (Daubenmire, 1943). Such local classifications define specific plant community types that have only one or two dominant (e.g., ecologically controlling) species, such as ponderosa pine forest or lodgepole pine forest. Identifying these relatively "fine-grained" and local community types is sometimes more interesting than using more generalized terms,

such as when evaluating the ecological affinities of birds and mammals in Wyoming (Maps 5 and 6). These local community types often spread out and extend over hundreds of square miles on relatively flat lands, but they are constrained on mountain slopes to fairly narrow horizontal zones that are stacked in altitudinal order, depending on their adaptations to local environmental variables such as temperature, wind, sunlight, substrate, slope, and the like.

Detailed descriptions of plant communities are available for the Rocky Mountains (Daubenmire, 1943; Habec, 1987) and for Wyoming (Porter, 1962; Green and Conner, 1989). Descriptions

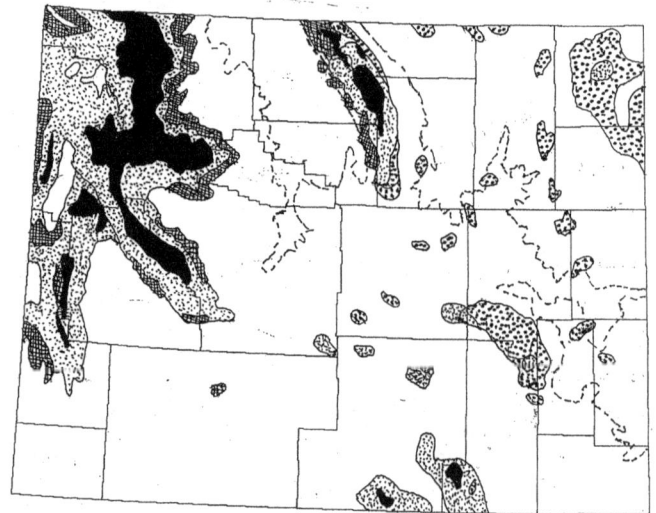

Map 5. Distributions of lodgepole pine (fine stippling), ponderosa pine (coarse stippling), Douglas-fir and spruce-fir forests (crosshatching), and alpine tundra (inked), along with the 5,000-foot-elevation contour (dashed line) in Wyoming. After Knight (1994).

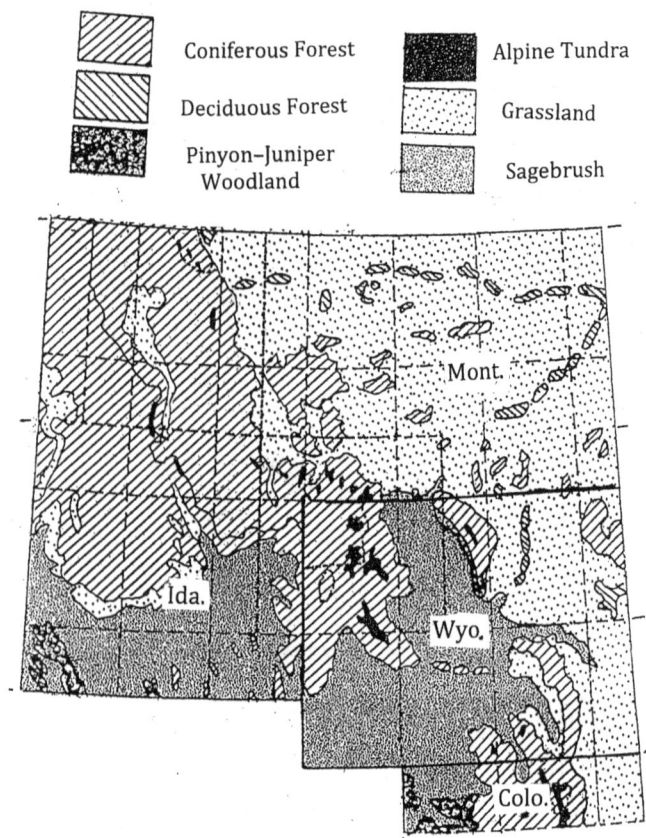

Map 4. Major native plant formations of Wyoming and adjacent Rocky Mountain states. Modified from Johnsgard (1986).

Map 6. Distribution of desert shrub vegetation (coarse stippling), grasslands (fine stippling), and altitudes above 9,000 feet (inked) in Wyoming. After Knight (1994).

are also available for forest communities of western Wyoming (Steele et al., 1983), the Wind River Mountains (Reed, 1969), the Medicine Bow Mountains (Alexander, Hoffman, and Wirsing, 1986), the Bighorn Mountains (Hoffman and Alexander, 1976), and the Black Hills (Hoffman and Alexander, 1987).

In Wyoming, the usual vertical sequence of major montane vegetation types (modified from Clark and Stromberg [1987] and Knight [1994]) that occur from alpine peaks downward is (A) alpine tundra; (B) spruce/fir (*Picea engelmannii, Abies lasiocarpa*) subalpine forest; (C) lodgepole pine (*Pinus contorta*) coniferous forest, typically after fire or on low-nutrient soils; (D) Douglas-fir (*Pseudotsuga menzeisii*) mixed coniferous forest; and (E) ponderosa pine (*Pinus ponderosa*) forest, often including limber pine (*Pinus flexilis*) or quaking aspen (*Populus tremuloides*) on lower or drier sites.

The usual foothills and lower-elevation nonforest sequence in Wyoming is (F) foothills woodlands of juniper (*Juniperus* spp.) and mountain mahogany (*Cercocarpus montanus*); (G) sagebrush (*Artemisia tridentata*) semidesert scrub; (H) grassland, including mixed-grass prairie (e.g., *Stipa, Andropogon, Schizachyrium*) on the more mesic eastern plains and shortgrass steppe (e.g., *Bouteloua, Buchloe, Poa*) in the western interior; and (I) desert and basin scrub, including xerophytes (e.g., *Grayia* and *Chrysothamnus*) on less alkaline soils and halophytes (*Sarcobatus, Atriplex, Kochia*, etc.) on highly alkaline substrates.

Similar altitudinal zonation sequences occur elsewhere in the Rockies. Naturalists should be aware of the major counterpart vegetational zones throughout the Rocky Mountains, both as a means of predicting the occurrence of particular plants and animals and also as a way of more fully appreciating the complex ecological interactions present in the region. Water-dependent communities also occur at various elevations throughout the Rockies and include both still-water (lotic) wetlands and flowing-water (lentic) habitats and associated riparian borders, which are often rich in species diversity.

Wyoming's Major Terrestrial Habitats and Their Associated Mammals

The major Wyoming montane and lowland plant communities, and some of their associated characteristic botanic and vertebrate components, are as follows.

Alpine Tundra
Tundra occurs above timberline where trees are absent or confined to exceptionally protected locations and is dominated by perennial herbs (grasses and sedges) and often some low-growing shrubs. In Wyoming, timberline typically occurs at 10,000 to 11,000 feet elevation (Daubenmire, 1943). Summers are very short at this elevation and wind effects are pronounced, with many of the woody species growing close to the ground where they remain under the snow cover during winter and avoid the desiccating effects of extreme temperatures and wind above snow level. Typical alpine mammals include the pika, mountain goat, dusky shrew, and dwarf shrew.

Spruce/Fir Forest (Timberline Zone)
This area of generally low and often twisted trees ("krummholz") is typically dominated by subalpine fir (*Abies lasiocarpa*) and Engelmann spruce (*Picea engelmannii*). In some areas of Wyoming, the whitebark pine (*Pinus albicaulis*) is primarily a timberline species, extending up to 11,000 feet, while in Yellowstone Park whitebark pine also grows at less than 9,000 feet, among stands of lodgepole pines. Somewhat farther south, limber pine (*Pinus flexilis*) is a characteristic timberline species at altitudes up to 11,000 feet, but it also occurs on drier foothill sites in the shortgrass plains near the Nebraska state line at elevations of less than 5,000 feet. The whitebark and limber pines are both small-stature, five-needled pines that are similar in appearance. Both of these pines also produce large and nutritious seeds that are relished by seed-eating birds such as Clark's nutcrackers and mammals ranging in size from mice to grizzly bears. The mature cones of whitebark pines often fall to the ground intact, where they may be quickly found and their seeds consumed by rodents, grouse, and many other birds. Typical mammals include the pygmy shrew, least chipmunks, golden-mantled ground squirrels, pine squirrels, snowshoe hares, lynx, wolverine, moose, and bighorn sheep. Wolverines are very rare denizens of these rich forests, and their populations seem to be slowly increasing.

Whitebark seedlings are shade-intolerant, so they may be unable to survive dense and mature spruce-fir forests. However, their heavy seeds are avidly collected and may be carried up to 15 miles by Clark's nutcrackers, which prefer to cache their seeds in bare or burned areas. Up to about 150 whitebark seeds can be carried by a Clark's nutcracker, and it can remember many of its cache sites for as long as nine months and recover the seeds in winter. The caching behavior by nutcrackers facilitates fairly rapid regrowth of trees after burns, as a result of germination by those seeds that the nutcrackers forget to recover. Some of these extremely slow-growing trees may reach many centuries of age (the apparent record is 1,270 years).

The seeds of limber pine are even larger than whitebark seeds, and their cones may remain attached to the tree for prolonged periods, where crossbills may seek them out and pry them open for their large seeds. Eventually, however, the cones open and drop their seeds to the ground, where they are often quickly found and consumed by rodents. The seeds of whitebark and limber pine are even a significant fall food for grizzly bears as well as a variety of small rodents. In the Rocky Mountain region, whitebark pines have been seriously affected by blister rust (*Cronartium*), bark beetles, and the parasite dwarf mistletoe (*Arceuthobium*). The blister rust came to America from Europe via Canada and has killed about 10 percent of Yellowstone's whitebark pines (and up to 100 percent of those in northwest Montana).

Lodgepole Pine Forest
Vast areas of the middle portions of the montane forest are covered by lodgepole pine (*Pinus contorta*) in the central and northern Rockies; for example, most of Yellowstone Park is dominated by such forests, which typically are regenerated only following

fires. In Wyoming, lodgepole pine extends from about 7,000 to 9,500 feet elevation. Lodgepole stands are usually very crowded and of nearly uniform ages, often as a result of having germinated after a forest fire, when their resin-sealed cones are burned and the long-held seeds are finally released. Because of their dense crowding and thin bark, lodgepole pines are easily killed by forest fires. However, they regenerate rapidly because of their fire-adapted seed-dispersal mechanism and fast growth, so that they might start producing cones when only 10 to 12 years old. Lodgepole pine seedlings are shade-intolerant but are able to grow on low-nutrient soils where few other trees can survive. In the Greater Yellowstone region, these are often volcanic-based (rhyolitic) soils that have few nutrients and can support little other vegetation, resulting in low plant species diversity and a very limited food supply for most birds and mammals. Lodgepole pine seeds are generally unavailable except to a few bird species that can pry open the tightly sealed cones, such as crossbills and probably also pine grosbeaks.

Mixed Coniferous Zone (Douglas-fir and Ponderosa Pine Forests)

The lower- to middle-altitude montane zone is dominated by Douglas-fir (*Pseudotsuga menziesii*), which sometimes forms dense, single-species stands but also often shares dominance in the northern Rockies with Engelmann spruce (*Picea engelmannii*) or white spruce (*Picea glauca*). Mature Douglas-firs have thick bark and are quite fire-resistant, sometimes forming magnificent stands several centuries old. These slow-growing trees and their seedlings are shade-tolerant and thus are able to survive in the densely shaded conditions of lodgepole pine forests, ultimately replacing them. This zone is best developed in Wyoming between 7,500 and 9,500 feet. Typical mammals of mid-montane coniferous forests include many seed-eating rodents such as the northern flying squirrel, least chipmunk, golden-mantled ground squirrel, red squirrel, and several mice and voles. Predators include the long-tailed weasel, ermine, least weasel, pine marten, bobcat, lynx, cougar, grizzly bear, and wolverine. Omnivores include the grizzly and black bears, raccoon, and skunk.

A slightly lower and sometimes open and savanna-like forest zone that is dominated by ponderosa pine (*Pinus ponderosa*) is extremely widespread throughout the Rockies, from Canada to central Mexico. In Wyoming, ponderosa pine extends from about 3,100 to 7,000 feet elevation. Ponderosa pines often form the lower edge of the montane coniferous forest and frequently also extend out into the high plains in scattered groves on mesas or other sites with favorable slope, soil, or moisture conditions. There the dominant, and sometimes only, tree species is ponderosa pine, which typically grows in fairly open rather than dense groves, with considerable grassy or shrubby cover between the trees. When mature, the ponderosa pine is fairly fire-resistant, and its thick bark can survive most of the ground fires that are common on the plains. Its cones initially point upward, but over time they bend downward and gradually release their seeds, which are consumed by many birds and mammals. The Abert's

squirrel is essentially confined to the ponderosa pine zone in southeastern Wyoming, where it eats the pine's inner bark, twigs, seeds, terminal buds, and flowers (Clark and Stromberg, 1987). Other seedeaters include the yellow-pine chipmunk as well as all the species mentioned for the Douglas-fir forest. In eastern Wyoming, the limber pine is an important additional component of this semiarid forest type, whereas in western Wyoming juniper (*Juniperus osteosperma*) replaces limber pine to some degree.

Aspen Woodland

Quaking aspen (*Populus tremuloides*) groves occur widely in the central and northern Rockies, either as a transitional ("seral") community following fire or logging, or as a long-term "climax" community on low hillsides too dry to support coniferous forests. Aspens are sensitive to fire and depend mostly on suckering for regeneration and local expansion; large areas in the Rockies are essentially clones of a single genetic type. Aspen woodlands often are rich in bird life, particularly woodpeckers, as well as various other cavity-nesting birds that often exploit old woodpecker excavations for their own nesting sites. Sapsuckers are particularly attracted to aspens, and their drillings attract many other sap-eating species of birds, mammals, and insects. Woodpeckers also eat many insects that live under the bark of various coniferous forest trees, such as bark and wood-boring beetles. Their bark-flaking activities may also expose the insects and their larvae or pupae to other predators or parasites. However, woodpecker-drilling activities may also expose the trees to infections by pathogens. Typical mammals of aspen stands include small rodents that use old woodpecker holes, such as white-footed and deer mice, least chipmunks, northern flying squirrels, and yellow-pine chipmunks.

Juniper Woodland

On foothills and other areas below the coniferous forest a "pygmy forest" of small, scattered, and arid-adapted trees occurs locally, especially in southwestern Wyoming. The woodland zone mainly occupies an elevation range of about 5,000 to 6,000 feet in Wyoming but can extend up to about 7,500 feet. It is composed of several rather low-growing species of junipers (mainly *J. scopulorum*) in Wyoming and, farther south, especially pinyon pine (*Pinus edulis*). Pinyon pines are very rare in Wyoming but produce nutlike seeds that are important foods for many rodents and birds. In Colorado, species diversity of mammals is very high in this habitat type, second only to diversity in riparian habitat. There, four species of *Peromyscus* (deer mice) occur sympatrically; they and other small rodents feed largely on the nutlike seeds of pinyon pine during years of high mast production and probably also on juniper "berries." Some typical pinyon-juniper mammals in Wyoming include the canyon mouse and bushy-tailed woodrat, and probably also bats such as the fringed myotis and Townsend's big-eared bat (Clark and Stromberg, 1987; Armstrong, Fitzgerald, and Meaney, 2011). The pinyon mouse, typical of this community type, once occurred in Sweetwater County but might be extirpated (Clark and Stromberg, 1987).

Oak–Mountain Mahogany Scrub

Like the juniper community type, the oak–mountain mahogany type is also an arid-adapted community that is better developed in the southern Rockies than in Wyoming. It is largely limited to lower montane slopes of Colorado and southern Idaho, and consists of woody shrubs such as mountain mahogany (*Cercocarpus* spp.), serviceberry (*Amelanchier* spp.), and farther south, several species of scrubby oaks (e.g., Gambel oak, *Quercus gambeli*). Rocky soils and exposed rock crevices are common in these scrublands, which provide hiding places for small rodents and foraging and roosting places for bats, such as the small-footed myotis. The trees and shrubs typically grow in clumps separated by grassy areas, forming a semidesert community similar to oak-dominated chaparral communities farther west. Typical mammals include the cougar, mule deer, bobcat, ringtail, western spotted skunk, bushy-tailed woodrat, and various mice such as the deer mouse.

Sagebrush Scrub and Sage-Steppe

Over more than half the land surface of Wyoming was historically dominated by sagebrush, especially big sagebrush (*Artemisia tridentata*), which ranges from 3,500 to 9,000 feet elevation. Silver sage (*A. cana*) reaches as low as 3,100 feet, whereas alpine sagewort (*A. scopulorum*) extends as high as 11,000 feet (Knight, 1994). Other species and varieties of sage are also often present. In some areas, various sage species share dominance with arid-adapted grasses (sage-steppe), which are more palatable if less nutritious than the leaves of *Artemisia*. There are two obligatory mammalian associates of sagebrush, the now-endangered pygmy rabbit and the sagebrush vole. Pygmy rabbits are entirely sage-dependent, with up to 99 percent of their food being sage (Clark and Stromberg, 1987). Sagebrush voles rely on eating sage leaves in the wintertime but consume mostly legume leaves during the growing season.

Nearly 80 species of mammals, including pronghorn, mule deer, elk, and bighorn sheep, are facultative obligatory associates of big sage. The Nuttall's (mountain) cottontail is notably common in this zone, where it eats mostly sagebrush leaves but also extends to the subalpine fir zone at elevations of nearly 9,000 feet. Desert cottontails also thrive in this zone. The pronghorn is nearly entirely dependent on sage as a winter food, and it is also an important winter food for mule deer. Wyoming ground squirrels, white-tailed jackrabbits, and white-tailed prairie dogs are also common in sage-steppe areas, where a wide diversity of grasses and forbs is seasonally available. Some common predators of sagebrush include the least, short-tailed (ermine), and long-tailed weasels as well as coyotes and swift foxes. Other reptilian predators, such as the bullsnake, prairie rattlesnake, and faded rattlesnake probably also eat rodents such as the plains pocket gophers and plains pocket mice, as well as Ord's kangaroo rats, grasshopper mice, and least chipmunks.

Grasslands

Within and to the west of the Rocky Mountains the grasses tend to be perennial bunchgrasses rather than the continuous sod-forming grasses typical of the Great Plains, with various wheatgrasses (*Agropyron*), needle-and-thread (*Stipa*), and fescues (*Festuca*) often dominant. The zone is best developed from about 3,500 to 6,000 feet, although mountain meadows can occur at much higher elevations. On higher plateaus, such as the Yellowstone Plateau and on midlevel mountains, perennial grasslands often dominate on fertile soils. For example, they cover about 20 percent of the Bighorn National Forest and are a major component of Yellowstone Park's northern range, the heart of the park's elk and bison summer habitat. The vegetation on Yellowstone's productive northern range consists mostly of perennial grasses, mainly Idaho fescue (*Festuca idahoensis*) and wheatgrasses, interspersed with open stands of big sagebrush. In the Bighorn Mountains, these grasslands are similarly dominated by wheatgrasses, pompelly brome (*Bromus pompellinaus*), Idaho fescue, and inland bluegrass (*Poa interior*). Bison, elk, coyote, prairie dogs, thirteen-lined ground squirrels, white-tailed jackrabbits, swift foxes, coyotes, badgers, cottontails, prairie voles, hispid pocket mice, and Ord's kangaroo rats are all common in these grasslands. Northern pocket gophers, northern grasshopper mice, olive-backed mice, plains mice, and silky pocket mice, kangaroo rats, and spotted ground squirrels are most common in grasses growing on sandy soils. Black-footed ferrets are by far the rarest inhabitants of grasslands and (through reintroductions) are found only where prairie dogs are abundant.

The vast plains of shortgrass lying to the east of the Rocky Mountains are less bunch-forming and instead are dominated by low, perennial grasses that produce continuous cover wherever there is adequate precipitation. These arid prairies have numerous species of grama grass (*Bouteloua* spp.) and buffalo grass (*Buchloe dactyloides*) as well as other taller species in protected or ungrazed areas. Heavy grazing or extended droughts favor invasion of the grass by cacti and other nonedible plants. Black-tailed prairie dogs (*Cynomys ludoviciana*) were historic keystone mammals over much of the Great Plains (Johnsgard, 2005), and such mammals as the Uinta ground squirrel, plains pocket gopher, prairie vole, and black-tailed jackrabbit are typical. Black-footed ferrets are by far the rarest inhabitants of Wyoming grasslands and (through introductions) are found only where prairie dogs are abundant.

Saltbush-Greasewood Scrub

In the Great Divide Basin area of Wyoming, and locally elsewhere, highly alkaline soils allow for the growth of an arid-land community type called saltbush-greasewood scrub. The Great Divide and also the Powder River and Bighorn basins range from about 4,000 to 6,400 feet elevation, and annual precipitation in the Great Divide Basin may be as little as six inches. The typical vegetation in these places is scattered, shrubby, and bunch-like, with the surprisingly dark green color of greasewood (*Sarcobatus vermiculatus*) contrasting with the more grayish and arid-looking shadscale (*Atriplex canescens*) and saltbush (*A. confertifolia*).

As in the sagebrush scrub community, edible foods in this arid environment are few and drinkable water is rare. Bird and mammal abundance and diversity are very low in this community type. The Wyoming pocket gopher and various voles and pocket mice are probably some of the resident mammals, and various predators such as coyotes and bobcats no doubt visit these habitats. Ringtails are known to occur in the lower Green River Basin, but they are too little studied to know much about their preferred habitats in Wyoming.

Streams and Riparian Shorelines

Flowing water (lentic) communities vary from aquifer seeps and tiny creeks to large rivers. The upper reaches of the Yellowstone, Missouri, North Platte, and other major rivers of the Great Plains bring west into the region an important biota that is especially rich in eastern bird life. Cottonwoods (*Populus angustifolia, P. acuminate*, and their hybrids) are important riparian-zone trees, and their large size provides many foraging niches and nesting sites for many songbirds. Other common trees are alders (*Alnus*) and willows (*Salix*), which sometimes form lower, dense shoreline thickets, depending on the amount and seasonality of water availability.

Riparian forests often extend for long distances along rivers otherwise passing through nonforested regions, which may allow for range expansion and dispersal across these barriers by forest-adapted bird species. Typical mammals include the white-tailed deer, porcupine, gray and red foxes, white-footed mouse, and fox and northern flying squirrels. High-elevation willow flats are used by moose, elk, snowshoe hares, water voles, and many other mammals.

Standing-Water Wetlands

Standing-water (lotic) ecosystems range from ephemeral spring ponds (playas) to semipermanent or permanent marshes and might also grade into larger and deeper lakes and impoundments. Typical mammals include the raccoon, water shrew, meadow and western jumping mice, muskrat, beaver, mink, river otter, Virginia opossum, Hayden's shrew, masked shrew, meadow voles, and many others.

The Birds of Wyoming's Terrestrial Habitats

Forest-adapted birds probably are affected by invisible or inconspicuous environmental aspects of the altitudinal and microclimatic gradient of montane habitats—such as weather and the availability of potential nesting sites—as much as they are affected by the specific botanic components of the habitat. The highest zone and the most climatically extreme conditions for life occur on alpine tundra, where the growing seasons are very short and breeding conditions are usually marginal. As a result, species diversity is very low in this zone but does include the American pipit, rosy-finches, and (at least formerly) the white-tailed ptarmigan. In the southern Rockies of southeastern Wyoming and Colorado, the breeding rosy-finch is the brown-capped; however, in the Rockies of northwestern Wyoming it is the black rosy-finch. Still farther north in the northern Rockies of Montana and

Canada, it is the gray-headed rosy-finch, which occurs in Wyoming only as a winter visitor. The white-crowned sparrow is a timberline nester, frequenting low willows and sedges, and nesting in low shrubs or on the ground. At least locally the Brewer's sparrow is also a timberline species.

The collective habitat requirements of a species most often are manifested by identifying the plant communities in which it is most often found, even if their specific components remain obscure. Hutto and Young (1999) did intensive fieldwork over three years in the montane forests of Idaho and western Montana. Using 566 sampling transects, they modeled 83 bird species as to their ecological affinities within 18 groups of unaltered and human-altered forest habitats. The unaltered habitat group included six undisturbed forest categories, and the altered communities included five types of post-lumbering habitats as well as post-fire sites. They found that post-fire sites tend to attract woodpeckers, including the hairy woodpecker, three-toed woodpecker, and northern flicker Among the various types of post-lumbering sites, open-county foragers, such as flycatchers, kestrels, ravens, and jays, were abundant as well as ground foragers such as spotted towhees, lazuli buntings, and dark-eyed juncos. Hein (1980) found that species diversity and breeding densities were somewhat higher in a burned Colorado coniferous habitat than in those that were unburned, and that the previously present hermit thrush, ruby-crowned kinglet, and the two chickadee species were replaced by the American robin, mountain bluebird, broad-tailed hummingbird, northern flicker, and Empidonax flycatchers.

In their studies of unaltered habitats, Hutto and Young reported that some important breeders in the subalpine spruce-fir forests are the brown creeper, ruby-crowned kinglet, hermit thrush, fox sparrow, Wilson's warbler, and pine grosbeak. The pine siskin, red crossbill, and white-winged crossbill are also frequent breeders. At somewhat lower altitudes where the trees grow taller, breeders include the northern pygmy-owl, black-backed woodpecker, American three-toed woodpecker, common raven, Steller's jay, and Clark's nutcracker. Golden-crowned and ruby-crowned kinglets also nest in these taller trees, as does the pine grosbeak.

Summarizing data from 12 Engelmann spruce–subalpine fir communities during the breeding season, Smith (1980) determined that the mountain chickadee and yellow-rumped warbler were observed in all 12; the hermit thrush, ruby-crowned kinglet, and pine siskin in 11; the Clark's nutcracker in 10; and the American robin in 9. In 8 studies the red-breasted nuthatch, brown creeper, Townsend's solitaire, pine grosbeak, and chipping sparrow were detected; in 7 the northern flicker, red crossbill, and dark-eyed junco; and in 6 the American three-toed woodpecker, dusky flycatcher, and Cassin's finch were present. The total number of species observed in all 12 studies ranged from 12 to 30; the 3 studies from the Rockies in Colorado, Wyoming, and Montana reported 12 to 19 species.

Sanderson, Bull, and Edgerton (1980) reported that 26 bird species of the interior Pacific Northwest breed only in mature or old-growth mixed-conifer forests. By comparison, a total of 10 species were reported to breed in the grass-forb stage, 30 in the shrub-seedling stage, 38 in the pole-sapling stage, 59 in the young forest stage, and 83 in the mature forest stage of succession. Among Wyoming's nonpasserine birds that breed in and are often associated with mature Douglas-fir forests are the Barrow's goldeneye, bufflehead, northern goshawk, great gray owl, boreal owl, and Williamson's sapsucker. Passerine species include the gray jay, Steller's jay, Clark's nutcracker, Wilson's warbler, yellow-rumped warbler, MacGillivray's warbler, golden-crowned kinglet, ruby-crowned kinglet, mountain chickadee, red-breasted nuthatch, Townsend's solitaire, hermit thrush, Swainson's thrush, evening grosbeak, pine grosbeak, and red crossbill.

The birds of Rocky Mountain lodgepole pine forests are in general much like those of the Douglas-fir and other mid-level coniferous forest communities. The hermit thrush, Cassin's finch, and red crossbill are also typical of this zone, and these same species often extend higher into the subalpine zone. In the Jackson Hole area, the common breeding birds of lodgepole pine forests are the northern flicker, downy woodpecker, gray jay, mountain chickadee, hermit thrush, yellow-rumped warbler, dark-eyed junco, and chipping sparrow (Anderson, 1980). Several woodpeckers are typical of lodgepole forests as well as some closely associated cavity-nesting songbirds. The most common breeding species in Colorado lodgepole forests are yellow-rumped warbler, ruby-crowned kinglet, dark-eyed junco, hermit thrush, black-capped and mountain chickadees, pine siskin, gray jay, and Townsend's solitaire (Hein, 1980). All these species tend to have broad niches and are also common in other coniferous communities. Hein noted that the number of breeding species ranged from 8 to 14 in four studies in lodgepole pine stands and from 10 to 20 in five studies of mixed lodgepole and other conifers. Species density was also substantially higher in the mixed-forest stands.

Diem and Zeveloff (1980) reported that at least 113 bird species utilize ponderosa pine forests. Insectivorous species that selectively nest in coniferous trees are the olive-sided flycatcher, ruby-crowned and golden-crowned kinglets, western tanager, and yellow-rumped warbler. Omnivorous species nesting selectively in coniferous trees include the Clark's nutcracker and the gray and pinyon jays. Tree-nesting species using either coniferous or deciduous trees are numerous (34 species) in these forests. Another 28 species are cavity nesters, 15 species nest in bushes or small trees, 15 nest on the ground, and 11 nest on cliffs, in caves, or on rocks or among talus. Many of these species also occur in other regional community types. Eleven ponderosa pine species (mostly woodpeckers) are self-excavating cavity-nesting species. Another 17 species use natural or preexcavated cavities, often those created by woodpeckers. The rare flammulated owl is often attracted to mature ponderosa pine stands with woodpecker holes.

Aspen forests typically have highly diversified breeding avifauna, including such Wyoming breeders as the ruffed grouse, northern pygmy-owl, warbling vireo, and about a half-dozen woodpeckers. Because of the excavating actions of woodpeckers

on soft-wooded trees such as aspens and birches, many cavity-nesting birds such as the tree and violet-green swallows, house wren, black-capped and mountain chickadees, and mountain bluebirds are attracted to this forest type. Canopy-nesters such as the warbling vireo, yellow-rumped warbler, and western wood-pewee are also common. Shrub-nesters include the yellow warbler, at least three Empidonax flycatchers (dusky, Hammond's, and cordilleran), black-headed grosbeak, and MacGillivray's warbler. Common ground-nesters include the dark-eyed junco, hermit thrush, Townsend's solitaire, and white-crowned sparrow (Flack, 1976). Salt (1957) found that 60 percent of the breeding bird species in aspen forests of the Jackson Hole region were cavity nesters, as compared with 54 percent for spruce-fir forests, and 19 to 31 percent for lodgepole or lodgepole and spruce or spruce-fir types. The aspen sites had substantially higher breeding bird densities (523 pairs per 100 acres) than did any of the other Rocky Mountain forest types analyzed by Scott, Whelan, and Svoboda (1980). Hutto and Young (1999) also found that cottonwood/aspen sites supported high densities of nearly a dozen species of birds.

Woodlands in Wyoming are dominated by junipers (*Juniperus scopulorum* in eastern Wyoming and *J. osteospermum* in the west), mountain mahogany (*Cercocarpus* spp.), and only rarely pinyon pine. Woodlands are the best developed in southwestern Wyoming. The woodland community brings into the state a distinctive group of birds, such as the common poor-will, gray flycatcher, pinyon jay, Woodhouse's ("western") scrub-jay, juniper ("plain") titmouse, blue-gray gnatcatcher, western bluebird, Bewick's wren, and black-throated gray warbler. Balda and Masters (1980) identified five "obligatory" species of the southwestern woodland community, of which the gray flycatcher, Woodhouse's scrub-jay, and juniper titmouse were included. Junipers provide important winter foods for Townsend's solitaires, waxwings, and many other seasonal berry-eaters such as thrushes, warblers, and woodpeckers. Pinyon-juniper plots of 10 to 15 acres may support 20 or so breeding species, a species diversity comparable to ponderosa pine or mixed conifer forests but having far lower breeding densities, namely 40 to 80 pairs versus 200 to 350 pairs per 100 acres (Johnson et al., 1980).

The birds of Wyoming's shortgrass prairies typically have wide breeding distributions on the Great Plains (Johnsgard, 1979, 2001), and some of these same species are also associated with the bunchgrass prairies of the Pacific Northwest. The mountain plover is one of the most typical, and now increasingly rare, breeding birds of shortgrass prairies. Others include such conspicuous species as the Swainson's and ferruginous hawks, prairie falcon, burrowing owl, sharp-tailed grouse, long-billed curlew, and upland sandpiper. Seed-eating birds are the most common breeders, such as the grasshopper, lark, and vesper sparrows, lark bunting, McCown's longspur, and chestnut-collared longspur. Nearly all of these rather inconspicuous species are declining nationally. Other mostly declining but more conspicuous birds are the Swainson's and ferruginous hawks, prairie falcon, burrowing owl, sharp-tailed grouse, long-billed curlew, and upland sandpiper as well as the loggerhead shrike and western meadowlark.

There are five obligatory bird species dependent on sage-brush communities, including the greater sage-grouse, sage thrasher, sagebrush sparrow, and Brewer's sparrow. Adult greater sage-grouse invest 90 to 100 percent of their diet in sage leaves during most of the year, and even juveniles eat similar proportions by the time they are three to four months old. Adult females most often nest under sage bushes (50–100 percent in various studies), and sage is often the most frequently used brood cover by late summer. Sagebrush sparrows and Brewer's sparrows most often place their nests on sage plants, and sage thrashers typically nest either directly under or on the plants. Paige and Ritter (1999) listed 90 bird species that are facultative sage associates—that is, species that benefit from sagebrush but are not dependent on it. In their review of birds associated with sage shrubland, they identified 17 species of national conservation concern. In addition to the five just-mentioned sage-obligate species, there are also three general shrub-adapted species: green-tailed towhee, lark sparrow, and black-throated sparrow (very rare in Wyoming). Five mixed shrubland and grassland species were also identified: Swainson's hawk, ferruginous hawk, prairie falcon, sharp-tailed grouse, and loggerhead shrike. Finally, four grassland-associated species were included: long-billed curlew, burrowing owl, short-eared owl, and vesper sparrow.

Paige and Ritter also identified eight habitat components of sage-steppe and the number of each that were utilized by the 17 species they classified as sage-scrubland birds. The most heavily used habitat component is open, patchy sagebrush, used by 15 species, followed by those using grass cover for nests, or grass-lands generally, by 10 species. Another high-usage component is short grass and bare ground, which is used by 9 species. The greater sage-grouse and sharp-tailed grouse use six components; the Swainson's hawk five; the ferruginous hawk, long-billed curlew, short-eared owl, sage thrasher, loggerhead shrike, Brewer's sparrow, and lark sparrow four; and the burrowing owl, prairie falcon, green-tailed towhee, sagebrush sparrow, and vesper sparrow three. It is thus clear that not all sage-associated species are responding equally to the same habitat elements, but the combined presence of sage and grassland plants most closely ties them together.

The still drier and lower alkaline-associated saltbush and greasewood communities have few breeding bird species and low bird populations. Plots of 10 to 15 acres of saltbush-greasewood habitat often support only 3 or 4 breeders and 15 to 20 pairs per 100 acres, whereas a similar plot of sagebrush might support 10 to 12 species and 50 to 100 pairs (Johnson et al., 1980).

Riparian woodlands that develop along montane drainages often have a mix of coniferous and deciduous species, and this plant diversity generates considerable bird diversity. Western riparian communities are not only fairly high in species diversity but population densities may reach 600 to 1,300 pairs per 100 acres (Johnson et al., 1980). In the Jackson Hole area, the common birds of riparian aspen groves are the northern flicker, downy and hairy woodpeckers, western wood-pewee, tree

swallow, black-capped chickadee, mountain bluebird, warbling vireo, yellow warbler, black-headed grosbeak, white-crowned sparrow, and Lincoln's sparrow (Anderson, 1980). The typical birds are a mix of eastern and western species and include broadly distributed species such as the willow flycatcher and warbling vireo, and such western birds as the spotted towhee, lazuli bunting, Bullock's oriole, and black-headed grosbeak. Eastern counterpart species of the deciduous forest, such as eastern towhee, indigo bunting, Baltimore oriole and rose-breasted grosbeak, or their interspecies hybrids, are often a part of the mix where their western counterparts also occur. Other eastern species that extend variably far westward along wooded river systems into Wyoming include the yellow-billed cuckoo, eastern screech-owl, red-headed and red-bellied woodpeckers, least flycatcher, eastern phoebe, eastern bluebird, brown thrasher, red-eyed vireo, and orchard oriole.

Recent Changes in Bird and Mammal Populations

Even in national parks not all the natural habitats are wholly pristine; historical and recent forest fires have placed much of the area of Yellowstone Park in various stages of vegetational transition ("succession") dominated by lodgepole pines. Invasion by bark beetles is rapidly changing the composition of coniferous forest tree species, and introducing lake trout into Yellowstone Lake has severely damaged the population of native cutthroat trout, affecting both bears and otters (Koel et al., 2005). Other human-caused effects on the environment are apparent everywhere throughout the Rocky Mountains, as grazing, lumbering, agriculture, mining, energy development, road building, and other familiar symbols of modern civilization have left their marks on the landscape.

Ranching activities in Grand Teton National Park and national forests in the Greater Yellowstone region have also influenced grassland and shrub succession in nonforested areas as a result of overgrazing. Damming of streams by beavers has resulted in the formation of unique beaver-pond communities, with an interesting and diverse association of plant, bird, and mammal life. Indeed, such species as the trumpeter swan and sandhill crane are largely dependent upon beaver activity in the Greater Yellowstone region for the production and maintenance of suitable breeding habitat. The reintroduction of wolves into Yellowstone Park has caused many beneficial changes to the vegetational ecology of that region, largely from the reduction of elk populations and changes in their distribution, which has allowed the redevelopment of aspens and stream bank vegetation that before were denuded by elk over-browsing, and resulting in improved water quality for aquatic life, increased food supplies for beavers, added nesting habitat for riparian-adapted birds, and the like.

A recent calamity has been the devastating effects of bark beetles (family Scolytidae) on the coniferous forests of the Rocky Mountains. Beetle populations have benefitted from global warming, which is causing increased aridity in the western mountains and reduced winter mortality to beetle populations. An associated increased vulnerability of several coniferous forest trees to beetle attack has resulted from the trees' increased water stress. About 6.4 million acres have been seriously affected by bark beetles in recent decades throughout the West, including 2.5 million acres in the US Forest Service Rocky Mountain Region, and especially in Wyoming, Colorado, and western South Dakota. The mountain pine beetle (*Dendroctonus brevicomus*) has had an impact on more than 1.5 million acres of coniferous forests in northern Colorado and southern Wyoming and has especially affected lodgepole pines and whitebark pines. In Wyoming, the infestation started in about 2000, with the mountain pine beetle, the spruce beetle (*D. engelmannii*), and the Douglas-fir beetle (*D. pseudotsuga*). Stephenson (2011) estimated that by 2012 nearly all the mature lodgepole pines in that region would be killed, and well over half of the whitebark pines in Yellowstone Park will have died. Of all the biological controls of these beetles, woodpeckers are perhaps the most effective. Woodpeckers such as the black-backed, American three-toed, and Lewis's are responding to the outbreaks of bark beetles in burned and water-stressed coniferous forests of the American West (Johnsgard, 2011a). The blister-rust fungus (*Cronartium*) has also long had a devastating impact on western pine forests (mainly affecting western white, ponderosa, and lodgepole), especially in Montana, Idaho, and Wyoming.

In general, the influence of humans is to reduce environmental diversity, either fortuitously or purposefully, by eliminating unwanted species in favor of more economically profitable uses for the land. As a result, some species have become extremely rare throughout the region, even in protected areas. These include several mammalian and avian predators, and many birds closely associated with native grasslands or shrublands, such as the sharp-tailed grouse and greater sage-grouse. However, some large predatory mammals, such as the gray wolf, grizzly bear, and wolverine, have responded to protection and management, and have gradually moved back into areas from which they have been absent for many decades. The same is true for several once-rare predatory birds such as the bald eagle, peregrine falcon, and osprey.

Still other species have benefited indirectly from human activities and have become extremely abundant in and around human activity centers. These include such introduced or expanding species as the house finch, European starling, and the explosively increasing Eurasian collared-dove. Several hummingbirds have altered their migration routes and even somewhat expanded their wintering ranges northward to take advantage of the increasing popularity of feeding them. Various once mostly country-living species, including the American crow, Cooper's hawk, merlin, American robin, common grackle, and great-tailed grackle, have benefited from human-caused habitat changes by adapting to urban parks and gardens.

Overleaf: Black-tailed prairie dog, adult

Chapter 2 • Mammals

Wyoming's mammals include some of the most spectacular of all North America's charismatic megafauna, with large populations of such memorable species as bison, elk, moose, bighorn sheep, mountain goat, pronghorn, two species of deer, and two bears. Additionally, it is the state where the black-footed ferret was making its last futile stand, until last-minute efforts at live-trapping, captive breeding, and reintroductions saved it from certain extinction.

The following 65 profile summaries include 70 of Wyoming's total mammal species. Wyoming currently supports 117 species of wild mammals (Buskirk, 2016), of which several groups—the shrews, moles, and pocket gophers—are unlikely to be seen by the average person during an entire lifetime, and the bats are likely to be seen only as flitting shadows in the moonlight. The largest group of Wyoming mammals, the rodents, includes nearly 50 species and comprises more than 40 percent of the total. Humans generally regard rodents with disdain, as they include some major pest species like the house mouse (*Mus musculus*) and Norway rat (*Rattus norvegicus*), which were both introduced from Europe and are not described in this book. However, one of Wyoming's rodents, the native beaver, is an extremely valuable manager of Wyoming mountain streams, preventing floods and establishing wet meadows, marshes, and other wetlands that are among the state's best wildlife habitats.

Nearly a dozen big game species are a major economic factor in the state. In 2017 the numbers of big game legally killed were: mule deer and white-tailed deer, 45,190; pronghorns, 42,294; elk, 24,530; moose, 296; mountain goats, 276; bighorn sheep, 175; black bears, 167; and bison, 70. Through hunting, these animals generated an estimated $243 million, or an average of about $2,100 per animal. In addition, 44 wolves were also legally killed in the pursuit of "sport," although the thrills of hearing and seeing wild wolves and grizzly bears have been, in the author's experience, a cherished lifetime memory. Current (2018) legally blocked efforts by the Wyoming Game and Fish Department to open the nationally endangered grizzly bear to limited trophy hunting might, if successful, bring out enough wealthy would-be killers of one of North America's most threatened and magnificent wild mammals to increase the average carcass value of Wyoming's big game to $2,200. In contrast, more than 4 million people visit Yellowstone National Park annually, many of them with the hope of seeing wolves or grizzly bears for the first time in their lives. Tourism benefits alone from wildlife viewing and other nature-related activities in the Yellowstone and Grand Teton national parks alone brings in an estimated $1 billion annually, making the economic value of hunting pale by comparison.

Of the 117 Wyoming mammal species listed in the synopsis (chapter 5), nearly 70 have been selected to be described here in limited detail. Those chosen are intended to be a sampling of the major mammalian groups, including representatives of all the genera that have well-represented Wyoming species. The profiles describe nearly all the very common (and a few of the rarest) species. These were also chosen to illustrate the amazing diversity and fascinating life histories of some of our nearest vertebrate relatives.

Measurements and weights shown here are mostly those of Clark and Stromberg (1987). Most of the life history information provided but not specifically attributed to a source is from Clark and Stromberg (1987) or Armstrong, Fitzgerald, and Meaney (2011). The Latin nomenclature used here follows Buskirk (2016), with older names inserted parenthetically, and the same is true for English vernacular names and any commonly used alternate names. See chapter 5 for definitions of the terms that follow each species' Latin name and that describe the species' general distribution pattern based on latilongs.

Family Didelphidae (Opossums)

Virginia Opossum. *Didelphis v. virginiana*. Localized.
Identification: Length: 643–900 mm (25–35 in); tail: 250–440 mm (9.8–17.3 in). Weight: 1.9–2.8 kg (4.2–6.2 lb). This is North America's only marsupial, but the female's pouch is unlikely to be visible in the wild even with young inside. However, the long, naked tail is visible, as are the large black ears and long pink nose. Opossums are as likely to be seen in a tree as on the ground and are active diurnally as well as nocturnally. They don't hibernate during winter but may retreat to a sheltered location during freezing weather. Wyoming is near the northern edge of their range, and the Virginia opossum is both the most northerly occurring opossum species, and the only member of the marsupial family to store fat.
Voice: Opossums hiss when threatened, and both sexes also utter clicking sounds in aggressive situations; females also use clicking sounds when communicating to offspring. These sounds stimulate ambulatory babies to follow her, either clinging to her back or belly, or by running along beside her.
Status: Opossums are local in southeastern Wyoming but are most common in cities, judging from the number of run-over carcasses that can be seen along streets. They have poor emergency responses to oncoming vehicles, often dazzled by headlights into a fatal "freezing" stance.
Habitats and ecology: Opossums are nearly omnivorous in their diets, selecting foods ranging from carrion to live animals. They often eat grain (such as corn) in the winter and gradually shift to insects, other invertebrates, bird eggs, small mammals, and plant materials during summer. During fall and early winter they also consume fruits and berries.
Breeding biology: Females become sexually mature during their first year, and in Wyoming have two litters per year, one in late January or February and another in May or June. The

gestation period is about 13 days, after which 4 to 23 embryo-like young emerge in rapid succession, either singly or in groups. Their forelegs are developed well enough for them to climb up and into the female's marsupium, where they try to find one of the approximately 13 teats, although some of these are nonfunctional. The average number of pouched young is about 8.5. As the young grow they are able to climb well, and their prehensile tail helps them maneuver in trees (but they can't hang upside-down from a branch by their tail alone, as charmingly portrayed in Pogo cartoons by Walt Kelly, at least not for more than a few seconds, as I have personally determined). Opossums are short-lived, with few surviving more than two years. Their tendency to "play possum" by becoming immobile and feigning death when threatened is of no survival value in the modern car-dominated world. Presidents and politicians might be able to survive indefinitely in the twenty-first century by dissembling but not opossums.

Suggested reading: Clark and Stromberg, 1987; Armstrong, Fitzgerald, and Meaney, 2011; Buskirk, 2016.

Family Soricidae (Shrews)

Cinereus Shrew (Masked Shrew). *Sorex cinereus*.
Widespread.

Identification: Length: 87–109 mm (3.4–4.3 in); tail 35–39 mm (1.4–1.5 in). Weight: 3.5–6 g (0.1–0.21 oz). This tiny shrew is uniformly brown above, becoming paler on the sides and underparts. Its nearly naked tail (maximum length 50 mm) is slightly bicolored and darkly tipped. It is extremely difficult to distinguish from its near relatives, such as the also common montane shrew, without examining their teeth, but the latter has a longer tail (50–70 mm) that is more clearly bicolored and with a tufted brown tip. The cinereus shrew is also extremely similar to the vagrant shrew, which occurs only in extreme western Wyoming (Lincoln County), and to the Hayden's shrew, which has a shorter (30–42 mm) brown-tipped tail and occurs only in extreme northeastern Wyoming (Crook and Weston Counties). In spite of its alternate common name, the cinereus shrew's face is not "masked," and its name cinereus (meaning "ashen") is scarcely better but might refer to the abdominal tint.

Voice: Although probably all shrews are highly vocal, the calls are weak and very high-pitched or of ultrasonic frequencies, and in some species, such as the vagrant shrew, are known to be used for echolocation (Gould, Negus, and Novicki, 1964).

Status: This shrew extends from eastern Siberia and Alaska south through Canada to New Mexico and Georgia. In Wyoming, it is one of the most widespread of Wyoming's shrew species, occurring almost everywhere except perhaps the northeast, where it is replaced by the almost identical Hayden's shrew. Both of these species avoid dry habitats, as do most of Wyoming's shrews.

Habitats and ecology: As for all shrews, the high metabolic rate of these tiny mammals forces them to be active and foraging almost constantly, and they consume any prey that they can kill, especially insects, spiders, and earthworms. In Wyoming, their habitats extend from subalpine spruce-fir downward through the mid-montane, lodgepole, and aspen zones, and even into sagebrush, although moist habitats are their primary and preferred environment. Their high metabolic rate prevents shrews from hibernating, and they remain active throughout the day, running nimbly, swimming, and jumping over distances as much as about eight inches. They live solitarily, with home ranges of less than 0.05 acre.

Breeding biology: Females have one or two litters per year, between March and September. The male remains with the female through pregnancy and for part of the post-partum period.

Suggested reading: Clark and Stromberg, 1987; Armstrong, Fitzgerald, and Meaney, 2011; Buskirk, 2016.

Western Water Shrew. *Sorex* (*palustris*) *navigator*.
Dispersed.

Identification: Total length 138–168 mm (5.4–6.6 in); tail 69–85 mm (2.7–3.3 in). Weight: 9–18 g (0.3–0.6 oz). This is the largest shrew in Wyoming. Its pelage is blackish dorsally with underparts that range from silvery white to gray, brownish, or blackish. The tail is strongly bicolored, and the large hind feet have unique fringes of stiff hairs and partially webbed toes.

Voice: Although water shrews emit continuous high-pitched sounds during their exploration, they evidently do not use echolocation. However, they can detect prey by smell, touch, shape, and movements, even without the aid of vision, which is generally poor in shrews.

Status: This shrew is found in Wyoming where clear, cold streams are present, especially if wet meadows are nearby and where large boulders and large tree roots provide crevices and ledges. It occurs in all the state's major mountain ranges except for the Bighorns and is notably common in the Yellowstone region.

Habitats and ecology: This shrew's ability to swim well and to skitter along the surface of the water for up to about five feet is achieved by buoyancy created from the trapping of small bubbles among the body fur and along the stiff hairs of the hind feet, along with foot webbing. When swimming, the animal appears to be a silvery bubble because of the countless small bubbles trapped among its water-resistant fur. In the water, the shrew catches and eats insect larvae, adult aquatic invertebrates, and even small fish. One observer (Streubel, 1995) found that in captivity a water shrew ate 1 ounce of food per 0.7 ounce of body weight per day, an amazing daily food intake, although other estimates have been substantially lower than this (Wilson and Ruff, 1999). In any case, the shrews most actively forage at night, especially shortly after dusk, with a second peak shortly before sunrise. They remain seasonally active through winter, under the cover of snow.

Breeding biology: Breeding occurs in Wyoming from January to August, with up to two litters produced per year. Most reproduction apparently occurs before June, with the young out foraging independently by July. Litters range from 5 to 8 young. Their maximum lifespan is about 18 months.

Suggested reading: Conaway, 1952; Clark and Stromberg, 1987; Carbyn, Fritts, and Seip, 1995; Armstrong, Fitzgerald, and Meaney, 2011; Catania, 2013; Buskirk, 2016.

Family Vespertilionidae (Vesper Bats)

Big Brown Bat. ***Eptesicus fuscus***. Widespread.

Identification: The big brown bat is the largest bat in Wyoming. It has uniformly brown fur and among North American bats is second only in size to the hoary bat. Length: 90–138 mm (3.5–5.4 in); wingspan 32–40 cm (12.5–15.7 in). Weight: 12–20 g (0.42–0.7 oz).

Voice: This is one of the many bats that use precise echolocation behavior for general orientation and also to locate aerial prey, which mostly consists of beetles but also includes ants, bees, and flies. Moths are infrequently taken, and it is known that some moths of the families Arctiidae and Noctuidae can create sounds that mimic and interfere with a bat's echolocation ability (Fullard and Fenton, 1979; Corcoran, Barber, and Conner, 2009). Moths might also fly in loops, make noises that might startle a bat, or dive to avoid capture. When foraging, the bats utter intense search calls and fly in stereotyped flight paths. They are able to simultaneously monitor background obstacles while also tracking small, often evasive, insects. One Maryland study documented that a big brown bat maternity colony of 150 bats ate 38,000 cucumber beetles, 16,000 June beetles, 19,000 stinkbugs, and 50,000 leafhoppers in a summer! Bats have a structure at the base of each ear called a tragus, which seals off the auditory canal each time a sound signal is sent out, and then quickly relaxes to allow for receiving the echo. Changes in the frequency and strength of the echo enable the bat to determine the prey's distance and direction of flight, and probably its identity.

Status: This is one of the most abundant bats in the state.

Habitats and ecology: Big brown bats are frequently found around buildings, rock crevices, caves, and hollow trees. Buildings

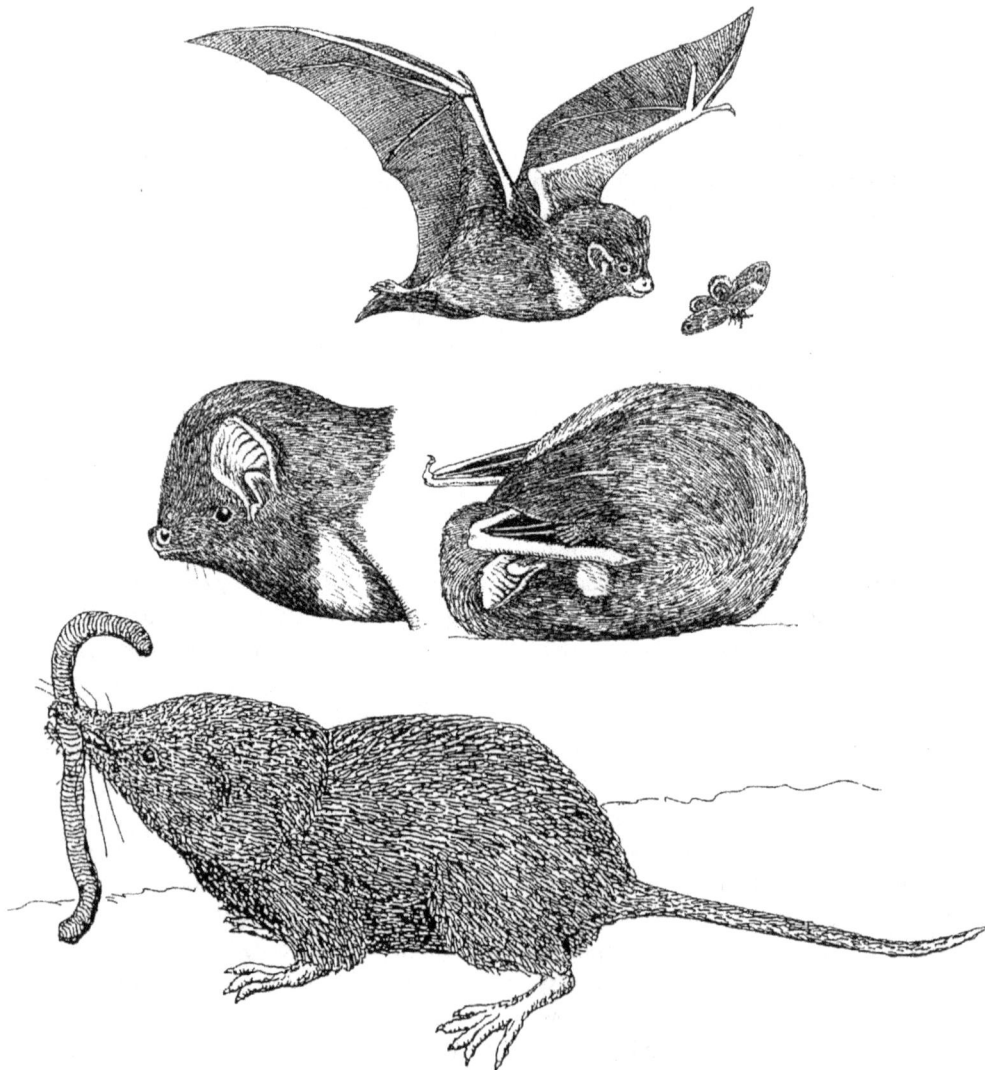

Eastern red bat: in flight, head detail, and sleeping (*upper*); cinereus shrew with earthworm (*below*)

serve both as summer daytime roosting sites and as hibernation sites. They are often seen around streetlights, indicating that visual hunting may also be important. They hibernate in winter, during which they might lose as much as a third of their body weight through metabolization. They have an amazing ability to navigate and, although they are not migratory, can return to their home roost over distances from as far away as 450 miles.

Breeding biology: Breeding behavior in this bat begins in the fall, with mating activities occurring as late as October, although ovulation and fertilization do not occur until early spring. Their young are born in May, when colonies of up to 300 nursing females might assemble. While the females are out foraging, their young are left hanging for up to an hour, although sometimes they lose their grip and fall fatally to the ground. Other females forage while carrying their young, which must cling tightly to their mother's teats to avoid the same fate. They are born at an unusually large size and can fly when they are only three to four weeks of age. Adults are known to have lived as long as 18 years.

Suggested reading: Barbour and Davis, 1969; Hill and Smith, 1984; Clark and Stromberg, 1987; Armstrong, Fitzgerald, and Meaney, 2011; Buskirk, 2016.

Hoary Bat. *Lasiurus cinereus*. Widespread.

Identification: This largest bat in Wyoming is easily recognized by its size and the overall "hoary" (frosted) appearance of white-tipped but otherwise dark brown fur. Length: 120–145 mm (4.7–5.7 in); wingspan 380–410 mm (14.9–16.5 in). Weight: 18–35 g (0.6–1.2 oz).

Voice: Hoary bats are usually audible when they are in flight uttering an audible chattering sound. However, Corcoran and Weller (2018) documented that these bats can fly within ten feet of sensitive microphones without producing detectable echolocation calls. Acoustic modeling indicated that hoary bats sometimes fly without using echolocation, probably to reduce echolocation output to avoid eavesdropping by conspecifics during the mating season. Their findings might help explain why tens of thousands of hoary bats are killed by wind turbines each year.

Status: These bats are among the most widespread of all North American bats but are solitary and rarely seen except during migration. Large swarms have been seen during August migration in eastern Wyoming. It is a strong flier and has reached Hawaii, where it is the only native land mammal and is recognized as an endemic subspecies.

Habitats and ecology: This large bat usually roosts in trees, using deciduous species in the eastern United States and favoring large ponderosa pines in the western states. They have been found from greasewood flats and the woodland zone to the Douglas-fir zone. They often roost in trees near the edges of clearings, where they are protected from above by leafy cover but can easily fly in from below. Their foods are mostly insects, especially moths, but they also take beetles,

flies, and wasps. They are also known to prey on several other species of bats. Hoary bats are strongly migratory, possibly flying distances in excess of 1,000 miles. Spring migration occurs in May and June, with females arriving on the breeding grounds about a month earlier than males (Armstrong, Fitzgerald, and Meaney, 2011).

Breeding biology: Little is known of the reproductive biology of this species, but it seems that mating probably occurs on the wintering grounds. Males evidently do not follow the females all the way to their northern breeding grounds. Twins are the usual litter size, but it can range from one to four. The young are carried by the female until they are 6 to 7 days old; they are able to fly by 34 days of age.

Suggested reading: Barbour and Davis, 1969; Hill and Smith, 1984; Clark and Stromberg, 1987; Armstrong, Fitzgerald, and Meaney, 2011; Buskirk, 2016; Corcoran and Weller, 2018.

Eastern Red Bat. *Lasiurus borealis*. Highly localized.

Identification: Male eastern red bats are easily identified by their unique bright orange-red pelage; females are reddish brown with white frosting. White patches are present at the shoulder and at the base of the thumb. Females are very slightly larger than males. Length: 107–128 mm (4.2–5.0 in); wingspan 28–33 cm (11.0–13.0 in). Weight: 5–16 g (0.2–0.6 oz).

Voice: Schmidt-French, Gillam, and Fenton (2006) found that echolocation calls of adults and subadults hunting flying prey differed in frequency components. Differences in the duration of their echolocation calls coincided with the environment (in open versus restricted spaces). While nursing, both mothers and pups make a vibrational humming sound, the components of which differed between mothers and young. However, calls produced by pups searching for their mothers' nipples showed little evidence of sonic individuality. In general, the female-young calls and calling behavior of eastern red bats, which are solitary and foliage-roosting, appear to differ from those of gregarious species that roost in more sheltered situations, since bats roosting alone in foliage may rely more on spatial memory than on acoustic cues to find their young.

Status: In Wyoming, red bats are mostly limited to the southeastern parts of the state. They are tree- or shrub-roosting, either as singles or in small female-litter groups; sometimes herbaceous plants such as sunflowers are also used for roosting.

Habitats and ecology: Red bats are solitary and occur at low densities of less than 0.2 per acre (Clark and Stromberg, 1987), and their home ranges are thought to be only about two acres in area. They migrate south in winter, where they hibernate lightly, emerging on warm days.

Breeding biology: Mating in this species occurs during fall, often while the pair is in flight. The young are not born until about mid-June, while the female is roosting in foliage. Red bats typically have three or four young and are among the few American bats having four nipples. However, they apparently do not try to fly with their young attached, as females have been found on the ground with attached young, apparently having fallen and being unable to take flight.

Suggested reading: Barbour and Davis, 1969; Hill and Smith, 1984; Clark and Stromberg, 1987; Armstrong, Fitzgerald, and Meaney, 2011; Buskirk, 2016.

Silver-haired Bat. *Lasionycteris noctivagans*.
Widespread.

Identification: The silver-haired bat is one of the easier species of small bats to identify. Its black fur is tipped with silver, except for its black face, and its ears are edged with yellowish. Length: 90–104 mm (3.5–4.1 in); wingspan 127–132 cm (5–5.2 in). Weight: 7–15 g (0.24–0.52 oz).

Voice: Little information. Its echolocation is still unstudied, but the sonograms resemble those of other species known to use echolocation. Given its tendency to forage late at night, its echolocation skills must be very good.

Status: These bats occur statewide in a variety of habitats and are probably abundant, but they are usually found in or near coniferous or deciduous forests close to water. In Colorado, they have been seen from about 4,500 to 9,500 feet, and in Wyoming they are often encountered during their fall migration during August and September.

Habitats and ecology: These bats forage late into the evening hours and nighttime, up to eight hours after sunset. Their flight is slow and erratic, and they forage mostly on moths but also on beetles, wasps, flies, and other winged insects. Summer roosts of this species are often trees, especially aspen and conifers. Mature trees or snags in the early stages of decay are preferred sites. The bats are migratory in Wyoming and perhaps winter in the Southwest; records in Colorado extend from March until early October. Silver-haired bats spend the winter as far south as northern Mexico, roosting or hibernating under loose bark, in rock crevices, leaf clumps, woodpecker holes, bird nests, sheds, garages, or other outbuildings.

Breeding biology: Like many of the Wyoming bats, silver-haired mating occurs in the fall, and the sperm remains viable through winter in the female's reproductive tract. Ovulation usually occurs in April and May, and gestation lasts 50 to 60 days. Typically two young are born and are able to fly by three or four weeks of age. The lactation period lasts about 36 days, and the precocial young are able to breed during their first summer of life. Silver-haired bats have been documented as living up to 12 years, but in one study the mean age of a population of wild individuals was 2 years.

Suggested reading: Barbour and Davis, 1969; Hill and Smith, 1984; Clark and Stromberg, 1987; Schmidt, 2003; Armstrong, Fitzgerald, and Meaney, 2011; Buskirk, 2016.

Little Brown Myotis. *Myotis lucifugus*. Ubiquitous.

Identification: This is one of the very large group (about 16 US species) of small-eared *Myotis* bat species (*Myotis* is Latin for "mouse-eared") that are probably impossible to identify without in-hand examination. It is very small with glossy yellow-brown fur and yellow-buff underparts. Length: 90–100 mm (3.5–3.9 in); wingspan 22–27 cm (8.7–10.1 in). Weight: 4–9 g (0.14–0.31 oz).

Voice: This species' vocalizations are well studied. Barclay, Fenton, and Thomas (1979) determined that at least ten vocalizations are used. Echolocation calls are used for orientation, avoiding near objects, and reconnoitering during first flights of young. Nonecholocation calls serve during agonistic encounters, for roost-space protection, in mother-infant interactions, and during mating behavior. Other studies (Kazial, Pacheco, and Zielinski, 2008) indicate that individual identities can be detected acoustically in flying bats, and individual identity, state of lactation, and age category can also be determined from calls by nonflying individuals.

Status: This species may be the most abundant bat in Wyoming and is widespread over nearly all of North America. It often roosts in rafters and is one of the common bats that enters houses when searching for a roosting site, the latter committed especially by inexperienced youngsters. During summer, nursing colonies of up to as many as 800 individuals have been found.

Habitats and ecology: The little brown bat often forages over water, flying slowly and erratically, catching insects. It especially preys on moths but also other insects as small as mosquitoes, the latter caught at a rate of up to 600 per hour! These bats forage through the night, and the chitin in the exoskeletons of the insects they eat is partially digested with the aid of enzymes produced by symbiotic bacteria in the bats' intestines. Roosting occurs in many different kinds of sites, such as in trees, buildings, and mines and under rocks.

Breeding biology: Mating occurs in late fall, and both sexes are promiscuous, with males sometimes copulating with torpid females. The sperm is stored, and ovulation might occur a few days after the female's emergence from hibernation. The gestation period is 50 to 60 days, with a single infant being born in June or July. The young can fly within three weeks and are at adult weight by a month. These tiny bats are remarkably long-lived, with many reaching 10 years of age; there is even a record of survival to 30 years.

Suggested reading: Barbour and Davis, 1969; Hill and Smith, 1984; Clark and Stromberg, 1987; Armstrong, Fitzgerald, and Meaney, 2011; Buskirk, 2016.

Townsend's Big-eared Bat. *Corynorhinus townsendii*. Widespread.

Identification: This easily identified pale brown bat has huge ears (30–39 mm long) and unique fleshy, glandular growths on each side of its snout. (It was once called the "lump-nosed bat," and its Latin name translates as "Townsend's club-shaped nose."). Length: 90–112 mm (3.5–4.4 in); wingspan 30–34 cm (11.8–13.3 in). Weight: 9–13 g (0.31–0.42 oz).

Voice: This species is a "whisper bat," meaning that it echolocates at much lower sound amplitudes than most bats, and thus it is difficult to obtain sound recordings. For humans

pinna

tragus

A

G

B

C

H

D

E

F

Heads of
(A) long-legged bat,
(B) long-eared bat,
(C) silver-haired bat,
(D) big brown bat,
(E) Townsend's
big-eared bat, and
(F) hoary bat.
Also shown are
(G) ear anatomy of
little brown bat and
(H) flight profile
of eastern red bat.
After sketches by
B. Siler (A–G) from
Armstrong (1975)
and (H) from
Johnsgard (2003).

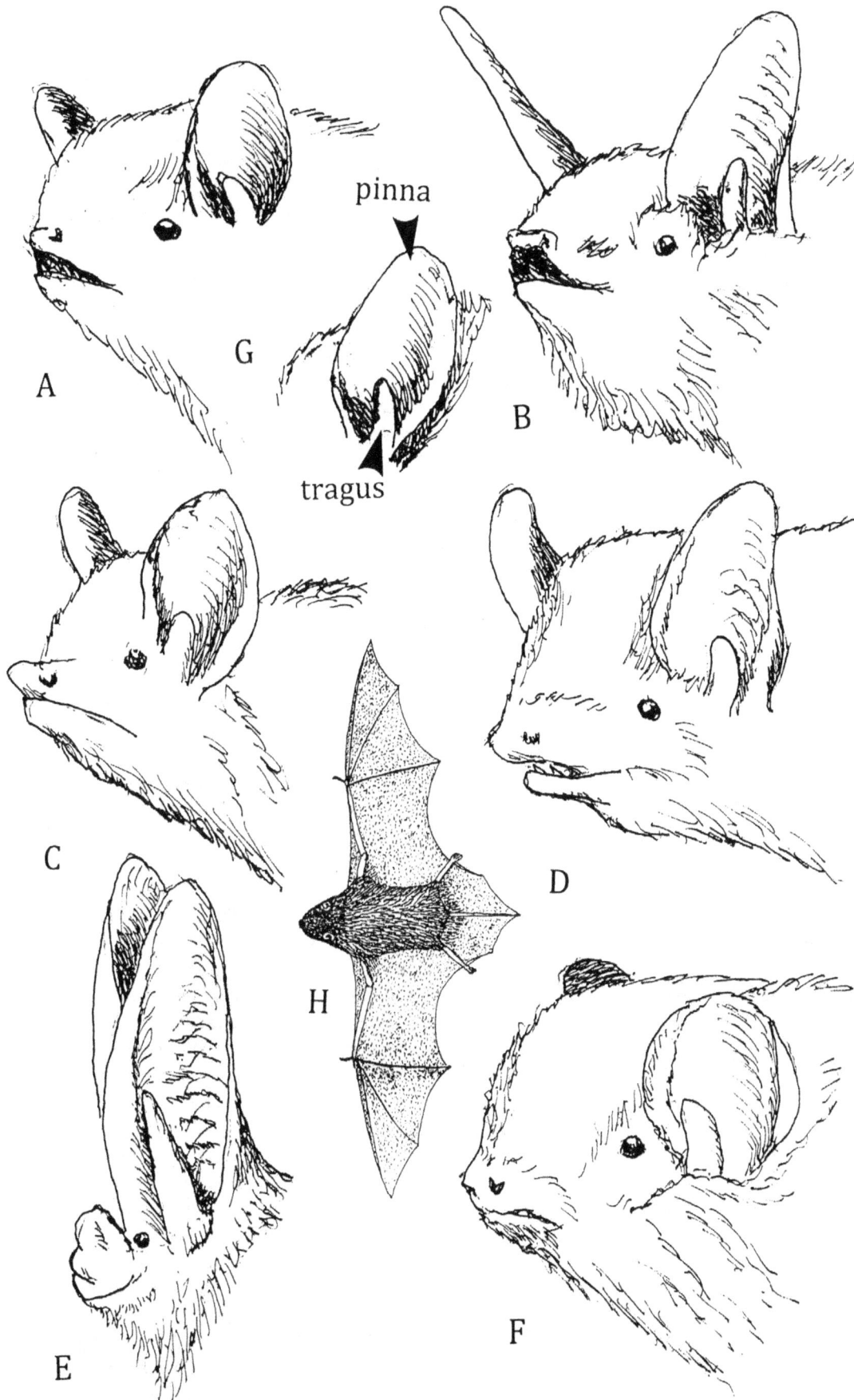

to hear the sounds these bats produce while echolocating, it is necessary to reduce the playback speed to one-tenth the recording speed. Their sound pulses range in duration from several to a few hundred per second. Ultrasonic frequencies above 20 kilohertz (kHz; 20,000 cycles per second) are used for prey detection and identification, since only extremely high-frequency sounds will be reflected back from tiny objects such as moths as small as 3 to 10 mm in length, whereas lower-frequency sounds serve for general in-flight orientation such as avoiding large objects. Vocalizations below 20 kHz are for social interactions such as adjusting spacing behavior in colonies, mother-offspring interactions, aggression, and warnings.

Status: This bat is common in Wyoming, especially in deserts, woodlands, and dry coniferous forests. The species might be active as early as late afternoon but usually begins after dark, with most foraging occurring four to five hours after sundown. They are a hibernating species, often choosing old mines or caves as hibernacula. They are also fairly sedentary, not moving far from their summer home ranges to their winter quarters. Home ranges have been estimated at 200 to 5,900 acres (0.8–24 km^2) (Armstrong, Fitzgerald, and Meaney, 2011).

Habitats and ecology: The big-eared bat's huge ears are used not only to transmit sound into its ear canal but possibly to impart lift during flight; when sleeping, the bat's ears are rolled back over its head and resemble ram horns. Rather than specializing entirely on moth eating (about 80 percent of their diet), these bats also eat caddisflies, flies, and other insects.

Breeding biology: The "lumps" on the side of the nose of this bat are glandular, and males scent-mark the females they are courting prior to copulation. Mating occurs during fall and winter while the bats are in their hibernacula, and sperm is stored until spring. Gestation lasts 56 to 100 days, with a single offspring being born in May or June. Initial flight occurs at about three to four weeks of age, and weaning is completed by six weeks.

Suggested reading: Barbour and Davis, 1969; Hill and Smith, 1984; Clark and Stromberg, 1987; Armstrong, Fitzgerald, and Meaney, 2011; Buskirk, 2016.

Family Ochotonidae (Pikas)

American Pika. *Ochotona princeps*. Dispersed.

Identification: The American pika is an easy species to identify—it is the only small rodent-like mammal at high altitudes that has large, rounded ears and no apparent tail. Whistled barking notes are frequently uttered in alarm or for other reasons and provide certain identification. Pikas are often found in talus rock piles, but in some areas they also live among old lava beds. The name pika is from a Siberian dialect, and the generic name is also a Latinized version of a Mongolian term for this group of mammals, which are notably common in Mongolia. Many species of pikas inhabit northeastern Asia,

Pika

where they live mostly in subarctic steppe habitats. Length: 157–203 mm (6.1–8.0 in). Weight: 119–176 g (4.2–6.2 oz).

Voice: Pikas are highly vocal. They utter short, high-pitched whistled calls—*hneee*—as alarm signals and also use similar but longer calls as sexual attraction signals. Such longer "songs" are most frequent during the spring breeding season, whereas a peak of short calls in late summer occurs during a period of late competition and aggression over food resources. Studies conducted from Wyoming to New Mexico have shown that regional dialects exist in pika vocalizations (Somers, 1973); additionally, individuals can be sonographically distinguished on the basis of their calls. Furthermore, males can also be individually identified by the secretions from their cheek glands, which are rubbed against rocks within the animal's territory.

Status: Pikas occur on the mountains of northwestern Wyoming as well as in the Bighorn, Sierra Madre, and Medicine Bow ranges from about 8,500 to 11,200 feet (2,500–3,600 m) elevation. With global warming there is concern that as pikas move upward in response to global warming, they will increasingly run out of suitable habitat on lower mountains that are losing their subalpine and alpine habitats. In the Great Basin of Nevada and adjacent states, 7 of 25 populations appear to have become extinct during the past century, and several western subspecies are now imperiled (Armstrong, Fitzgerald, and Meaney, 2011).

Habitats and ecology: Territories range in size seasonally from 0.7 to 0.17 acre (300–680 m^2), and home ranges vary from 0.2 to about 0.5 acres (860 to about 2,000 m^2). Populations in Colorado range from 0.5 to 3.9 animals per acre (1.4–9.8 per hectare) (Armstrong, Fitzgerald, and Meaney, 2011). Within territories, storage places for hay piles of leaves and stems of forbs, sedges, and shrubs are hidden under rocks. These are strongly guarded and defended, although some thefts of materials among neighbors are frequent. Territorial disputes reach their highest levels in fall, as the animals must acquire sufficient food reserves to get them through winter. Pikas cannot digest the cellulose in their totally plant-based diet, but intestinal bacteria are able to; then the animals re-ingest their own droppings and are able to use the enzymes within these excreted materials to digest some of their cellulose intake.

Breeding biology: Breeding behavior occurs in spring, from April through June in Colorado. Gestation lasts 30 days and the young are born in Wyoming as early as April. From one to five babies compose a litter, and two litters might be produced during the short montane summer. Females become sexually mature during the first summer of their lives. Pikas typically live about three years in the wild but have survived up to seven years in captivity.

Suggested reading: Orr, 1977; Clark and Stromberg, 1987; Armstrong, Fitzgerald, and Meaney, 2011; Buskirk, 2016.

Family Leporidae (Hares and Rabbits)

Snowshoe Hare. *Lepus americanus*. Dispersed.

Identification: Snowshoe hares are closer to cottontails than to jackrabbits in size. They have short ears like those of the cottontail, long hind legs like those of the jackrabbit, and mostly brown (summer) or white (winter) tails and body pelage. During about two and a half months in spring (March to May) and for nearly three months in fall the animals are rather bicolored, with their bodies partly in their white winter pelage and the remainder in summer brown. Head and body length: 36–52 cm (16 in); tail 25–55 mm (1–2.2 in). Weight: 1.4–1.8 kg (3–4 lb).

Voice: Normally silent, snowshoe hares sometimes thump on the ground with their hind feet, as do many other rabbits and hares.

Status: Snowshoe hares are common in the coniferous woodlands and forests of Wyoming (except in the Black Hills), especially at higher elevations up to 11,000 feet. Populations vary greatly in abundance over time, undergoing cycles of about 8 to 11 years duration; during years of high abundance they might rarely reach densities of more than 100 animals per acre, but typically they range from about 1 to 3 per acre. These cycles are of widespread occurrence and tend to be synchronized over broad areas, and their fundamental causes are still not understood. Certainly predation is influential, as these hares are important prey for mammalian predators such as lynx, bobcats, foxes, minks, and coyotes and larger owls such as great horned owls.

Habitats and ecology: Snowshoe hares favor shrubby habitats, with dense herbaceous and woody vegetation that provides both food and protective cover. Twigs and bark are eaten during winter, but succulent plants such as clovers are favored as well as the buds and leaves of willows, aspens, and poplars.

They typically forage at night and rest in heavy cover during the day. Home ranges may vary from about 5 to 37 acres for males and about 3 to 4 acres for females. There is no evidence that the animals are territorial. Male home ranges overlap with those of several females, and sometimes females are courted simultaneously by several males, which jump over the females and drum their feet in competitive attempts to attract her attention.

Breeding biology: Courtship begins in March and continues through August, with females producing two to four litters per breeding season. Females become sexually mature in their second year. The gestation period is 38 days, and the litter size varies greatly from one to six young with litters averaging the largest in the northern parts of the species' range. Only the female tends the young, which, as in all hares, are born well furred and with their eyes open. Within a week they are moving about and by three weeks are ready to depart from their natal nest, traditionally called a "form." When fully grown they can run as fast as 32 miles per hour and can change course quickly, so few predators can run them down in the open field. Nonetheless, up to 40 percent of the population might be taken by predators during winter, so the life expectancy of snowshoe hares is very short.

Suggested reading: Aldus, 1937; Lawrence, 1955; Clark and Stromberg, 1987; Hodges, Mills, and Murphy, 2009; Armstrong, Fitzgerald, and Meaney, 2011; Buskirk, 2016.

White-tailed Jackrabbit. *Lepus townsendii*.
Ubiquitous.
Black-tailed Jackrabbit. *Lepus californicus*. Slightly
dispersed.

Identification: The white-tailed jackrabbit is the largest jackrabbit in North America. I have sometimes thought I'd flushed a white-tailed deer fawn after being suddenly startled by a bounding white-tailed jackrabbit. The two species of jackrabbits differ most obviously in their tail color, the black-tailed's being black on the tail's upper half rather than entirely white. They also differ in ear length, with the white-tailed's being relatively longer, albeit very slightly (white-tailed, 110–130 mm versus black-tailed, 96–114 mm). The white-tailed jackrabbit has a total length of 22 to 26 inches (55–65 cm) and weighs 6.6 to 13 pounds (3–6 kg), whereas the black-tailed has a total length of 18 to 25 inches (46–63 cm) and weighs 5.3 to 8.8 pounds (2.4–4 kg).

Voice: Like all rabbits and hares, these two species are normally silent but do scream in pain or distress, as when captured by a predator, and the young similarly scream in distress or when frightened. Females also vocalize softly to their young. Probably all members of the rabbit-hare group have glands under the chin, in the groin area, and in the anal area, the secretions of which are presumably as important during close social interactions as vocalizations.

Status: The black-tailed jackrabbit is confined to the eastern third of Wyoming and favors shortgrass steppe but also uses

White-tailed jackrabbit (*left*), and black-tailed jackrabbit (*right*)

sage-steppe and greasewood scrub desert. The white-tailed jackrabbit has a statewide distribution and is common in mixed grass and shrub communities, especially sage-steppe habitat, and also occurs in meadow-like forest openings from the foothills to as high as alpine tundra. In Colorado, its altitudinal range is from about 4,000 feet (1,200 m) to more than 14,000 feet (2,400 m).

Habitats and ecology: Although the two jackrabbits are similar in many ecological respects, the white-tailed jackrabbit is more closely tied to native vegetation than is the black-tailed, even though it seems to have a greater capacity to live in many different habitat types. Shortgrass prairie is a commonly used habitat in Wyoming, but it also does well in tallgrass prairies of the eastern Great Plains, bunchgrass prairies of the Pacific Northwest, annual grasslands of California, and the desert grasslands of the American Southwest and northern Mexico. Both of these species consume a wide diversity of grasses, forbs, and shrubs, with shrubs being major winter survival foods, grasses favored in spring, and forbs important during summer. Like other rabbits and hares, they also consume some of the own feces, which allows them to digest the nutrients they had earlier eaten and were subsequently transformed by intestinal bacteria into digestible forms. Densities of black-tailed jackrabbits in Colorado remarkably range from as few as 2.5 to an amazing 85 animals per acre (0.1–34.6 per hectare), whereas white-tailed jackrabbits were estimated in one study to have densities of 0.009 animal per acre (2.2 per km²). Home ranges of black-tailed jackrabbits were judged to be 39 to 451 acres (16–183 hectares) in Colorado (Armstrong, Fitzgerald, and Meaney, 2011), which is similar to an estimate of 220 acres (89 hectares) for the white-tailed jackrabbit (Forsyth, 1999).

Breeding biology: Both species breed in spring, the white-tailed starting during March in Wyoming, with young born as late as

August. At least two litters per breeding season have been reported for the white-tailed species, and up to four litters might be produced in some parts of the black-tailed species' range (Armstrong, Fitzgerald, and Meaney, 2011). Litter sizes vary greatly, with one to nine young reported for the white-tailed and one to eight for the black-tailed. The largest litters occur in the middle of the breeding season, and litters average larger in northern parts of the species' ranges, although fewer litters per season are possible in northern regions. Gestation periods for these two species have been reported as about 40 to 45 days. The young are born fully furred and are weaned by only two to four weeks of age. Black-tailed young reportedly reach full size by two months of age, and at least in southwestern Wyoming young white-tailed jackrabbits are independent by two months but do not reach full body size until about December. Female black-tailed jackrabbits might breed in their first year, whereas white-tailed females reportedly do not breed until their second year. Perhaps these reproductive traits vary from year to year or regionally, but like several other hares both species undergo marked population cycles of about eight to ten years, for reasons that still remain elusive. Possibly annual changes in litter size and number of litters per breeding season under different ecological conditions could have marked influences on short-term population trends.

Suggested reading: Clark and Stromberg, 1987; Armstrong, Fitzgerald, and Meaney, 2011; Buskirk, 2016.

Desert Cottontail. *Sylvilagus audubonii*. Ubiquitous.

Mountain (Nuttall's) Cottontail. *Sylvilagus nuttallii*. Widespread.

Identification: Wyoming has three species of cottontail rabbits, of which the desert cottontail is intermediate in mass; it is pale gray and has the longest ears with sparse hair on the inner surfaces. The mountain cottontail is smallest in mass and is grayish brown; its short ears are densely lined with white hair. The eastern cottontail is the heaviest, with a grayish, grizzled pelage and the shortest ears; it is limited in range to extreme southeastern Wyoming (Larimer County). *Mountain cottontail:* Weight: 638–871 g (22.5–30.7 oz); ear length 55–65 mm (2.2–2.5 in). *Desert cottontail:* Weight: 755–1,250 g (26.6–44.1 oz); ear length 62–78 mm (2.4–3.0 in). *Eastern cottontail:* Weight 801–1,533 g (28.2–54.1 oz); ear length 48–60 mm (1.9–2.4 in).

Voice: Like other rabbits, cottontails are mostly rather silent except when they are in distress or pain, when a loud child-like scream is produced. However, they also utter clucking, purring, and humming sounds, the former two when eating or content, and the last by males during courtship. During aggressive situations cottontails may growl, snort, or hiss.

Status: The desert and mountain cottontails are both widespread and common. Together with jackrabbits they are commonly hunted for sport, but they don't undergo the massive changes in population densities that are common in snowshoe hares and jackrabbits. Home ranges in the desert cottontail have been estimated at 7.4 to 9.9 acres (3–4 hectares) (Mark and Stromberg, 1987), and also at 1.2 to 19.8 acres (0.5–6.0 hectares), with densities of up 6.5 animals per acres (16 per hectare) (Armstrong, Fitzgerald, and Meaney, 2011).

Habitats and ecology: The mountain cottontail favors rocky, sagebrush habitats at higher elevations among rock crevices and burrows, but it ranges from the juniper woodland zone to forest-edge situations in subalpine habitats from about 3,900 to 8,900 feet (1,200–2,700 m) elevation. It strongly favors habitats with sagebrush, which along with juniper and rabbitbrush are important winter foods, while summer foods are a variety of grasses and forbs. Populations in Oregon's juniper scrubland varied in one study from as many as 1.0 animal per acre (2.5 animals per hectare) in summer to as few as 0.02 per acre (0.06 per hectare) in winter, while in British Colombia densities ranged from 0.08 to 0.16 animals per acre (0.2–0.4 per hectare). The desert cottontail is found in lower altitude desert valleys as well as on the shortgrass plains, where they often occur with prairie dogs and use their burrows for escape hatches. They also occur in riparian woodlands, montane scrublands, juniper woodlands, and other habitats with little vegetation but having places where burrows, crevices, or other places offering escape cover are present.

Breeding biology: Desert cottontails breed from late winter through the summer months, while mountain cottontails begin breeding in early spring. Gestation in both species is 28 to 30 days, with one to eight (mountain) or three to four (desert) young born per litter. The young leave their nests at about two to three weeks of age and are independent by four to five weeks (mountain) or three to four weeks (desert). There are three to five (mountain) or two to four (desert) litters per breeding season (Mark and Stromberg, 1987). The average litter size is about four.

Suggested reading: Clark and Stromberg, 1987; Armstrong, Fitzgerald, and Meaney, 2011; Buskirk, 2016.

Pygmy Rabbit. *Brachylagus idahoensis.* Highly localized.

Identification: The pygmy rabbit is the smallest of North American rabbits and one of the rarest. It has notably small ears, a nearly invisible brown tail, and a pelage that varies in color from dark brown (summer) to grayish brown (winter). Length: 22–28 cm (9.5 in); tail: 15–24 mm (0.6–0.9 in). Weight: 373–458 g (13.1–16.2 oz). It is never found far from sagebrush.

Voice: Little information. Vocalizations have been described as including squeals, squeaks, and chuckles. They also have a pika-like "chirp" alarm call, a feature that is unusual for rabbits but that probably reflects their semicolonial behavior.

Status: This species is limited in Wyoming to Lincoln and Uinta Counties. It is considered endangered in Washington where it is known from only a single location, and it is a species of concern in Idaho. It also occurs in eastern Oregon, northern Nevada, and western Utah. It currently has federal endangered status.

Habitats and ecology: The pygmy rabbit is a sagebrush obligate species, with up to 99 percent of its food being sage, mostly or entirely big sagebrush (*Artemisia tridentata*). The animals live in small groups among sage thickets, in which they have a system of burrows and runways. Their home ranges are very small; individuals generally remain within about 100 feet of their burrow. Densities average only 0.3 to 0.6 rabbit per acre (0.7–1.4 per hectare), but these rabbits have also been reported to live in colonies of up to as many as 16 per acre (45 per hectare).

Breeding biology: Breeding behavior extends from late December through March. Gestation lasts 27 to 30 days, and litters of four to eight young are typical, averaging six. There may be as many as three litters born per year, but mortality rates are very high, as much as 88 percent annually. High rates of mortality occur during the first five weeks of life and also among adults during late winter and early spring. Captive animals have survived for as long as two years, even with only periodic access to sagebrush (Mark and Stromberg, 1987; Wilson and Ruff, 1999).

Suggested reading: Clark and Stromberg, 1987; Purcell, 2006; Armstrong, Fitzgerald, and Meaney, 2011; Buskirk, 2016.

Family Sciuridae (Squirrels, Marmots, and Prairie Dogs)

Yellow-bellied Marmot. *Marmota flaviventris.* Widespread.

Identification: This large, heavy-bodied mammal is a very large ground squirrel easily identified by its yellowish-orange underpart coloration that extends up the sides of the neck and becomes russet brown on the upperparts. Often some white to yellow markings appear on the face and over or around the nose. However, in the Teton range a melanistic variant occurs, which is almost entirely blackish, except for limited white facial markings. Other intermediate pelage color variants sometimes also occur. Length: 470–680 mm (18.5– 26.3 in). Weight: 1.59–3.97 kg (3.6–10.9 lb). Males are generally larger and heavier than females.

Voice: Whistling is well developed in marmots and serves alert, alarm, and threat functions. An undulating scream indicates fear, and tooth chattering accompanies aggression. Reportedly marmots can vocally differentiate among threats from wolves, coyotes, and eagles in their alarm calls; pikas and golden-mantled ground squirrels evidently respond to marmot alarm calls. Marmots can also reportedly differentiate the scents of various predators from nonpredators such as moose and elk (Armstrong, Fitzgerald, and Meaney, 2011). Marmot and *Marmota* are from the French "marmotte" and are derived from the Latin "mus-montanus," or "mountain mouse."

Status: The yellow-bellied marmot is a generally common species in Wyoming mountains, and it is present in all the major ranges. In Colorado, it occurs at altitudes from 5,400 to 14,000 feet (1,650–4,270 m).

Habitats and ecology: Mountain meadows at the base of talus slopes are favored habitats. There the marmots can dig a complex system of tunnels leading away from a primary burrow that is usually situated among and under rocks and boulders. These rocks and multiple short tunnels leading from the burrow give the marmots at least temporary safety from highly efficient digging predators such as grizzly bears and badgers. The burrow also provides a place for winter hibernation, which might begin as early as August in high mountains. During summer the male defends a territory that includes the territories of several females, with the most-fit males having the largest harems. These groups hibernate communally, the largest males becoming dormant first, followed by the females, and lastly the young of the year. Hibernation lasts seven to eight months, depending in part upon the late summer food supply and the time of spring snowmelt. Home ranges vary greatly in size, averaging about 2.5 acres (1 hectare), as determined by the density of males and the distribution of females, but many marmots live in semi-colonial communities. Some females live monogamously with solitary, low-ranking males, and these females produce a larger number of offspring than those in harems but lack the advantages of protection from predators that is provided by group living (Forsyth, 1999).

Breeding biology: Shortly after emerging from hibernation mating occurs, although yearling females do not breed, nor do the majority of two-year-old females. Gestation lasts four to five weeks and is followed by a lactation period of another three to four weeks. The average litter size is about four. The young emerge at about a month of age and grow rapidly. Littermates den together their first winter and probably disperse as yearlings. Yearling males are forced to leave their natal home and face attack from older males if they fail to do so. Some females also disperse, but others remain behind and become part of the breeding colony. In one population study of 1,000 animals, the oldest male was 9 years old and the oldest female was 15, but mean life expectancy at birth was less than two years for both sexes (Armstrong, Fitzgerald, and Meaney, 2011).

Suggested reading: Armitage, 1962, 1998; Clark and Stromberg, 1987; Armstrong, Fitzgerald, and Meaney, 2011; Buskirk, 2016.

Least Chipmunk. *Tamias minimus*. Ubiquitous.

Identification: The least chipmunk is the smallest chipmunk in Wyoming and is easily recognized by its long tail (almost as long as the head plus body length) and the prominent striping on its head and back. Five dark and four light stripes occur on the back and sides—the middle dark stripe extends to the tip of the tail, and the outermost pair of dark and white stripes are very distinct. It has a white belly and holds its tail vertically when it runs. Total length: 190–212 mm (7.5–8.3 in); tail: 74–91 mm (2.9–3.6 in). Weight: 29–55 g (1–1.9 oz).

Voice: Apart from it chipping calls, little has been written on the vocalizations of this species, perhaps because it so closely resembles and co-occurs with about ten other similar western chipmunks. Its calls are known to advertise territory, indicate alarm, and attract mates. The vocalizations are uttered in synchrony with body posturing and rhythmic movement of its tail. Apart from the chip, which is uttered almost constantly, there are also chipper and chuck calls as well as a bird-like trill. Youngsters squeal and so do adults occasionally (Wilson and Ruff, 1999).

Status: The least chipmunk is a common species statewide but is most common in sagebrush and on dry rocky slopes. Its altitudinal range extends in the central Rocky Mountains from 5,000 to 12,000 feet (1,525–3,600 m), and from lowland semidesert and foothill woodlands through montane shrublands, montane forest edges, aspen groves, and even to alpine tundra, where few if any other chipmunks are to be found. Densities no doubt vary greatly, but in a provisioned population in Rocky Mountain National Park there were 22 per acre (54.3 per hectare) (Armstrong, Fitzgerald, and Meaney, 2011).

Habitats and ecology: All chipmunks are seedeaters, and the least chipmunk especially favors conifer seeds; secondarily it also consumes grass and forb seeds. It also eats flowers, leaf buds, leaves, mushrooms, other fungi, and even insects, birds' eggs, and carrion. It is solitary and territorial but is able to coexist with other larger chipmunk species. Least chipmunks are diurnal in activity, infrequently climb trees or shrubs, and spend much of their lives in underground burrows and rock crevices. They dig and use hibernacula, provisioned with a good supply of seeds they can eat when they periodically arouse. Hibernation might extend from September until April, a time period of reduced metabolic rate that perhaps allows them to live longer lives than some other nonhibernating mammals of similar size (Wilson and Ruff, 1999).

Breeding biology: Least chipmunks have a single litter per season, with breeding extending in Wyoming from mid-March through mid-May. Females enter estrus almost immediately after emergence from hibernation, and mating soon follows. Gestation lasts four weeks, and the litter size is four to six. Within two weeks the eyes of the young have opened and they are well furred. By 30 days they are weaned and emerging from their natal burrow. Lifespans are relatively long for these tiny animals, with some individuals known to have survived for as long as six or seven years (Wilson and Ruff, 1999).

Suggested reading: Skryja, 1970, 1974; Clark and Stromberg, 1987; Armstrong, Fitzgerald, and Meaney, 2011; Buskirk, 2016.

Uinta Chipmunk. *Tamias umbrinus*. Slightly dispersed.

Identification: The Uinta chipmunk is Wyoming's largest widespread chipmunk. It differs from all the others in that its outermost lateral stripe is white rather than blackish brown, and immediately below that the fur is rufous brown. Also, its dark dorsal stripes are brown rather than blackish. It has white underparts and a uniformly brown tail. Total length: 200–243 mm (7.9–9.6 in), tail: 90–115 mm (3.5–4.5 in). Weight: 55–80 g (1.9–2.8 oz).

Voice: The vocalizations of this species are reportedly very similar to those of the least chipmunk.

Status: In Wyoming, this species exists as three isolated populations that are considered separate subspecies and are associated with three montane regions, the Greater Yellowstone ranges of northwestern Wyoming, the northern perimeter of the Uinta Mountains in Sweetwater County, and the Sierra Madre–Medicine Bow ranges in Carbon County. There the animals occupy forested mountains from the ponderosa pine zone to subalpine spruce forests. In Colorado, its range in elevation is from 7,000 to 11,200 feet (2,135–3,600 m).

Habitats and ecology: This species is most closely associated with ponderosa pine forests, but it also occupies lower-elevation juniper woodlands and higher-elevation subalpine forests. They are not common in lodgepole pine forests. Rocky areas and talus slopes are also favorite habitats. They eat the usual assortment of conifer seeds as well as the seeds of many shrubby species, such as junipers and chokecherries, along with the buds, pollen, and fruits of various shrubs and trees. They also eat fungi, insects, and birds' eggs. They spend a greater amount of time in trees than do most chipmunks and more aggressively defend territories than most. They are diurnal and probably are partial hibernators, with supplies of winter foods stored in their hibernacula. Home ranges in Colorado's Front Range were found to be 5 to 12 acres (2–5 hectares) (Armstrong, Fitzgerald, and Meaney, 2011).

Breeding biology: In Colorado, these chipmunks hibernate by early November (probably as early as October in Wyoming), but breeding begins shortly after spring arousal, which in Wyoming occurs by May. Gestation requires about 30 days, and a single litter of three to five young are produced each breeding season. Sexual maturity probably is attained in the year following birth, but little specific information on breeding biology is available from Wyoming (Clark and Stromberg, 1987).

Suggested reading: Clark and Stromberg, 1987; Armstrong, Fitzgerald, and Meaney, 2011; Buskirk, 2016.

Northern Flying Squirrel. *Glaucomys sabrinus*.
Slightly dispersed.

Identification: The northern flying squirrel is a very easy species to identify, but it is nocturnal and shy, so chances of seeing it in the wild are slight to nil. At rest its enormous ebony-black eyes, its sharply divided brown upperparts and white underparts separated by a loose skin fold, and its flattened bushy tail are unique. Total length: 290–315 mm (11.4–12.4 in); tail: 129–142 mm (5.1– 5.6 in). Weight: 105–170 g (3.7–6.0 oz).

Voice: In addition to some audible vocalizations, including chucks and chattering notes, flying squirrels also emit ultrasonic calls ranging between 51 and 38 kHz (Murrant et al., 2013). These consist of frequency-modulated (FM) calls of several types, although their possible functions are still not well understood. So far there is no evidence they are used for navigation by echolocation while gliding, although this seems a valid possibility. There is also the possibility that they serve for cryptic

Northern flying squirrel, adult

communication above the hearing range of owls, which are among their major predators.

Status: In Wyoming, flying squirrels occur only as isolated populations in the Black Hills, Greater Yellowstone region, and Sweetwater County. In Utah, they occur at elevations of 7,900 to 10,200 feet (2,400–3,100 m) in mixed coniferous forest and cottonwoods. Probably tall, mature forests are preferred for providing the best gliding opportunities. Almost nothing is known of regional population densities or home ranges. Home ranges are generally believed to average about 2 to 3 acres (0.8–1.2 hectares) but may be as large as 37 to 50 acres (15–20 hectares). Other home range estimates have been as large as several square kilometers, and density estimates have ranged from less than 2.5 to more than 25 per acre (less than 1 to more than 10 per hectare).

Habitats and ecology: The gliding abilities of flying squirrels are remarkable; they can readily glide over distances in excess of 65 feet (20 m), and there are reports of flights as long as 300 feet (90 m) (Wilson and Ruff, 1999). One of their major sources of food is fungi such as mushrooms (Dubay, 2000); in the Pacific Northwest they are believed to play an important role in the dispersal of fungus spores that symbiotically help forest trees absorb nutrients. They also eat seeds, nuts, fruit, insects, and other invertebrates as well as birds' eggs and some carrion.

Breeding biology: Flying squirrels do not hibernate and are able to be active at temperatures as low as −4°F (−20°C). They breed during spring, in Wyoming from late March to May, and perhaps produce a second litter during summer. The gestation period is 37 to 42 days, and the litter size varies from two to six. Newborns weigh only five to six grams, but by day 32 their eyes are open, and they begin explore by day 40. They are weaned by 60 days of age but remain with their mother for several additional months. Typical longevity is three to four years (Wilson and Ruff, 1999).

Suggested reading: Woods, 1980; Clark and Stromberg, 1987; Goldingay and Scheibe, 2000; Armstrong, Fitzgerald, and Meaney, 2011; Murrant et al., 2013; Buskirk, 2016.

Red Squirrel (Chickaree). *Tamiasciurus hudsonicus*.
Widespread.

Identification: The red squirrel, a small squirrel of western Wyoming, is a rich orange-brown dorsally, which contrasts with white underparts and, in summer, a blackish flank line that separates the brown from the white. It also has a white eyering and white cheeks and throat. Total length: 270–385 mm (10.6–15.6 in); tail: 95–158 mm (3.7–6.2 in). Weight: 145–260 g (5.1–9.1 oz).

Voice: Highly vocal, the common call is an alarm signal of endless and rapid series of barking cherks or chuckles, and a rapidly trilled "rattle" call that is territorial in function. When recordings of the territorial rattle are broadcast, a squirrel is more likely to respond with its own rattle if it hears the voice of a non-kin squirrel than if it hears that of a kin individual, proving that the rattle call can be used for kin recognition. Other research suggests that the alarm call does not encode information that identifies the species of predator that has been detected, nor is it intended to alert neighbors or kin but rather is directed at the predator, to announce its detection and possibly help in trying to deter it. Growls, screeches, and other vocalizations are also produced.

Status: This squirrel is a common and conspicuous resident of coniferous forests (pines, spruces, and firs) in western Wyoming and is often called a chickaree (from its vocalizations) or a pine squirrel. It is common in such forests, often at densities of up to 7 to 8 squirrels per acre (17–20 per hectare), and where they are very common they might consume up to two-thirds of the forest's entire cone crop (Armstrong, Fitzgerald, and Meaney, 2011).

Habitats and ecology: The conifer cones holding the edible seeds that are the squirrel's primary diet are gathered just before the seeds ripen. Spruce cones are highly preferred, perhaps because of their large size and because cones are produced most years. Lodgepole cones are also favored because the cones remain on the trees unopened for several years. The red squirrels strip away the cones' protective scales when the seeds are to be eaten, producing piles of cone scales called middens. These piles are then used to store more cones and are jealously guarded. Such piles can grow to be very large over time (up to about 20 feet in diameter) and might persist over several years or even decades. The squirrels also gather and store mushrooms for eating later. Tree twigs are also clipped off conifer branches and are gathered and used for making spherical nests that are usually placed in trees or sometimes taken into underground burrows.

Breeding biology: Single litters are produced in Wyoming, usually between March and May. Estrus females briefly allow males to visit them, their receptivity lasting a single day, after which the females again expel all squirrels from their territories. Gestation lasts 36 days, and an average of four young are born. Weaning occurs about mid-July when the young are 9 to 11 weeks old, but they remain with their mother until they are approximately a third grown. About 5 percent of wild squirrels survive at least five years, and a captive individual once lived for nine years (Clark and Stromberg, 1987).

Suggested reading: Woods, 1980; Clark and Stromberg, 1987; Armstrong, Fitzgerald, and Meaney, 2011; Buskirk, 2016.

Abert's Squirrel. *Sciurus aberti*. Highly localized.

Identification: This large, rare, and beautiful squirrel is limited in distribution to ponderosa pine forests near Harriman in Albany County. Its long ears are conspicuously tufted from fall to spring, and the species is then easily distinguished from all other Wyoming squirrels. Its pelage is mostly dark gray to black, with varying amounts of white flecking and white fringes on its broad tail. There may be a middorsal brownish stripe, a black flank stripe separating the dark upperparts from the lighter underparts. Total length: 463–584 mm (18.2–23 in); tail: 195–280 mm (7.6–11 in). Weight 540–971 g (19–34.2 oz).

Voice: Nothing specifically unique to this species' vocalizations has been noted.

Status: The Abert's squirrel is limited in Wyoming to a single ponderosa pine community near Harriman, its northernmost range limit. Elsewhere it occurs at elevations from 5,900 to 9,800 feet (1,800–3,000 m) within the range limits of ponderosa pine.

Habitats and ecology: This squirrel is a near-obligate of ponderosa pine, which it uses for its primary food (the seeds, buds, and outer cambium layer of twigs and small branches) and nest sites. The squirrels also feed on ground fungi that have symbiotic relationships with the pine's roots, so the dissemination of fungal spores through the squirrels' feces may be an important aspect of the tree community's health. However, the squirrels' harvesting of the trees' seeds and cambium exert pressures on the health of the pines too, and in at least some populations the trees have evolved secondary defenses in the form of producing distasteful compounds. The squirrels evidently seek those trees that are lower in such chemicals for their foraging and placement of their nests. Their twig nests are usually placed in large trees well up in the tree canopy or in "witches' brooms," the balls of twigs caused by dwarf mistletoe. Territories are apparently not defended, and home ranges are relatively large, 10 to 25 acres (4–10 hectares) (Armstrong, Fitzgerald, and Meaney, 2011). Densities of 12 to 295 animals per square mile (5–114 per square kilometer) have been estimated.

Breeding biology: This squirrel breeds in spring, from late March to May, and females are pursued by many males, several of which might mate with her. A gestation period of 40 to 46 days produces a litter of two to five young (averaging about 3.5). After about 15 to 16 weeks they are fully weaned and leave the nest. They reach adult size by 25 weeks of age. Their estimated longevity is seven to eight years.

Suggested reading: Farentino, 1972; Clark and Stromberg, 1987; Allred, 2010; Armstrong, Fitzgerald, and Meaney, 2011; Buskirk, 2016.

Thirteen-lined Ground Squirrel. *Ictidomys* (*Spermophilus*) *tridecemlineatus*. Widespread.

Identification: The thirteen-line ground squirrel is the familiar "gopher" that is often seen at the edge of a gravel road, where its striped and spotted body is easily observed. Its upperparts consist of a series of about 13 alternating dark brownish and whitish stripes, the dark stripes marked with regularly spaced whitish spots. The legs are short and the ears are small, and the tail is fairly long but not bushy. Adults weigh about 4 to 10 ounces and are much heavier prior to fall hibernation than during spring. The spotted ground squirrel (*S. spilosoma*) of southeastern Wyoming's sandy grasslands is similar in size and appearance, but its spots are not arranged in such a linear manner and are not separated by buffy stripes.

Voice: When frightened, and as a warning call, this ground squirrel utters a loud birdlike trill or tremulous whistle.

Status: This species is a common ground squirrel in eastern Wyoming, and population densities in the northern plains are even higher, ranging from about one to ten per acre, with still higher numbers occurring late in the breeding season as young emerge. In one Great Plains study, the home ranges of males averaged 11 to 12 acres and was largest during the breeding season, while the ranges of females averaged 3 to 4 acres and was largest during pregnancy and lactation.

Habitats and ecology: Associated with shortgrass and mixedgrass habitats, this species is common statewide in grazed pastures, golf courses, and roadsides where the soil is well drained and can be easily excavated. Burrows may be up to 20 feet long and vary in depth from a few inches to as much as 4 feet. There are also short escape burrows scattered through the animal's home range. Except during spring, the entrances of burrows are not revealed by piles of soil.

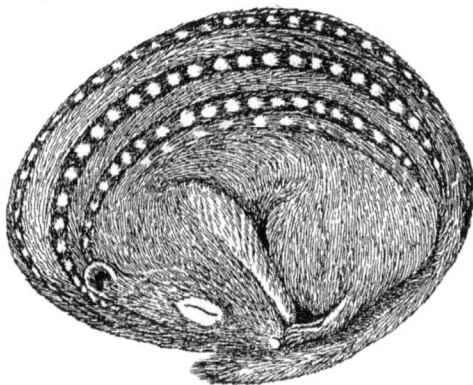

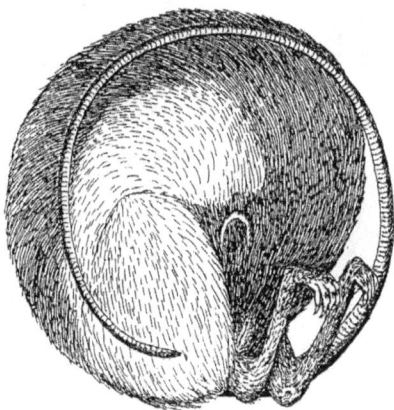

Hibernating adults of thirteen-lined ground-squirrel (*upper left*) and western harvest mouse (*lower left*), and adult Wyoming ground squirrel, alert posture (*right*)

Above ground are grassy runways that the ground squirrels use to travel from one burrow to another or out on foraging excursions.

Breeding biology: This species is among the least social of the ground squirrels, but adults do have a greeting ceremony during which they touch noses, and scent markings are made by rubbing their faces over objects in their environment. Males emerge from their hibernation in late March or April, ready to mate with females when they emerge a week or more later. After mating, a gestation period of 27 or 28 days follows, and litters of 5 to 13 young are born in May or early June. In Wyoming, a single litter per year is produced. Hibernation begins in late September or early October.

Suggested reading: Clark and Stromberg, 1987; Armstrong, Fitzgerald, and Meaney, 2011; Buskirk, 2016.

Golden-mantled Ground Squirrel. *Callospermophilus* (*Spermophilus*) *lateralis*. Dispersed.

Identification: This chipmunk-like rodent is common, relatively tame, and often seen in parks and preserves. Its single broad white lateral stripe, bounded above and below with black stripes, and the orange-brown "mantle" around its hind neck are obvious. Unlike chipmunks, it has no facial stripes. Length: 245–295 mm (9.6–11.7 in); tail: 70–120 mm (2.7–4.7 in). Weight: 175–350 g (6.2–12.3 oz).

Voice: The best-studied aspect of this species' vocalizations is its alarm call, uttered when confronted with a canid such as a dog or coyote. It consists of what has been described as a trill call, which is a multinote vocalization consisting of a single high note followed by a series of lower notes. In one study (Eller and Banack, 2004), dialect differences and geographic variations were found within this call. Vocalizations extending above 20,000 hertz, into the ultrasonic range, have also been detected. Eller and Banack suggested that within-group variations of the alarm call promote group recognition. Largely asocial, this species probably has a limited range of vocalizations associated with maintaining social cohesiveness.

Status: This species occurs throughout the Greater Yellowstone ecoregion and across much of southern Wyoming, including the Sierra Madre and Medicine Bow ranges. It also extends into the Green River Basin and along the Colorado border on the northern edges of the Uinta Mountains.

Habitats and ecology: It inhabits forest edges, rocky outcrops and slopes, roadsides, logged or burned areas, mountain meadows, and other generally open habitats from the juniper woodland zone to alpine tundra. They mostly forage on the stems and leaves of herbaceous plants but also eat seeds, berries, and fungi. To a lesser degree they eat insects, birds' eggs, carrion, and sometimes even young birds. Home ranges vary from 2.5 to 30 acres (1–12 hectares) and densities of 2 to 3.2 animals per acres (5–8.7 per hectare) have been observed. The animals are mostly solitary and territorial near their dens but might congregate near reliable food sources, such as at tourist sites. They are deep hibernators, their dormancy period

lasting from about early October to March.

Breeding biology: Breeding begins soon after the animals emerge from hibernation, with females coming into estrus about two weeks after they emerge. Gestation lasts 28 to 30 days, and a litter of two to eight (averaging five) young is produced. There is only one breeding cycle per season, and young are nearly full-grown by 2.5 months of age. They begin breeding as yearlings (Armstrong, Fitzgerald, and Meaney, 2011).

Suggested reading: McKeever, 1964; Clark and Stromberg, 1987; Armstrong, Fitzgerald, and Meaney, 2011; Buskirk, 2016.

Black-tailed Prairie Dog. *Cynomys ludovicianus*.
Slightly dispersed.

White-tailed Prairie Dog. *Cynomys leucurus*.
Widespread.

Identification: Prairie dogs are the most gregarious of Wyoming's rodents and are always found within colonies that are marked by scattered mounds that indicate burrow entrances. The two Wyoming species are very similar except for the color of their tail tips: black or white. The white-tailed species also differs in having blackish smudges above the eyes and on the cheeks. The two species occupy almost entirely complementary ranges, the black-tailed occurring mostly east of a line from the eastern foothills of the Bighorn Mountains southeast to western Laramie County, and the white-tailed west of a line from the western foothills of the Bighorn Mountains southeast to eastern Albany County. *White-tailed prairie dog:* Length: 340–370 mm (13.4–14.6 in); tail: 40–65 mm (1.6–2.6

White-tailed prairie dog, adult alert posture (*left*), and black-tailed prairie dog, jump-yip call (*right*)

in). Weight: 650–1130 g (22.9–39.8 oz). *Black-tailed prairie dog:* Length: 312–410 mm (12.3–16.1 in); tail: 72–90 mm (2.8–3.5 in). Weight: 680–1500 g (24–52.9 oz).

Voice: Both species are highly vocal. The common call of the black-tailed species is a sharp, whistled *wee-ooo* call, or "jump-yip" display, which reputedly serves as a territorial call and also as an "all-clear" signal. It certainly has other meanings, as it is highly contagious among colony and family members and seems to function largely in group-cohesion or social familiarization (Waring, 1970). The white-tailed species' territorial call is a "laughing bark," uttered without an associated jump but with a throwing back of the head. Many of the other calls are similar between the species; the black-tailed species has an estimated eight vocalizations plus tooth-chattering whereas the considerably less social white-tailed species has six detected vocalizations (Waring, 1970).

Status: The black-tailed prairie dog has suffered greatly at the hands of humans; its total US population has probably been reduced by 99 percent during historic times (Johnsgard, 2005), and in Wyoming by more than 80 percent (Campbell and Clark, 1981). The bacillus plague has also had significant population effects on prairie dogs.

Habitats and ecology: Home ranges of the white-tailed species have been estimated at 12.3 to 17 acres (5.9–6.9 hectares) and densities of 0.3 to 5 animals per acres (0.8–12.6 per hectare). Home ranges and densities of black-tailed prairie dogs are made complex by the highly social nature the species, with "towns" divided into "wards," and wards divided into individual family units called coteries. Densities vary greatly by season, but averages ranging from 4 to 20 animals per acres (10–50 per hectare) have been reported. In one study, a ward was estimated to occupy 7.4 acres (3 hectares), within which eight coteries occupied 0.5 acre (0.21 hectare) each (Armstrong, Fitzgerald, and Meaney, 2011). Prairie dogs carry with them many ancillary species that variously benefit from their presence. The black-footed ferret cannot survive in the absence of large populations of prairie dogs, and others such as burrowing owls, prairie rattlesnakes, grasshopper mice, and thirteen-lined ground squirrels also benefit. Nearly 150 species have been found to be facultative associates of prairie dogs (Clark et al., 1982; Johnsgard, 2005).

Breeding biology: Both prairie dog species breed in early spring (March and April), and in both the gestation period lasts 28 to 32 days. Litter sizes vary greatly from two to eight (white-tailed) to ten (black-tailed), averaging about five. Young appear in early June at about 40 days of age. By 90 days they appear to be fully grown, although adult weight may not be attained until about 120 days of age. Prairie dogs might hibernate from 30 to 120 days in various colder regions, although in some parts of their ranges they exhibit little or no true dormancy.

Suggested reading: Clark and Stromberg, 1987; Hoogland, 1995, 2006; Johnsgard, 2005; Slobodchikoff, Perla, and Verdolin, 2009; Armstrong, Fitzgerald, and Meaney, 2011; Buskirk, 2016.

Family Geomyidae (Pocket Gophers)

Northern Pocket Gopher. *Thomomys talpoides.*

Ubiquitous.

Identification: The presence of pocket gophers is best recognized by the occurrence of the surface trails they make under snow. After snowmelt these consist of raised ridges on the ground, which are casts of tunnels filled with soil that had been excavated elsewhere. Their burrows are marked by somewhat conical mounds of soil that have been pushed out along a ramp. The entrance to the burrow is kept plugged with soil. The animals are rarely if ever seen above ground but have small eyes and ears, and strong forelegs with large claws. Like pocket mice, they have fur-lined cheek pouches. All four of Wyoming's species are very similar in appearance, with brownish or tan pelage, but the northern pocket gopher is the only one that occurs statewide. Length: 164–263 mm (6.4–10.4 in); tail: 40–75 mm (1.7–3.0 in). Weight: 75–180 g (2.6– 6.3 oz).

Voice: No information on Wyoming species. A related southern species (*Geomys breviceps*) was found to have four audible signals, two of which were uttered only during close contact with other individuals. Considering the solitary nature of pocket gophers, the limited diversity of pocket gopher vocalizations is not surprising.

Status: Pocket gophers are generally little studied because nearly their entire lives are spent underground. They emerge only long enough to clip off the stems of herbaceous forbs and pull them into their burrows. In Colorado, this pocket gopher occurs from lower-altitude grasslands to mountain meadows and alpine tundra but mainly occurs at altitudes of 9,000 to 11,000 feet (2,750–3,350 m).

Habitats and ecology: It is extremely difficult, if not impossible, to estimate population densities and home ranges, but density estimates of 6 to 30 animals per acres (15–74 per hectare) have been made. One estimate of 74 animals per hectare was judged to have moved up to 74 metric tons of soil per hectare (30 tons per acre) annually, thus markedly affecting soil structure, carbon content, and erosion rate as well as increasing plant productivity and driving the rate of plant community succession (Armstrong, Fitzgerald, and Meaney, 2011).

Breeding biology: In Wyoming, animals in breeding condition have been found from March to April, with pregnant females obtained as late as July. Gestation lasts 18 to 19 days, and a single litter of four to six young are born. Weighing only about three grams at birth, they attain almost adult weight within six months. Their usual longevity in the wild is 1.5 to 2.5 years, but some wild individuals have lived for more than 5 years (Clark and Stromberg, 1987; Armstrong, Fitzgerald, and Meaney, 2011).

Suggested reading: Clark and Stromberg, 1987; Armstrong, Fitzgerald, and Meaney, 2011; Buskirk, 2016.

Family Heteromyidae (Pocket Mice and Kangaroo Rats)

Olive-backed Pocket Mouse. *Perognathus fasciatus*.
Widespread.

Identification: Of Wyoming's five species of pocket mice, the olive-backed is the most widespread, occupying the eastern four-fifths of the state. All five are similar in size and appearance, and all have fur-lined cheek pouches. These allow them to carry seeds back to their burrows without getting them wet and thereby subject to mold during storage. The pouches can be everted for emptying out their contents and for cleaning, using the forepaws. All of the pocket mice also have much larger hind legs than forelegs, large heads, large eyes, and long, bicolored tails. The olive-backed species has a distinctive olive-brown, almost greenish, upper body pelage that is separated from the white underparts by a yellowish line (which is sometimes lined with black) along its flanks. Three other species of pocket mice occur in Wyoming that are quite similar in appearance, but only one these, the hispid pocket mouse (*P. hispidus*) is substantially larger (minimum length 200 mm, or 7.9 in) and has a coarse dorsal pelage. Length: 127–137 mm (5–5.4 in); tail: 57–68 mm (2.2–2.7 in). Weight: 8–14 g (0.3–0.5 oz).

Voice: No specific information is available on the vocalizations of this or other Wyoming pocket mice.

Status: This pocket mouse is widespread in Wyoming, but its status and biology are little known. It is especially adapted to shrub-steppe and shortgrass prairies having sandy to gravelly soils at elevations below 4,500 feet (1,400 m) in the Black Hills, but in Colorado it extends into the ponderosa pine zone over elevations of 5,000 to 7,000 feet (1,525–2,135 m). Sites with loamy sand to clay soils with low vegetation and a good deal of bare ground, which allows bipedal hopping when escaping danger, are preferred.

Habitats and ecology: All pocket mice are heavily dependent on seeds for their diets. A wide variety of grass and forb seeds are eaten, and it is likely that some insect materials are too, based on studies of captive animals. In the very closely related plains pocket mouse, seeds were found to make up 81 percent of the stomach contents in one Utah study, while in a New Mexico study 91 percent of the diet consisted of seeds, and the other 9 percent was arthropods (Armstrong, Fitzgerald, and Meaney, 2011). It is known that in several pocket mice, such as the plains pocket mouse, access to drinking water is not needed, as moisture is generated from the carbohydrates that they consume (Clark and Stromberg, 1987). They evidently do not hibernate but undergo periods of torpor, periodically waking to eat some of their stored seeds. The mice dig burrow systems, often under a yucca or shrub, with side tunnels for food storage. Summer nests are about 12 to 15 inches (300–375 mm) below ground, but winter or summer dormancy (estivation) burrows may be as much as 6 feet (2 m) deep. Densities on the plains of Montana and North Dakota have ranged from 0.2 to 1.6 animals per acre (0.6–4.0 per hectare) (Armstrong, Fitzgerald, and Meaney, 2011).

Breeding biology: Little studied in Wyoming, breeding in this species probably occurs in late fall or winter. Gestation lasts 30 days. In Colorado, one or two litters are probably produced during spring and early summer, with litters typically of four to six young (six teats are present on females).

Suggested reading: Eisenberg, 1963; Clark and Stromberg, 1987; Armstrong, Fitzgerald, and Meaney, 2011; Buskirk, 2016.

Ord's Kangaroo Rat. *Dipodomys ordii*. Dispersed.

Identification: Wyoming's only kangaroo rat, this well-named mammal has highly developed hind limbs that allow for long horizontal and high vertical leaps. Its forelimbs are correspondingly small and mostly used for holding and manipulating foods. Its tail is long (longer than the head-body length), is doubly striped with black and white, and is noticeably tufted. Its dark eyes are large, with whitish "eyebrow" markings. The upperparts are otherwise rather uniformly tan. Length: 249–280 mm (9.8–11.0 in); tail: 138–163 mm (5.4–6.4 in). Weight: 45–100 g (1.6–3.6 oz).

Voice: Most observers claim that kangaroo rats do not vocalize but rather generate noises only by tapping or scratching with their hind feet. Although kangaroo rats vocalize only rarely, foot-drumming and tooth-chattering are common during aggressive encounters. However, there are a few accounts of squeaks or squeals uttered by animals under duress, and purring growls, clucks, and grunts produced by captive animals (Eisenberg, 1963). The large ears and specialized middle ear structures of the species in this mostly desert-dwelling group reflect the importance of hearing in these nocturnal mammals. Their large outer ears (pinnae) help localize sound sources, and their enlarged tympanic bullae (bony chambers enclosing the middle ears) allow for the detection of extremely weak and very low-frequency sounds, such as those that might be made by approaching predators. One

Ord's kangaroo rat, adult

study of hearing in a kangaroo rat (Heffner, 2005) found that the animals could respond to frequencies ranging from 50 to 65,000 Hz and could locate the source of very brief sounds.

Status: Kangaroo rats are highly mobile, and home ranges of from 5.7 to 8.1 acres (2.3–3.3 hectares) and a density of about 14 animals per acre (32 per hectare) have been reported (Clark and Stromberg, 1987). Other estimates in Texas have ranged from 0.01 to 6.3 animals per acre (0.04–15.6 per hectare); in Wyoming, Maxell and Brown (1968) found the highest densities to occur in sand dunes, with yucca grasslands and sage grasslands having progressively lower densities.

Habitats and ecology: Because of their efficient seed-gathering abilities, kangaroo rats are among the rodent species that are possibly important seed dispersers. One study in Utah suggested that 68 percent of foods taken there were seeds, 25 percent was green vegetation, and 7 percent was arthropods, while in the Pawnee National Grasslands 85 to 95 percent of the winter diet consists of seeds.

Breeding biology: The breeding season is extended, with activity in all months except December noted in some regions, but in Wyoming there are only two litters per year, with one born in late winter or early spring and another in late summer. Gestation lasts about 30 days, and litter sizes range from one to six. The eyes of the young open at about two weeks, and by two months of age they are independent of their mother. However, they might remain with their mother for several months, until the young males become aggressive. The animals remain active year-round and breed as yearlings (Clark and Stromberg, 1987; Armstrong, Fitzgerald, and Meaney, 2011).

Suggested reading: Eisenberg, 1963; Garner, 1974; Clark and Stromberg, 1987; Armstrong, Fitzgerald, and Meaney, 2011; Buskirk, 2016.

Family Castoridae (Beavers)

Beaver. *Castor canadensis*. Ubiquitous.

Identification: America's largest native rodent is easily recognized by its size, aquatic behavior, and flattened tail. The muskrat is much smaller and has a rounded tail. Length: 850–1,000 mm (33–40 in); tail: 258–325 mm (10–13 in). Weight: 11–35 kg (24–77 lb).

Voice: Although generally thought to lack vocalizations, beavers do have a limited vocabulary (Novakowski, 1969). Adults utter burps, whines, and hisses and also produce gnawing and chewing sounds. Beavers hiss when confronted by other animals and when defending their territory. Young beaver are particularly vocal and often produce a soft repetitive whine when soliciting food or when in uncomfortable social situations.

Status: Beavers are common in rivers, ponds, and other aquatic habitats across Wyoming, as a result of reintroductions in the 1900s following earlier extirpations. Their densities are hard to judge, but in Colorado they have been found to have colony territories of about 1 to 20 acres (0.4–8 hectares) and nearest-neighbor distances of 0.3 to 0.9 mile (0.7–1.5 km). Distances between colonies vary greatly and in part are related to the abundance of willows along the river or stream. Aspens and cottonwoods are important winter foods, and alder, birch, and conifers are used to a limited extent (Armstrong, Fitzgerald, and Meaney, 2011).

Habitats and ecology: Beavers are master water engineers; they build dams and cut canals to regulate water flows and provide channels for moving foods. They also construct wood and earthen lodges and dig bank burrows for safety, food storage, and reproduction. The dams may be as much as several hundred feet long, and up to about eight feet high, and may last for decades, owing to constant maintenance and repair. Most dam construction is done during spring and fall, and during fall a colony might fell a mature aspen every other night while assembling a winter food supply. Beavers are monogamous, a family typically consisting of an adult pair plus their yearlings and juvenile offspring, totaling from about four to eight animals. Both sexes have scent glands (castor glands) that are spread on mud mounds and used to mark the boundaries of their territory. Pair bonds last an average of 2.5 years, with the death of a mate the most common reason for mate turnover (Armstrong, Fitzgerald, and Meaney, 2011).

Breeding biology: Mating behavior in Wyoming occurs from January to February, and gestation lasts 104 to 111 days. In Colorado, litter sizes vary with altitude; females living at elevations above 5,000 feet have smaller litters than those living at lower elevations (average 4.4 vs. 2.7). The young ("kits") are born fully furred and are weaned by about two months. They become sexually mature by their second winter and disperse, the males being more likely to disperse than females. Maximum longevity is about 20 years, but average lifespans are 8 to 10 years (Clark and Stromberg, 1987; Armstrong, Fitzgerald, and Meaney, 2011).

Suggested reading: Novakowski, 1969, Clark and Stromberg, 1987; Armstrong, Fitzgerald, Meaney, 2011; Müller-Schwarze, 2011; Buskirk, 2016.

Family Cricetidae (New World Mice)

Northern Grasshopper Mouse. *Onychomys leucogaster*. Ubiquitous.

Identification: Grasshopper mice generally resemble deer mice but have much shorter tails (less than half the length of the head and body) and are noticeably stockier in build. They have large blackish eyes, furred ears that often have a patch of white at the base, and very long whiskers. Length: 120–155 mm (4.7–6.1 in); tail: 29–50 mm (1.1–2.0 in). Weight: 35–45 g (1.2–1.6 oz).

Voice: The grasshopper mouse is the most vocal of Wyoming mice. Adults "sing" in an upright posture with an open mouth; the vocalization is an easily heard wail of about 12 kHz that lasts about a second and identifies the individual, its sex, location, and probably other information. These are highly mobile mice, with large home ranges of up to 7.4 acres (3 hectares), so long-distance sound communication is probably highly adaptive.

Status: This mouse occurs statewide except for in the higher mountains and is most common in overgrazed pastures, shortgrass prairies, and grassy sandhills. Although mostly nocturnal, grasshopper mice are carnivorous and sometimes actively hunt grasshoppers (a favorite prey), other insects, and small vertebrates. They have a powerful bite for their size, as my young son Scott once discovered when, seeing one hiding in sandhill grasses, impulsively grabbed it and stuffed it in a pocket. He soon yelped in pain, quickly pulled it from his pocket, and dropped it back into the grass. Grasshopper mice readily kill other mice and even have been known to kill mammals up to the size of hispid cotton rats (*Sigmodon hispidus*), which average three to four times their own weight (Armstrong, Fitzgerald, and Meaney, 2011).

Habitats and ecology: As noted, grasshopper mice have large home ranges that average about 5.7 acres (2.3 hectares) and include scent-marked and defended territories. In one Colorado study, six male home ranges averaged 4.2 acres (1.7 hectares) and overlapped with the home ranges of several females. Like many carnivores, they are monogamous, with both parents tending the young. They do not hibernate, but in winter they shift to a diet in which seeds predominate (Armstrong, Fitzgerald, and Meaney, 2011).

Breeding biology: Grasshopper mice breed in spring, with young being born from March to September in Colorado, where three or four litters per season might be produced. Gestation lasts from 27 to 47 days, being longer for lactating females than those that are not lactating. Typically three or four young are born, but litters of up to six might occur. By a month of age the young are weaned, and they reach sexual maturity at three to four months. In captivity they have lived for more than four years, but most wild animals do not survive much longer than two years (Clark and Stromberg, 1987; Armstrong, Fitzgerald, and Meaney, 2011).

Suggested reading: Ruffer, 1965; Clark and Stromberg, 1987; Armstrong, Fitzgerald, and Meaney, 2011; Buskirk, 2016.

Western Harvest Mouse. *Reithrodontomys megalotis*. Dispersed.

Identification: Harvest mice closely resemble several other grassland mouse species but are very small. The more widespread western harvest mouse has a long bicolored tail (as long or longer than the head-body length) and a dark brown back that is grizzled with black hairs. The plains harvest mouse is limited to the eastern fourth of the state, has a shorter tail (less than the head-body length), and lacks a black grizzled dorsal appearance. Length: 122–155 mm (4.8–6.1 in); tail: 56–73 mm (2.2–2.9 in). Weight: 11–17 g (0.4–0.6 oz).

Voice: No specific information on this species is available, but studies of several other harvest mice indicate that their

vocalizing includes both audible and ultrasonic components and probably represents announcement calls (Miller, 2010).

Status: This widespread but inconspicuous mouse is common in shortgrass and scrub habitats as well as in riparian and taller grassy environments and in weedy sites. Densities vary greatly, from about 1 to 24 animals per acre (2.5–60 per hectare), and reportedly even up to 50 per acre (123 per hectare) (Clark and Stromberg, 1987; Armstrong, Fitzgerald, and Meaney, 2011).

Habitats and ecology: Home ranges of these small mice are also rather small, from 0.08 to 0.2 acre (0.2–0.56 hectare). Western harvest mice are active all year, nocturnally eating a wide variety of seeds, insects, and various vegetable matter. They are unusual in that they construct nests that are spherical, baseball-sized structures composed of woven grasses and other fibrous vegetation and lined with soft plant materials. These might be placed on the ground or elevated in vegetation, with a small entrance hole at the bottom. The mice huddle socially in these grassy balls during cold weather, and in some regions might even hibernate, or at least become torpid for short periods (Armstrong, Fitzgerald, and Meaney, 2011).

Breeding biology: Harvest mice breed from spring to fall, with many litters produced per season. In the wild these probably total about three or four, and gestation takes 24 days. From three to nine young are born, with five being the typical number. By five weeks they are fully grown, and they can breed by two months of age (Clark and Stromberg, 1987).

Suggested reading: Clark and Stromberg, 1987; Miller, 2010; Armstrong, Fitzgerald, and Meaney, 2011; Buskirk, 2016.

Deer Mouse. *Peromyscus maniculatus*. Ubiquitous.
White-footed Mouse. *Peromyscus leucopus*. Localized.

Identification: Deer mice and white-footed mice are so similar in appearance that an in-hand inspection is needed to tell them apart. Both are variable geographically by season and age in their upperpart colors, which vary from a gray to a rich brown color, but they invariably are a sharply contrasting white underneath. This bicolored pattern includes even the long, sparsely haired tail. The ears are very large, and the eyes are similarly large and protruding. The white-footed mouse can often be separated from the deer mouse by the white-footed mouse's slightly longer tail (60–100 mm vs. 55–69 mm). The white-footed mouse's tail is also less strongly bicolored than that of the deer mouse, and it is restricted in range to the Bighorn Mountains, northeastern Wyoming, and the eastern tier of counties.

Voice: These mice are not notably vocal (at least to the human ear), but young deer mice of two different subspecies were found to produce distress calls at frequencies of 3,600 to 26,500 Hz in a pulsed series of two to five notes lasting 0.1 to 0.2 second, the pulses separated by shorter pauses (Hart and King, 1966). It is also known that ultrasonic calls (35,000 Hz) are used by male deer mice prior to and after copulation (Pomerantz and Clemens, 1981). An alarmed white-footed mouse will rapidly drum its forefeet.

White-footed mouse, adult dozing (*upper*), adult nose-to-nose encounter (*upper middle*); northern grasshopper mouse, male howling (*lower middle*), and adult crouching (*below*)

Status: Only the deer mouse is extremely common in Wyoming. It is the most widespread and probably most common North American rodent, and this is probably also true in Wyoming.

Habitats and ecology: In Wyoming, white-footed mice are almost exclusively found in wooded areas, such as riparian woods, shelterbelts, and fencerows, whereas the deer mouse is more adaptable, and it occurs in open meadows, prairies, badlands, croplands, shelterbelts, hedgerows, and coniferous woodlands. Where white-footed mice are present in woodlands, deer mice are rare or absent, but deer mice are found in woodlands that lack white-footed mice. Both species live in burrows, the deer mouse's often under rocks, among debris, or under fallen logs, where they make vegetation-lined nests that are for sleeping, protection from cold, and rearing their young. Both are mostly nocturnal. The home ranges of white-footed mice vary greatly but rarely are larger than an acre, and females usually have smaller ranges than males.

Deer mouse home ranges in the northern Great Plains vary widely, from 0.1 to 10 acres (0.04–4 hectares). Both species eat a variety of berries, seeds, fruits, nuts, and other vegetation as well as insects, other invertebrates, and occasionally small vertebrates.

Breeding biology: Both species are active year-round, but the period of reproduction is mostly during spring and fall. The gestation period of both species is 22 or 23 days, and litter sizes range up to nine but average about four. The babies' eyes open at about two weeks and the youngsters are weaned at about four weeks. Both sexes are mature by about seven to eight weeks of age, so by fall several generations may be present in the population.

Suggested reading: Pomerantz and Clemens, 1981; Clark and Stromberg, 1987; Armstrong, Fitzgerald, and Meaney, 2011; Buskirk, 2016.

Muskrat. *Ondatra zibethicus.* Ubiquitous.

Identification: Muskrats are easily distinguished from much larger beavers (body 24 inches long exclusive of tail), the only other large swimming mammal in Wyoming. Muskrats are about the size of a prairie dog (12 to 13 inches long exclusive of tail), and their tail is long and ratlike but laterally flattened. They are never found far from water, and usually there are associated domelike lodges or "muskrat houses" of cattails built in shallow water where muskrats occur, although muskrats may also dig shoreline burrows. Length: 420–620 mm (16–24 in); tail: 200–290 mm (7.9–11.8 in). Weight: 700–1500 g (24–53 oz).

Voice: Undescribed vocalizations have been heard among captive muskrats huddling in cold weather, but no detailed accounts of their vocal communications are available.

Status: Muskrats probably occur in every Wyoming county wherever standing water and emergent vegetation are found, from low elevations to the edge of alpine tundra. The state's muskrat population probably numbers in the millions. In good habitats, populations might exceed 22 animals per acres (55 per hectare). In recent years, about 5,000 muskrats have been trapped annually, compared with about 2,700 beavers, 1,800 bobcats, and 1,200 pine martens.

Habitats and ecology: Like beavers, muskrats are herbivorous and eat mostly aquatic vegetation, especially cattails; however, they also eat crayfish, fish, and mollusks. They are active throughout the year and are largely nocturnal but often can be seen during late afternoon hours, and sometimes even near midday. Home ranges of females and young are small and center on a lodge or burrow. Both sexes are territorial and strongly defend breeding territories, which are scent-marked.

Breeding biology: Muskrats are promiscuous or loosely monogamous. The male will take over care of the young if the female dies. Females are polyestrous, with cycles of about 30 days, and two or perhaps three breeding cycles might be completed within a single season. Gestation lasts 25 to 30 days, six to seven young are the usual litter, and the young are independent by about two months of age. They become sexually mature during their second spring. Typical lifespans are three to four years (Clark and Stromberg, 1987).

Suggested reading: Clark and Stromberg, 1987; Armstrong, Fitzgerald, and Meaney, 2011; Buskirk, 2016.

Southern Red-backed Vole. *Clethrionomys (Myodes) gapperi*. Widespread.

Western Heather Vole. *Phenacomys intermedius*. Slightly dispersed.

Identification: Of the eight Wyoming species of rather small, short-legged, short-tailed, stocky-bodied mice collectively called voles (from a Norwegian name for field mice), the red-backed vole is one of the easiest to identify. They have a reddish-brown tint on the back and shoulder pelage, white feet, and short tails. They are common in midaltitude lodgepole pine forests, although they have been found as high as alpine tundra. The similar western heather vole is also found in high-elevation lodgepole pine and spruce-fir forests, and it has an even shorter tail but lacks rufous pelage tints. The five *Microtus* vole species are similar but are mostly found at lower elevations, except for the very similar montane vole. *Red-backed vole:* Length: 125–172 mm (4.9–6.8 in); tail: 30–50 mm (1.2–2.0 in). Weight: 18–42 g (0.6–1.5 oz). *Heather vole:* Length: 115–145 mm (4.5–5.7 in); tail: 22–33 mm (0.9–1.3 in). Weight: 25–40 g (0.8–1.4 oz).

Voice: When they are disturbed, red-backed voles are known to make chirplike barking sounds that can be heard up to about six feet away. They also make chattering or gnashing sounds with their teeth. No specific information on the heather vole's vocalizations is available.

Status: Red-backed and heather voles are nocturnal and not often encountered in their typical habitat of upper montane coniferous forest that has a ground cover of heaths and other evergreen woody vegetation. The heather vole occurs at altitudes from 7,000 feet (2,130 m) to above timberline in Colorado. The red-backed vole range extends up to at least the timberline (krummholz) zone.

Habitats and ecology: Old-growth spruce and mature lodgepole pine forests with abundant surface litter, woody debris, and understory vegetation are the favorite habitats of red-backed voles in Wyoming. Heather voles in Colorado similarly favor spruce and lodgepole forests with a good groundcover of heathers. Red-backed voles also favor undisturbed sites with downed and decaying trees as well as coarse woody debris. Probably both species are attracted to forests with abundant fungi, an important food source that can be easily dried and stored for later consumption (Dubay, 2000). Both species are also active year-round and are mostly nocturnal. They are not social, but during winter they may become more tolerant, and at least the heather vole is known to engage in winter huddling. Home ranges for the red-backed voles have been estimated in Colorado at 0.02 to 1.2 acres (0.01–0.5 hectare), and densities of individuals from 0.8 to 19 per acre (2 to 48 per hectare), while heather vole densities in the northern Rockies have been estimated at 0.2 to 4 per acre (0.5–10 per hectare) (Armstrong, Fitzgerald, and Meaney, 2011).

Breeding biology: Females of both species are polyestrous, having multiple breeding cycles per year, and mating is promiscuous. The gestation period is about 18 days in the red-backed vole and 19 to 24 days for the heather vole, with about five or six young being the usual litter size. The heather vole might have as many as three litters per year, with females reaching breeding age at four to six weeks, and males initially breeding the following year (Armstrong, Fitzgerald, and Meaney, 2011).

Suggested Reading: Clark, 1973c; Clark and Stromberg, 1987; Armstrong, Fitzgerald, and Meaney, 2011; Buskirk, 2016.

Long-tailed Vole. *Microtus longicaudus*. Widespread.

Prairie Vole. *Microtus ochrogaster*. Dispersed.

Identification: These typical *Microtus* voles (*Microtus* translates as small-eared, but these voles simply have most of their moderately large, rounded ears hidden by long body fur) are stocky grassland- and meadow-dwelling mice with short legs, large heads, and relatively small eyes. These two species are very similar, being uniformly grayish brown to reddish brown, with variably grayish underparts. They are most easily separated by tail-length differences. *Long-tailed vole:* Length: 175–220 mm (6.8–8.7 in); tail: 55–90 mm (2.2–3.5 in). Weight: 40–55 g (1.4–1.9 oz). *Prairie vole:* Length: 144–180 mm (5.7–7.1 in); tail: 33–45 mm (1.3–1.8 in). Weight: 30–70 g (1.1–2.5 oz).

Voice: Both of these species are known to emit ultrasonic vocalizations as infants, and at least the prairie vole also produces ultrasonic vocalizations as adults (Lepri, Theodorides, and Wysocki, 1988). The functions of these calls are still poorly understood, but in the case of infants, the calls appear to be related to begging behavior associated with sibling competition for access to their mother's nipples (females have four pairs of mammae) or when the infant is cooled and in distress.

Status: The prairie vole is a common mammal over all of Wyoming and overlaps with the long-tailed vole in grasslands over most of the state. Prairie voles in Wyoming occur in drier areas where shrubs are not co-dominants, but they favor relict mixed-grass and tallgrass prairies. They also extend into riparian habitats of willows and cottonwoods and into shrub-woodland habitats. Long-tailed voles favor areas close to water, such as riparian willow and alder thickets, but they also occur in semiarid sagebrush, subalpine forests and meadows, and alpine tundra.

Habitats and ecology: Typical densities of long-tailed voles in good habitats are 4 to 8 animals per acre (10–20 per hectare) but may exceed 20 per hectare during high populations. Their home ranges average 0.5 acre (0.19 hectare) for females and 0.7 acre (0.27 hectare) for males. Home ranges of the prairie vole average about 0.04 to 0.6 acre (0.1–0.25 hectare).

Unlike the long-tailed vole and nearly all other *Microtus*, the prairie vole is rather social, and adults form strong monogamous pair bonds. Both parents tend to their young and help maintain trails and burrows. Social groups form by the addition of offspring and nonrelated individuals; communal nests of 20 or more have been found in some areas (Clark and Stromberg, 1987; Armstrong, Fitzgerald, and Meaney, 2011).

Breeding biology: Both of these species are polyestrous, the long-tailed vole breeding from spring to fall (April to October in Colorado), and the prairie vole throughout the year. Gestation lasts about 21 days, and litter sizes are typically three or four but range up to seven. Young are weaned by three weeks of age and can breed by 30 days of age. Longevity in both species is very short, usually averaging well under a year (Armstrong, Fitzgerald, and Meaney, 2011).

Suggested reading: Clark, 1973c; Clark and Stromberg, 1987; Armstrong, Fitzgerald, and Meaney, 2011; Buskirk, 2016.

Sagebrush Vole. *Lemmiscus curtatus*. Dispersed.

Identification: This is the only Wyoming vole that is uniformly pale gray, except for a whitish belly, and has a very short tail. The animals are rarely found near sagebrush, although they consume a wide variety of plants and also live in greasewood- and rabbitbrush-dominated semidesert. Length: 110–140 mm (4.3–5.5 in); tail: 16–25 mm (0.6–1.0 in). Weight: 17–35 g (0.6–1.2 oz).

Voice: No information is available.

Status: Because of the widespread occurrence of sagebrush, this species' status seems to be secure, even if the sage-dependent birds are declining. Population densities seem to vary greatly. Those estimated in Sweetwater County have ranged from 0.2 to 0.4 per acre (0.3–0.6 per hectare), but in Idaho densities of from 1.6 to 6.5 per acre (4–16 per hectare) have been reported (Clark and Stromberg, 1987; Armstrong, Fitzgerald, and Meaney, 2011). Nationally, the species' elevational range has been reported as 1,000 to 12,140 feet (305–3,700 m) (Wilson and Ruff, 1999).

Habitats and ecology: This species' overall range corresponds fairly well with that of big sagebrush, making it the nearest thing to a sagebrush obligate among small mammals. At least during winter it is the prime food for sagebrush voles, which eat its bark, twigs, and flowers. It is entirely vegetarian and does not eat seeds or insects, nor does it store food but rather consumes a wide diversity of green vegetation. This species is quite social, living in colonies that share collective burrows. These burrows might have up to 30 entrances that are well hidden. The burrows are often paved with clippings of sage and grasses. The voles are active year-round and are mostly crepuscular.

Breeding biology: These voles probably breed throughout the year, with a peak in spring and autumn, when succulent vegetation is readily available. The gestation period lasts 25 days, and estrus begins again within a day of giving birth. The litter size varies from 2 to 13 young, usually 4 to 6, and weaning

Bushy-tailed woodrat, adult

occurs by 21 days. Females are sexually mature by 60 days and the males by 70, although they are not full grown until they are about 90 days of age (Armstrong, Fitzgerald, and Meaney, 2011).

Suggested reading: Clark and Stromberg, 1987; Armstrong, Fitzgerald, and Meaney, 2011; Buskirk, 2016.

Bushy-tailed Woodrat. *Neotoma cinerea*. Ubiquitous.

Identification: This rat, with its light brown to blackish body pelage, white underparts, and well-furred, bicolored tail resembles an overgrown white-footed mouse rather than a Norway rat (*Rattus norvegicus*). They have huge eyes and large ears, and the tail is about three-fourths the length of the head and body. Length: 350–470 mm (13.8–18.5 in); tail: 135–223 mm (5.3–8.7 in). Weight 240–290 g (8.5–10.3 oz).

Voice: In addition to aggressive foot thumping and tooth chattering, several vocalizations are used. These include a buzzing sound uttered by the male prior to copulation and sometimes also by females in sexual situations. A loud scream is uttered by animals in distress or during fights, and a short squeal is emitted in less extreme situations (Escherich, 1981a).

Status: This species is nearly ubiquitous in Wyoming, especially in rocky habitats, from the lowest dry basins to mountainsides as high as 11,500 feet (3,500 m), and from semidesert scrubs and arid grasslands to forests. They are also attracted to abandoned buildings and occupied cabins, where their tendency to chew on isolated wires can be troublesome. They are entirely vegetarian and consume a very wide variety of leaves, twigs, berries, bark, and other plant materials. Foods are left out to dry and then stored for winter consumption.

Habitats and ecology: This well-known "packrat" is famous for finding and carrying off small items to its storehouse of both edible and inedible materials in and around its nest. While camping in rocky scablands of Washington state while a graduate student, I often lost coins that I happened to leave in my tent, and I was in constant fear that a woodrat might find my hidden car keys and leave me stranded, as its huge nest was located in an inaccessible rock crevice. Both sexes scent-mark and strongly defend individual territories, although males will allow females into their territories. No pair-bonding occurs. Densities in southwestern Wyoming have been estimated at 0.3–1.9 animals per acre (0.8–4.6 per hectare) (Belitsky, 1981; Clark and Stromberg, 1987).

Breeding biology: These woodrats breed over several months, from April through August in Colorado, and possibly produce two litters per season. Gestation lasts 27 to 35 days, and litter sizes vary from two to five young, usually three or four. Only females tend the offspring. The young are independent by two months and begin to breed at two years. Average longevity in the wild is about three to four years.

Suggested reading: Escherich, 1981a; Clark and Stromberg, 1987; Armstrong, Fitzgerald, and Meaney, 2011; Buskirk, 2016.

Family Zapodidae (Jumping Mice)

Western Jumping Mouse. *Zapus princeps*.
Widespread.

Meadow Jumping Mouse. *Zapus hudsonicus*. Highly localized.

Identification: These two species are too similar to allow for separation except by experts, but the two probably overlap only in southeastern Wyoming. Except for that region, jumping mice in northeastern Wyoming are probably meadow jumping mice, and those over the rest of the state are probably western jumping mice. Size differences exist but overlap extensively. *Western jumping mouse:* Length: 210–250 mm (8.3–9.8 in); tail: 125–150 mm (4.9–5.9 in). Weight 18–25 g (0.6–0.9 oz). *Meadow jumping mouse:* Length: 180–220 mm (7.1–8.7 in); tail: 115–136 mm (4.5–5.3 in). Weight 12–22 g (0.4–0.8 oz).

Voice: No information is available.

Status: The western jumping mouse has a widespread distribution in Wyoming. In Colorado, it ranges from about 6,000 feet (1,830 m) to 8,200 feet (2,500 m) and is especially associated with montane riparian habitats. By comparison, the meadow jumping mouse is a grassland species that is most common in taller grasses near water and in herbaceous understory vegetation of wooded areas.

Habitats and ecology: The jumping mice are well named—they can make long jumps of up to about 12 feet and hops of up to about 10 feet when frightened. They can also perform a series of short jumps and then hide motionless in tall grass. Altering course while in the air is possible by using their long tail as a rudder. They are deep hibernators, going dormant for nearly half of the year and spending their period of activity in nocturnal foraging, accumulating fat in preparation for their next hibernation. Home ranges for female meadow jumping mice average about 1.5 acres (0.6 hectare) and those of males about 2.5 acres (one animal per hectare), while densities range from 2.3 to 19.5 per acres (7.4–48 per hectare) (Clark and Stromberg, 1987). The mice eat a variety of seeds, insects, and fungi but do not store food. They must accumulate about six grams of fat to survive hibernation, sometimes doubling their post-hibernation spring body weight; heavier animals have the best chance of winter survival (Armstrong, Fitzgerald, and Meaney, 2011).

Breeding biology: Jumping mice begin to breed shortly after emerging from hibernation, probably in May, and might have two or three litters during their summer breeding period. Gestation lasts 17 to 20 days, and litter sizes usually number four to six. The young are weaned by four weeks and reach adult size by three months of age. They begin breeding the following spring and have average longevities of about three years (Clark and Stromberg, 1987).

Suggested reading: Quimby, 1951; Clark, 1971b; Clark and Stromberg, 1987; Armstrong, Fitzgerald, and Meaney, 2011; Buskirk, 2016.

Family Erethizontidae (Porcupines)

Porcupine. *Erethizon dorsatum*. Ubiquitous.

Identification: The porcupine is easily identified by its array of sharp, barbed quills, which are modified hairs and mostly obscured by long body fur. The quills vary in length up to several inches long and become longer as the animal matures. They are raised during defensive threat, exposing their white bases, which results in a white skunk-like stripe appearing on the animal's otherwise dark back, perhaps helping to deter further aggression. During an attack, the tail is lashed from side to side, which deeply embeds the quills into any enemy's body. As many as 30,000 quills may be present, not all of which are barbed. They are hollow, and the body heat of the victim causes embedded quills to expand and become harder to extract. Cutting off the end of an imbedded quill releases the air pressure inside and can make it somewhat easier to remove. Length: 79–103 cm (31–40 in); tail: 145–300 mm (5.7–11.8 in). Weight: 3.5–18 kg (7.7–39 lb).

Voice: Normally silent, porcupines do perform prolonged tooth-chattering prior to a defensive attack. They also emit a pungent odor as a secondary warning. Their generic name translates as "one who rises in anger." Vocalizations that have been mentioned by various observers include squeaking, moaning, grunting, hooting, sobbing, wailing, shrieking, and howling. A detailed inventory of their vocabulary apparently remains to be done.

Status: Porcupines are common statewide but are rarely found far from trees, which they use as a refuge, as a den if there is a cavity, and as a source of food, eating the tree's inner bark, leaves, needles, buds, and twigs, as well as other vegetable materials. Maples, aspens, beeches, basswoods, ashes,

Porcupine, adult

and apples are among their favorite food sources. Cellulose-digesting bacteria in an intestinal caecum allow the porcupine to exploit this energy source (Whittaker, 1997).

Habitats and ecology: Porcupines are nocturnal, but sometimes they can be seen in daytime resting or sleeping in trees. They move slowly and are unlikely to retreat or flee from any source of possible danger. Among predators, only the fisher is efficient at flipping a porcupine onto its back, so that its vulnerable belly is exposed to a lethal attack. In the wild, longevity up to 10 years is known, and in captivity individuals have lived up to 20 years. They are reported to make intelligent, if potentially dangerous, pets.

Breeding biology: The answer to the most common question asked about porcupine reproduction is "Very carefully." The actual answer is that the female relaxes and lowers her quills before raising her tail to the male. Prior to that there is an extended courtship, and before mating the male squirts a spray of urine over the female. The breeding season occurs during autumn, and the gestation period lasts 29 to 31 weeks. Typically only a single, relatively precocial offspring is born,

whose soft spines at birth quickly harden and in less than an hour are functional weapons. Nursing lasts two or three weeks, and the youngster remains with its mother for several months (Clark and Stromberg, 1987; Whittaker, 1996).

Suggested reading: Clark and Stromberg, 1987; Roze, 1989; Armstrong, Fitzgerald, and Meaney, 2011; Buskirk, 2016.

Family Canidae (Dogs)

Coyote. *Canis latrans*. Ubiquitous.

Identification: Coyotes are very doglike (rarely they hybridize with large dogs to produce "coydogs") and are uniformly grizzled gray overall, except for a black-tipped tail and rusty to yellowish legs. The ears are prominent, and the tail is bushy and held almost between the legs when the animal is running. Length: 110–122 cm (44–48 in); tail: 330–400 mm (13–20 in). Weight: 10–20 kg (22–44 lb).

Voice: The communal howling calls of coyotes are familiar to farmers, ranchers, and outdoors-lovers. They are prolonged vocalizations, usually uttered during evening or early

Overleaf: Coyote, adult

morning hours. The call consists of a few sharp barks followed by a prolonged mournful howl, ending with several short, sharp yips. Barking calls are used in threat situations. Other vocalizations and social posturing are essentially like those of domestic dogs. The coyote's Latin name means "barking dog"; its English name is from the Aztec Nahuatl language, *coyotl*.

Status: In spite of constant human persecution, coyotes have managed to survive across most of their historic range and no doubt still occur in every Wyoming county. Mange has affected their heath and reduced their populations in some areas.

Habitats and ecology: Coyotes are highly adaptable, occurring in habitats ranging from desert to woodlands and even extending into tundra and tropical forests. They are mobile daytime predators, ranging long-term over distances up to 400 miles. They can trot at speeds of 25 to 30 miles per hour, run at speeds up to 40 miles per hour, and can leap as far as 14 feet. Although most ranchers hate them, coyotes are highly beneficial to them because they eat primarily rodents. The livestock that are eaten are mostly scavenged or involve otherwise weakened animals.

Breeding biology: Coyotes form monogamous pair-bonds, and females come into estrus for only a few days once per year, between February and March, following a prolonged courtship period. The gestation period is 63 days, and the average litter size is six pups. The young are born blind and helpless, nursed for six or seven weeks while the male brings food to the den, and then weaned. The pair-bond lasts through the period of reproduction, after which family ties are the basis for autumn and winter social hunting parties, as in wolf packs. Lifespans in the wild rarely exceed ten years; given Wyoming's gun-addicted culture and contempt for all predators, most Wyoming coyotes are probably very lucky if they reach five years of age.

Suggested reading: Ryden, 1977; Pringle, 1977; Camenzind, 1978; Clark and Stromberg, 1987; Laydet, 1988; Dobie, 2006; Armstrong, Fitzgerald, and Meaney, 2011; Buskirk, 2016.

Gray Wolf. *Canis lupus*. Highly localized.

Identification: This species is very similar to a large German shepherd but has a tuft of hair projecting downward from each ear, usually has white fur around the mouth, and holds its tail straight rather than curled upward when active. Length: 130–185 cm (51–73 in); tail: 300–450 mm (11.8–17.8 in). Weight: males: 20–80 kg (44–176 lb); females: 18–55 kg (40–121 lb).

Voice: The prolonged, far-carrying howl of wolves is unforgettable and the perfect symbol of wilderness. Pack howling serves to make a territorial claim, may be a call to assembly, a statement of solidarity. Pups also engage in pack howls. Howls by a lone individual may be uttered by a lost animal or out of grief. Barking is rarer in wolves than in dogs and is used as an alarm signal. Whining and whimpering serve as signals of submission, anxiety, or frustration but can also be uttered during periods of friendly interaction.

Status: Wolves are federally protected in most states but were

Gray wolf, adult

removed from the Endangered Species list in Wyoming in 2017, so their degree of protection there is under the control of the Wyoming Game and Fish Department. State control is also true in Montana and Idaho. After the reintroduction of 13 animals in 1995, the population in Yellowstone National Park slowly increased, reaching 111 by the spring of 1999, and 171 in 19 packs by 2007. Since then there has been a slow decline and stabilization, and by the end of 2016 the park population totaled 108 in 11 packs. During winter, each wolf might kill and eat 1.8 elk per month. Sarcoptic mange, canine distemper, and perhaps other diseases have contributed to wolf mortality. In 2015 an estimated 528 wolves were present in the Greater Yellowstone ecosystem, of which nearly 400 were in Wyoming. At that time an estimated 1,704 wolves, including 96 breeding pairs, were estimated to be present in the US Fish and Wildlife Service's Northern Rocky Mountain Population Segment.

Habitats and ecology: In the Yellowstone region, elk have been the primary large prey animal of wolves. Between 1996 and 2001, the pioneering wolves in Yellowstone killed 77 elk, as compared with 14 bison (through 1999) and fewer than 10 moose. Based on 20 years of study as of 2017, it was estimated that wolves kill up to 2,156 elk in the park each year. Wolves have their greatest success with young elk (aged 8–9 months) and animals 11 years of age and older. The average age of adult cows taken was 14 years, or well past their reproductive prime (Halfpenny, 2003).

Breeding biology: Wolves reach sexual maturity in their second year but may not breed until they are three years old. Females become sexually receptive between January and April, and have an estrus period of five to seven days. They probably mate for life, and the alpha male of the pack is the sole father of the pack's young. Gestation lasts 63 days, and an average of six pups compose a litter. The pups emerge from the den during their third week and are protected and fed by the parents and the pack members. By ten months of age the young are participating in hunts (Clark and Stromberg, 1987).

Suggested reading: Mech, 1970; Clark and Stromberg, 1987; Mech and Boitani, 2003; Smith, 2005; Halfpenny, 2007; Ripple and Beschta, 2007; Armstrong, Fitzgerald, and Meaney, 2011; Buskirk, 2016.

Red Fox. *Vulpes vulpes*. Ubiquitous.

Identification: This canid is easily recognized by its long ears, pointed snout, and large, bushy tail tipped with white. Usually it is rufous brown, but variations often occur in the pelage color. Melanistic (black) pelage occurs commonly in western populations. A silver variant is also melanistic, but white tips to the long blackish guard hairs give the overall pelage a silvery cast. There is also a cross fox variant that is yellowish dorsally but has a cross-like pattern of darker fur along the dorsal midline and down the shoulders. These variants are most common in Canadian and Alaskan populations. Length: 83–101 cm (32.5–40 in); tail: 291–461 mm (11.4–18.1 in). Weight 3–7 kg (6.6–15.4 lb).

Voice: The most common red fox vocalization is a series of rapid, high-pitched, yip-like barks, generally thought to be an identification signal, as it is known that individual foxes can recognize one another by this barking. Another vocalization is a loud "screamy howl," uttered by vixens in estrus to attract males to them; it has been described as sounding like a human baby in distress. Other vocalizations are much softer and used in close-quarter communication. One is "gekkering," a rapid chattering series of notes with occasional inserted howls that is used during aggressive encounters between adults and during play-fighting by juveniles. Parents chortle to their young when bringing food to them.

Status: Red foxes occur statewide, mostly at lower elevations and in open country, such as open woodlands, riparian corridors in grasslands, farmlands, pastures, and brushlands, Occasionally they are found in towns, where they gain some protection from coyotes and might rob food put out for pet dogs. They eat a wide variety of vegetable foods, including grass, berries, acorns, and fruits. Grasshoppers and other insects plus crayfish compose about a quarter of the normal diet, and during winter mice, rabbits, squirrels, and other small mammals are important foods. Red fox hearing is highly sensitive to low-frequency sounds such as underground or below-snow noises of digging mammals, which they can pounce upon with a surprising degree of accuracy. Caching of excess food is typical.

Habitats and ecology: In Colorado, red foxes occur upward in mountains to at least 11,400 feet (3.475 m), where they forage in subalpine meadows and alpine edges, and remain above 8,000 feet (2,440 m) during winter. Home ranges are highly variable and have been estimated to be as small as 116 acres (57 hectares) in urban areas to 14,800 acres (6,000 hectares) in less food-rich habitats. Daily movements of an individual might exceed six miles (10 km). Territories of a mated pair are defended during the breeding season and are scent-marked with urine and feces (Armstrong, Fitzgerald, and Meaney, 2011).

Breeding biology: Breeding occurs from December to March (in Colorado, primarily during January and February). Estrus lasts one to six days, and gestation requires about 52 days. The average litter size is 5.5 but ranges widely from 1 to 17. Both members of the pair provision the young, which remain in the den for about a month. They are fully grown by six months of age, and by late September the juveniles disperse. Females can breed by about ten months of age, although about 10 percent of young females might not breed (Clark and Stromberg, 1987; Armstrong, Fitzgerald, and Meaney, 2011).

Suggested reading: Clark and Stromberg, 1987; Gese, Stotts, and Grothe, 1996; Armstrong, Fitzgerald, and Meaney, 2011; Buskirk, 2016.

Swift Fox. *Vulpes velox*. Dispersed.

Identification: This tiny (Yorkshire terrier–sized) fox closely resembles the gray fox, which occurs sympatrically (has overlapping ranges) with it in eastern Wyoming. Like gray foxes, swift foxes have black-tipped tails, but the top of the tail is not black. They are also generally paler than the gray fox and are only slightly cinnamon-tinted along the flanks and legs. Length: 735–880 mm (28.9–34.6 in); tail: 240–350 mm (9.4–13.8 in). Weight: male: 2.2–2.9 kg (4.8–6.4 lb); female: 1.8–2.3 kg (4.0–5.1 lb).

Voice: Swift foxes utter rapid sequences of 4 to 25 barks, mainly during the breeding season. The barks are thought to be contact calls for social units and associated with mating and territoriality. They have been found to be acoustically unique to individuals (Darden, Dabelsteen, and Pedersen, 2003).

Status: The swift fox is limited to eastern Wyoming and is generally declining across its range. It was a candidate for federal listing as endangered or threatened until 2002, when that status was terminated. In Wyoming, it is considered a Species of Greatest Conservation Need. Surveys by the Wyoming Game and Fish Department in 2013 detected the swift fox at 14 locations in Campbell, Weston, Niobrara, Converse, Albany, Platte, and Laramie Counties, suggesting it is still widespread if rare in the state.

Habitats and ecology: Associated with shortgrass prairies, swift foxes were reported by Clark and Stromberg (1987) to be locally abundant in Laramie County with densities for a pair at 3 to 5 square miles (5–8 km²) in wetter shortgrass prairies. These authors noted that swift foxes have suffered from predator poisoning, over-trapping, and habitat losses, and that cars often kill them. Coyotes also commonly kill them and at times have been judged to be the species' most significant source of mortality. Based on studies in five states, their primary food sources by number of items eaten are arthropods (mainly insects), although by volume mammals such as jackrabbits and cottontails are probably most important (Johnsgard, 2005).

Breeding biology: Swift foxes mate monogamously, although in some areas males have been found paired with two females that share the same den. They breed from early March through July, and the gestation period is 51 days. The litter size ranges from one to eight young (eight mammae are present on females), but typically the number is four to five. The eyes of the young open by two weeks of age, and they begin to leave their den by the fourth or fifth week. Females begin to breed during their first year of life. Annual survival rates in Wyoming have been estimated at 40 to 69 percent (Olson and Lindzey, 2002a), so few wild animals survive beyond their third year. However, captive females have bred at six years of age, and one male bred at ten years of age (Armstrong, Fitzgerald, and Meaney, 2011).

Suggested reading: Clark and Stromberg, 1987; Covell, 1992; Schauster, Gese, and Kitchen, 2002; Dark-Smiley and Keinath, 2003; Armstrong, Fitzgerald, and Meaney, 2011; Buskirk, 2016.

Gray Fox. *Urocyon cinereoargenteus*. Slightly dispersed.

Identification: This fox closely resembles the substantially smaller swift fox but has a black stripe along the top of its bushy tail. A cinnamon-orange tint usually extends along the bottom of the tail and underparts up over the sides of the shoulders and neck to the back of the ears but sometimes these areas are whitish. Gray foxes also somewhat resemble coyotes, which are substantially larger and lack the gray fox's strong cinnamon-orange tints in the pelage. Length: 80–112 cm (31.5–44.3 in); tail: 275–343 mm (10.8–13.5 in). Weight: 3–7 kg (6.6–15.4 lb).

Voice: Gray foxes utter loud barks or yips during the breeding season and produce a variety of other chuckles, growls, and squeals. Probably their vocabulary is much like that of the red fox, but no detailed comparisons seem to be available.

Status: Gray foxes are rare and have a limited range in eastern Wyoming. They are often found along stream courses lined with deciduous riparian woods as well as in brushy woodlands among areas of hilly or irregular topography. In some areas, they have also occupied places close to human habitation. They den in rocky outcrops, brush piles, burrows, and hollow trees, and unlike other foxes are able to climb trees with ease by grasping with their forefeet and long front claws and using their hind legs for propulsion. Tree dens up to 25 feet (7.6 m) above the ground have been found (Clark and Stromberg, 1987). In the western states, gray foxes have large home ranges, estimated to range from about 74 to 864 acres (30–350 hectares), and in the east their home ranges

are sometimes even larger, up to and exceeding 1,235 acres (500 hectares) (Armstrong, Fitzgerald, and Meaney, 2011).

Habitats and ecology: Gray foxes are too rare in Wyoming for researchers to know much about their ecology, but in general it is known that they eat a wide variety of plant and animals, much like other foxes. Their prey includes many species of mammals up to the size of muskrats and jackrabbits as well as (presumably small) domestic dogs and cats, and other animals as small as mice. Birds and nestling birds of species such as crows, ducks, and songbirds and bird eggs are known to be eaten, as well as insects. All sorts of vegetable matter is also consumed. Gray fox predators include eagles, mountain lions, bobcats, and coyotes. In some areas, they have been affected by rabies (Armstrong, Fitzgerald, and Meaney, 2011).

Breeding biology: Gray foxes are monogamous and breed from January to April over much of their range. The gestation period is about 59 days, with from one to seven young being born, the average litter size being about four. The young remain in their den for the first month, but begin to disperse that fall. Sexual maturity among females occurs at ten months, but many females probably do not breed during their first year (Armstrong, Fitzgerald, and Meaney, 2011).

Suggested reading: Clark and Stromberg, 1987; Armstrong, Fitzgerald, and Meaney, 2011; Buskirk, 2016.

Family Ursidae (Bears)

(American) Black Bear. *Ursus americanus*. Dispersed.

Identification: Black bears differ in general appearance from grizzly bears by their smaller size, the lack of a pronounced shoulder hump, larger ears, and a straight or convex muzzle profile. They are usually uniformly black except for a brown muzzle but sometimes are brown or cinnamon-colored; they may show some white on the chest and neck. Their front claws are not remarkably long, and their adult paw prints are usually four to five inches wide, rather than up to more than six inches across, as in grizzlies. Length: 137–188 cm (54–74 in); tail: 85–130 mm (3.3–5.1 in). Weight: 40–300 kg (88–660 lb).

Voice: Black bears have a variety of vocalizations. The threat call is a pulsing sound, whereas a high-pitched moan indicates submission or fear. Jaw-popping and woofing sounds are also uttered in antagonistic situations. Nursing or contented cubs make motor-like sounds similar to a cat's purring, and they utter bawling sounds when in distress.

Status: Black bears are mostly limited to the wooded mountainous regions of Wyoming, especially the Greater Yellowstone ecoregion, and are most common where understory vegetation and thick, berry- and acorn-producing plants are present, and where there is abundant food in the form of grasses, forbs, berries, roots, twigs, nuts, and buds as well a pine nuts (Kendall, 1983). They are quite mobile, with home ranges up to 77 square files (200 km²), and may travel over distances of as much as 100 miles (160 km) (Clark and Stromberg, 1987).

Black bear, adult

Habitats and ecology: In Colorado, densities of one bear per 1.9 to 12 square miles (5–31 km²) have been reported, and home ranges of up to 174 square miles (451 km²). Densities of as high as a bear per 1.1 square miles (2.77 km²) were reported in an area of mixed oak–brush–aspen–dry meadow. In one study, resident adult females were found to have annual home ranges of 5.4 to 76 sqaure miles (14–199 km²), whereas those of adult males were from 12 to 56 square miles (31–145 km²) (Armstrong, Fitzgerald, and Meaney, 2011).

Breeding biology: Females begin breeding at an average age of five years, but the range is from three to seven years. Breeding occurs in early summer, from June through July and perhaps into August. Mating stimulates ovulation, and the gestation period lasts a surprisingly long seven to eight months, owing to delayed implanting of the fertilized egg until November or December. The cubs are born in January or February, while the female is hibernating in her winter den. The litter size is two, occasionally three. In Colorado, the intervals of producing litters range from one to four years, with most females having offspring every other year. The young remain with their mothers through the second winter of their lives. Survival rates vary between hunted or trapped and protected populations, but most black bears in wild populations do not live more than eight to ten years (Armstrong, Fitzgerald, and Meaney, 2011).

Suggested reading: Clark and Stromberg, 1987; Schullery, 1992; Lynch, 1993; Gasson, Grogan, and Kruckenberg, 2003; Halfpenny, 2007; Armstrong, Fitzgerald, Meaney, 2011; Buskirk, 2016.

Grizzly Bear. *Ursus arctos horribilis*. Local.

Identification: This huge bear has a distinctive hump on its shoulders (from a mane of long hair), a slightly dished-in muzzle profile, and a variable but often brown to grayish black pelage that is usually "grizzled" with silvery tips on the long upper-body hair. Other color variations occur regionally. Their very long front claws are often evident, and their somewhat human footprint–like paw prints exceed six inches in width. *Yellowstone population:* Length: males avg. 1.64 m (5.4 ft), females avg. 1.53 m (5.0 ft); tail: males avg. 43 mm (1.7 in), females avg. 40 mm (1.6 in). Weight: dominant males avg. 575 pounds (260 kg), maximum 714 pounds (323 kg) (Halfpenny, 2007). *Note:* The Alaska population of this species, locally called the brown bear, is substantially larger and heavier, with a length of 1 to 2.8 meters (3.3–9.2 ft) and weight of more than 600 kg (1,322 lb).

Voice: Like black bears, grizzly bears have many vocalizations. Threatening animals make huffing sounds and repeated jaw-popping noises (about 2–3 per second). Snorting, tooth clacking, and blowing air out through the nostrils also occur during aggressive encounters. While communicating with her young, females make moaning, grunting, or growling sounds. Males make similar noises as well as bellowing during the breeding season.

Grizzly bear, adult

Status: The grizzly bear has been a federally listed endangered species since the 1973 Endangered Species Act was enacted. In 1975 an estimated 136 bears inhabited the Greater Yellowstone ecoregion, a number that had increased to 571 by 2007. More recently the Yellowstone population has been at about 700 animals, probably at or near the region's carrying capacity, and the reason in 2017 for removing the species' endangered status in the three affected states—Montana, Idaho, and Wyoming. Wyoming's effort to initiate a limited trophy hunt in the fall of 2018 was blocked by a federal judge before the season could begin, but it is likely continuing efforts will be made to overturn this ruling in the three states involved.

Habitats and ecology: Like the black bear, the grizzly is catholic in its food selection. Studies in Yellowstone Park of grizzly specimens obtained since the 1970s that used blood, bone, and hair samples indicate that the proportion of meat consumed by the grizzlies was 79 percent for adult males, 37 percent for subadult males, and 45 percent for adult and subadult females. Corresponding data for black bears indicate that they consume a higher proportion of vegetable foods than grizzlies (Halfpenny, 2007). The most important single plant food source for grizzly bears in Yellowstone are the "pine nuts" of the whitebark pine (Kendall, 1983), a subalpine tree whose nut-like seeds are also major foods for Clark's nutcrackers and many small rodents. These hard seeds, like those of limber pine and pinyon pine, are large and contain triacylglycerol, which is a high-energy lipid. The bears eat them in great quantities during autumn while building up fat reserves, prior to their entering hibernation (Halfpenny, 2007). Surprisingly, moth larvae living in the litter of mountain meadows and talus slopes are also an important fall food.

Grizzly bear, adult male

Breeding biology: Male grizzlies, like black bear males, are polygynous, and females likewise might mate with more than one male. Females are sexually mature and reproduce at an average age of 5.9 years; they have reproductive cycles that produce litters at intervals averaging 2.9 years. Breeding extends from May to July with a mating peak in June. Estrus lasts 27 days. Implantation of the fertilized egg is delayed, and the cubs are born in the winter den during late January. Litter sizes range from one to four, with two cubs being the norm. The cubs remain with the mother for one or two years. Studies in Yellowstone Park indicated that 13 percent of that population were yearlings, 25 percent two- to four-year-olds, and 44 percent adults. Females breed throughout their entire adult lifetime and have been found to survive for as long as 25 years. Even in protected populations such as Yellowstone Park up to 90 percent of their mortality is human caused, often by careless drivers when bear families are trying to cross the heavily trafficked roads of the park (Clark and Stromberg, 1987; Armstrong, Fitzgerald, and Meaney, 2011; Wilkinson, 2013).

Suggested reading: Clark and Stromberg, 1987; Knight, Blanchard, and Eberhardt, 1988; Schullery, 1992; Lynch, 1993; Craighead, Sumner, and Mitchell, 1995; Reed-Eckert, Meaney, and Beauvais, 2004; Schwartz, Haroldson, and White, 2006; Halfpenny, 2007; Armstrong, Fitzgerald, and Meaney, 2011; Wilkinson, 2013; Wyoming Game and Fish Department, 2016; Buskirk, 2016.

Family Procyonidae (Raccoons)

Northern Raccoon. *Procyon lotor*. Ubiquitous.

Identification: Another easily recognized mammal, raccoons have a white-bordered black face mask and a tail with four to seven brown and black rings. They also have conspicuous white-bordered black patches behind their ears but are otherwise rather uniformly brown, grayish, or reddish brown, varying somewhat with the season and the regional population. Northern populations are the largest and are able store body fat for winter survival. Length: 60–1.05 cm (23.6–41 in); tail: 200–400 mm (7.8–15.7 in). Weight: 3.6–9 kg (8–19.8 lb).

Voice: Highly vocal, raccoons produce a wide variety of sounds, including whimpering, purring, screaming, growling, hissing, snarling, and chuckling as well as a shrill, tremulous whistle that is often uttered by adults during autumn.

Status: Raccoons are widespread and highly adaptable, with the ability to survive almost as well in cities as in the countryside. They are most common where water is present, such as marshes and riverine forests. Their foods are diverse, depending on their habitats. Raccoons vary from being carnivores to vegetarians and consume such items as insects, fish, crustaceans, bird eggs, mollusks, earthworms, grain, fruit, and discarded kitchen wastes. Their front feet are relatively handlike, adapted for holding and manipulating objects.

Habitats and ecology: Raccoons are probably most abundant in riverine woodlands, especially those with oaks, elms, and sycamores, where they use tree hollows, squirrel nests, stumps, or rotten logs for dens. Individual raccoons might have several dens within their home range, which for males might exceed 4,800 acres and for females up to 2,400 acres. Mostly nocturnal, raccoons can move at speeds of up to 15 miles per hour, and long-term movements of more than 175 miles have been reported.

Breeding biology: Raccoons do not establish territories but often move in family groups of up to about six animals. Males become sexually active in January and February, moving from den to den in search of receptive females. Gestation requires 63 to 65 days, so most litters of three to four young are born in late April or early May; rarely births occur as late as October. Blind at birth, the eyes of newborns open at 18 to 29 days of age. The youngsters are weaned by eight to ten weeks and soon begin to follow their mother on foraging searches. They become sexually adult as yearlings. Average longevity in the wild is usually only a few years, although lifespans of up to 12 years have been reported (Clark and Stromberg, 1987).

Suggested reading: Clark and Stromberg, 1987; Armstrong, Fitzgerald, and Meaney, 2011; Buskirk, 2016.

Ringtail. *Bassariscus astutus*. Highly localized.

Identification: The ringtail is about the size of a domestic cat and has a similar body form but with a bushy, black-and-white ringed tail (6–9 rings) that is as long as the body and head combined. Highly agile, it can climb trees and navigate steep rocky slopes and nearly vertical cliff walls with ease. Length: 630–810 mm (24.8–31.8 in); tail: 305–438 mm (12.0–17.2 in). Weight: 0.7–1.1 kg (1.5–2.4 lb).

Voice: Ringtails have a wide variety of vocalizations associated with their sociality, parental care, and the fact that as many as three generations might share a communal territory (Willey and Richards, 1981). Aggressive calls include growls, barks, and (under extreme duress) loud wavering screams. Calls associated with social interactions include rhythmic whistle-grunts, which are often uttered by several animals in a chorus-like manner; prolonged chitters by subadults; and mewing by nursing young. Estrus females also utter a chittering call similar to that of subadults, but it is lower in pitch.

Status: Ringtails are rare and highly restricted in their Wyoming distribution to the lower Green River Basin. They have also been reported along a short stretch of the North Platte River south of Seminoe Reservoir (Clark and Stromberg, 1987).

Habitats and ecology: This nocturnal omnivore is associated with a variety of habitats, including riparian brushland, oak chaparral, and pinyon-juniper woodland, especially in rocky topography. In Colorado, it occupies roughlands, rocky canyons, and foothills at elevations up to about 9,200 feet (2,800 m). A variety of small mammals are eaten, as well as birds and lizards, but insects and fruits such as juniper "berries" probably constitute the majority of their diet. They evidently do not require an external source of water, and they den in rock crevices, hollow logs, old buildings, and trees.

Home-range studies in various regions indicate marked variations, from as little as an average of 50 acres (20 hectares) for females and 106 acres (43 hectares) for males to as much as a monthly average of 336 acres (136 hectares). Woodland densities of 3.9 to 7.5 animals per square mile (1.5–2.9 per km²) have been estimated (Armstrong, Fitzgerald, and Meaney, 2011).

Breeding biology: Breeding occurs in spring, with the precise month uncertain for Wyoming but usually during March in Utah. Young there are born from late April to early June. Litters range from one to five, averaging about three or four. Likely, only the female tends to the young, and possibly the male is excluded from the den. The young are weaned by three to five months and may begin foraging with their mother when only two months old. Both sexes are sexually active by ten months of age (Armstrong, Fitzgerald, and Meaney, 2011).

Suggested reading: Willey and Richards,1981; Clark and Stromberg, 1987; Armstrong, Fitzgerald, and Meaney, 2011; Buskirk, 2016.

Family Mustelidae (Weasels)

American Badger. *Taxidea taxus*. Ubiquitous.

Identification: The badger is another of the grassland mammals that is instantly recognizable. It has a distinctively black-patterned face with a white line up the midline of the head and a broader buffy stripe extending from the mouth diagonally back below the eye to the ear. Badgers have long, grizzled, grayish brown dorsal fur and a low-slung body profile. Males are larger than females. Badgers have powerful forelegs with partially webbed front toes and long claws, which provide for highly efficient digging. Length: 600–730 mm (23.6–28.7 in); tail: 105–135 mm (4.1–5.3 in). Weight: 6.4–11.5 kg (14–25 lb).

Voice: Various hissing and grunting noises are uttered when a badger is threatened, but these are otherwise quite silent animals.

Status: Badgers have a widespread occurrence in Wyoming but are most common on the open plains, where rodent populations are high. Badgers are sometimes shot for "sport," and at times bounties have been paid for them, but most are probably killed in traps set for other mammals. Their digging activities help with soil development, and the holes they produce are used for escape or as dens by many other animals. They are effective killers of venomous snakes and are relatively immune to rattlesnake venom.

Habitats and ecology: Badgers range across many habitats and ecosystems but favor open areas of meadows, prairies, steppe grasslands, or other places where subterranean dens for breeding and semihibernation during winter can be dug. They are extremely adept at digging out ground squirrel and prairie dog tunnels; sometimes two badgers will collaborate, with one digging and the other waiting at another tunnel entrance to catch any escaping inhabitants.

Breeding biology: Badgers breed in the summer or fall, but the pair-bond lasts only until the female is fertilized. After conception she takes on all further responsibilities for reproduction, including protecting the offspring. The young are not born until March or April because of delayed implantation of the embryos. Two to four babies typically compose a litter; they are born helpless and blind but are fur-covered at birth. The young are weaned when they are about two-thirds grown, and females may become sexually mature within a year of birth. Both sexes are highly mobile as adults, males having enormous home ranges of 600 to 4,000 acres (240–1,600 hectares). Home ranges of males often overlap those of several females; anal scent glands are present that may help with social communication and coordinating sexual activities. Badgers can potentially live up to 14 years, but there is high mortality among the young, probably owing to starvation and being killed by humans.

Suggested reading: Clark and Stromberg, 1987; Neal, 1996; Goodrich and Buskirk, 1998; Armstrong, Fitzgerald, and Meaney, 2011; Buskirk, 2016.

Pacific (American) Marten. *Martes americana*. Slightly dispersed.

Identification: Martens (often called pine martens) are weasel-like mammals with almost entirely chocolate brown bodies but golden yellow on the throat, lower neck, and chest. They have a long bushy tail that is somewhat darker brown than the body. Martens are highly agile, climbing and running among tree branches and vegetation fast enough to catch squirrels, one of their many prey species. Length: 465–659 mm (15.7–25.9 in); tail: 135–160 mm (5.3–6.3 in). Weight: 400–1,400 g (0.8–3.1 lb).

Voice: Belan, Lehner, and Clark (1978) described some of the marten's numerous vocalizations. They include huffing, which might be associated with fear or threat; a single hisslike exhalation; a quite variable chuckle; a high-pitched scream of fear or anger; and a whine, also associated with fear or discomfort. Another was a high-pitched tone associated with the last two vocalizations. These calls were all heard from a wild captive individual. Vocalizations associated with courtship and mating are apparently still undescribed.

Status: Martens are almost entirely limited in Wyoming to the Greater Yellowstone ecoregion and are probably rare except in the national parks. Under such protection, they might become fairly tame; I frequently and delightedly watched them for hours on end near a den at Mardy Murie's cabin during the mid-1970s (Johnsgard, 1982). Often I would put out scraps of meat along their runways to see how long it would take for the martens to find them; most often a sharp-eyed raven would get there first. Marten pelts are valuable, and very regrettably they are still trapped in Wyoming. In recent years, about 8,000 martens have been killed annually for their fur.

Overleaf: Northern river otter, family

Habitats and ecology: Martens are mostly associated with montane coniferous forests, especially mature spruce-fir forests, but they also use lodgepole pine, Douglas-fir, and cottonwood riparian corridor communities. They also extend into alpine tundra but do not tolerate more than about 25 to 30 percent of their home range in open areas. Large-diameter trees, logs, and snags are important aspects of their habitat, as are elements that enhance their prey base, such as shrubs and berry sources. In some areas, voles and other mice make up a large majority of their diet, but pine squirrels are also locally important, as are snowshoe hares in Manitoba (Armstrong, Fitzgerald, and Meaney, 2011).

Breeding biology: Martens become sexually mature at 12 to 15 months of age. Most mating occurs between late July and early September, with males being polygynous and females selectively polyandrous. Because of delayed implantation, the gestation period lasts 220 to 276 days, with the young born in spring or summer. Litter sizes range from one to five, averaging about three. By five to six weeks of age the young leave the nest and become very active, attaining adult size by about three months of age (Armstrong, Fitzgerald, and Meaney, 2011).

Suggested reading: Clark and Stromberg, 1987; Ruggiero, 1994; Armstrong, Fitzgerald, and Meaney, 2011; Aubry et al., 2012; Buskirk, 2016.

(American) Mink. *Neovison vison*. Ubiquitous.

Identification: The mink is the largest of Wyoming's typical weasels. It has the most uniformly dark brown pelage, except for some white on the chin and chest. Its tail is long, up to one half the animal's total length, but is not so bushy as the similar-sized marten's. Mink are almost always found close to water. Length: 460–700 mm (18.1–27.6 in); tail: 150–230 mm (5.9–9.0 in). Weight: 0.9–1.6 kg (1.98–3.5 lb).

Voice: Mink have a wide variety of vocalizations that include aggressive hissing, screaming, chuckling, squeaking, barking, and purring. Screaming is used as a defensive threat, and chuckling is uttered by both sexes during the breeding season and is associated with sexual stimulation. Squeaking is used by both sexes when in pain or fearful (Gilbert, 1969). No detailed analysis of the mink's other calls and their functions is apparently available.

Status: Mink have statewide distribution in Wyoming but are always found close to streams, rivers, lakes, or ponds. Their home ranges tend to be linear along river or stream courses, and in one study averaged 5,350 feet (1,630 m) (Clark and Stromberg, 1987).

Habitats and ecology: Mink are active throughout the year, becoming more diurnal during winter months and increasing their use of fish as prey. However, mink are also opportunistic predators, often feeding on rodents—they are evidently able to hear the ultrasonic sounds often made by them (and also some of the ultrasonic sounds made by their own young). Compared with river otters, mink are more likely to take terrestrial prey, such as mammals and birds, and less likely to take fish. In a Colorado study, their density was estimated as 1.7 animals per square mile (Armstrong, Fitzgerald, and Meaney, 2011).

Breeding biology: Breeding occurs from late February to April. Males are polygynous, and females may also mate with multiple males, so that a single litter might have more than one father. The gestation period is highly variable, from 40 to 74 days, owing to variably delayed implantation. Litter sizes range from one to eight but average five. The young are born in late April or May, and they reach sexual maturity at about ten months. Mink might remain reproductively active for about seven years (Armstrong, Fitzgerald, and Meaney, 2011).

Suggested reading: Clark and Stromberg, 1987; Armstrong, Fitzgerald, and Meaney, 2011; Buskirk, 2016.

Northern River Otter. *Lontra canadensis*. Dispersed.

Identification: River otters are well named; they are mostly found along rivers, are substantially larger than mink, and have a stout, gradually tapering tail. They swim very rapidly, often while totally submerged, and are dark brown overall, except for lighter underparts. Length: 915–1,346 mm (36–53 in); tail: 352–510 mm (13.8–20 in). Weight: 5–13.7 kg (11–30 lb).

Voice: Like other members of the weasel family, river otters are highly vocal. Their utterances are said to include buzzes, whistles, chirps, twitters, growls, and a staccato chuckle as well as a hair-raising scream that is audible for as far as 1.5 miles over water. As highly social animals, it is not surprising that otters use a wide diversity of vocalizations, which have seemingly not yet been fully analyzed.

Status: Otters are largely limited to western Wyoming and are concentrated in the Greater Yellowstone ecoregion, especially Yellowstone National Park and the Snake River. Water quality and quantity are important features of the habitat that influence otter numbers, as are an abundance of fish and crustaceans, and the presence of ice-free stretches of river during winter. Crayfish and slow-moving fish such as suckers are important foods, but amphibians, birds, and insects are also consumed (Clark and Stromberg, 1987; Armstrong, Fitzgerald, and Meaney, 2011).

Habitats and ecology: Home ranges of otters vary greatly and have been reported in Colorado to be from 1.2 to 48 miles (2–78 km) in length with a mean of 20 miles (32 km). However, in Alberta male home ranges were sometimes greater than 124 miles (200 km), whereas female home ranges were typically about 43 miles (70 km). Males not only have larger home ranges than females but also tend to form social groups. Groups vary from families composed of females and their young to groups of bachelor males and mixed-sex congregations of up to 30 animals (Armstrong, Fitzgerald, and Meaney, 2011).

Breeding biology: Otters reach sexual maturity at two years of age, but some males might not breed successfully until they are at least five years old. Mating occurs immediately after the birth of a female's litter, during March and April in Colorado. A female's estrus period may last more than 40 days, during which scent tracks are produced that can be followed by males. Apparently ovulation is induced by copulation, and because of delayed implantation the gestation period might last 290 to 375 days. The average litter size is about three, ranging from one to six. The young begin to leave the den at two months and are weaned by about three months. They remain with their mother for about seven months, and siblings might remain together for more than a year (Armstrong, Fitzgerald, and Meaney, 2011).

Suggested reading: Clark and Stromberg, 1987; Ruggiero, 1994; Kruuk, 2006; Wengeler, Kelt, and Johnson, 2010; Armstrong, Fitzgerald, and Meaney, 2011; Buskirk, 2016.

Long-tailed Weasel. *Mustela frenata*. Ubiquitous.

Identification: The long-tailed weasel is the largest of the typical (*Mustela*) weasels and very probably the most common. It is almost as large as a marten, but its long tail is less bushy and is tipped with black. This weasel is slightly larger than the very similar ermine (short-tailed weasel), which in summer has whitish to buff rather than tawny underparts. During winter these two species are both entirely white, except for a black-tipped tail. Length: 300–450 mm (13.7–17.7 in); tail: 110–175 mm (4.3–6.9 in). Weight: 130–316 g (0.2–0.7 lb).

Voice: Svenden (1976) described three basic vocalizations: the trill, screech, and squeal. The trill occurs under calm conditions, such as when the animal is playing or hunting. The screech occurs when it is disturbed, or it is used as a defensive signal. A squeal is uttered when the animal is in distress.

Status: Judging from the number and distribution of records, this species is probably the most common and widespread of the three weasels of Wyoming. All three weasels subsist primarily on small mammals, including species up to the size of rabbits, and also take birds at least as large as grouse. They also eat the eggs of waterfowl and grouse. The animals have been reported to consume 20 to 40 percent of their body weight daily, indicating a very high metabolic rate for an animal of that size. Long-tailed weasels are generally solitary except when breeding, probably becoming more social when food sources are abundant and more solitary when they are rare.

Habitats and ecology: During summer these weasels are mostly diurnal and spend much of the time hunting and searching for mates, primarily during early morning and late afternoon hours. Home ranges of males are larger than those of females, often of about 30 to 40 acres (12–16 hectares) in favorable habitats, but in areas of fragmented agricultural land may be as large as 126 acres (51 hectares) for females and 445 acres (180 hectares) for males. In Colorado habitats from sagebrush to alpine tundra, a weasel density of 2.1 per square mile (0.8 per km²) was estimated, but in the eastern United States estimates of as many are 98 animals per square mile (38 per km²) have been made (Clark and Stromberg, 1987; Armstrong, Fitzgerald, and Meaney, 2011).

Breeding biology: Breeding occurs during summer, mostly in July and August, with the female mating with the male whose home range includes hers. Males might form transitory pair-bonds with several females but play no part in parental care. The gestation period lasts 220 to 237 days as a result of delayed implantation, with young mostly being born in April and May. A single litter is produced per year, which averages about seven young. By six weeks of age the young are weaned. Females reach sexual maturity at about 3 months of age, whereas males begin to reproduce at about 15 months of age (Armstrong, Fitzgerald, and Meaney, 2011).

Suggested reading: Svenden, 1976; Clark and Stromberg, 1987; Armstrong, Fitzgerald, and Meaney, 2011; Buskirk, 2016.

Black-footed Ferret. *Mustela nigripes*. Localized.

Identification: This rare weasel has a distinctive facial pattern of a black mask on a mostly white face, a black-tipped tail, and black legs on an otherwise yellowish tan body. The species

Black-footed ferret: adult head (*upper*), inside prairie dog burrow (*middle*), and on a prairie dog mound (*below*)

Overleaf: Black-footed ferret, adult

is essentially limited in distribution to large prairie dog colonies. Length: 480–567 mm (18.9–22.3 in); tail: 114–127 mm (4.5–5.0 in). Weight: 530–1,300 g (1.2–2.9 lb).

Voice: Clark et al. (1986) listed the vocalizations they observed in this species, which they identified as the bark, chattering bark, bluff-hiss, growl, and "ungh." These were contextually classified as variously associated with threat, defense, greeting, and mating. Some additional calls are uttered by the young. A few other vocalizations have been mentioned in the literature, but no comprehensive analysis of the species' vocalizations and their situational functions seems yet to be available.

Status: This species is perhaps America's rarest wild mammal and was among the first species selected for listing as endangered under the Endangered Species Act of 1973. At that time, this ferret was generally feared to already be extinct, until a small central Wyoming population was discovered near Meeteetse in the Bighorn Basin during 1981. That population was immediately placed under the strictest protection but was nearly lost in 1985 as the result of a canine distemper outbreak. However, 17 still-surviving animals were live-trapped and placed in captivity. This small group became the reproductive nucleus of the population that is alive today, having produced a genetic line of several hundred relatively inbred animals. Interestingly, the similarly endangered North American whooping crane population of about 600 birds has likewise descended from a nucleus of only about 16 to 18 wild birds that existed in 1941. Reintroduction efforts for both species have proven to be extremely difficult, and it is impossible to judge their long-term prospects for success.

Habitats and ecology: Black-footed ferrets are almost wholly dependent on prairie dogs for their food intake, about 90 percent of their food intake is prairie dogs. To sustain life, adults must eat 1 prairie dog every two to six days, or about 100 per year. Assuming typical prairie dog densities, that means each adult ferret needs 47 to 94 acres (19–38 hectares) of prairie dog habitat to survive, if no other foods are consumed (Armstrong, Fitzgerald, and Meaney, 2011). For a reintroduced South Dakota ferret population, the prairie dog colonies averaged 47 acres (19 hectares) in area and were 0.9 mile (1.5 km) apart, while in Wyoming they averaged 795 acres (322 hectares) and were 0.4 mile (0.6 km) apart (Clark and Stromberg, 1987).

Breeding biology: Like other weasels, male ferrets are probably polygynous, and females probably mate with males sharing their overlapping home ranges. They mate during spring (mid-March to early April), and the gestation period is about 42 to 45 days. Most young are born in May. Litter sizes average 3.5 young but range from 1 to 5. A single litter is produced per breeding season. The young emerge from the burrow when they are about three-fourths grown and begin to disperse by September or October. They become sexually mature in their first year (Clark and Stromberg, 1987; Armstrong, Fitzgerald, and Meaney, 2011).

Suggested reading: Clark, 1986b; Clark and Stromberg, 1987; Clark, 1989; Seal et al., 1989; Miller, Forrest, and Reading, 1996; Armstrong, Fitzgerald, and Meaney, 2011; Buskirk, 2016.

Striped Skunk. *Mephitis mephitis*. Ubiquitous.

Identification: Striped skunks are easily (thankfully) identified by their black-and-white-striped pelage pattern. A narrow white stripe extends from the nose to the crown, and two black-bordered white stripes extend from the nape or forehead along the lower back to the tail. The rest of the pelage is black, unlike that of the more extensively white-patterned spotted skunk. Length: 520–770 mm (20.4–30.3 in); tail: 170–400 mm (6.7–15.7 in). Weight: 1.8–4.5 kg (3.9–9.9 lb).

Voice: Skunks produce a variety of calls, ranging from hissing, whining, and churring to squealing and screaming. They also stamp their feet loudly when confronted with a threat.

Status: Skunks occur throughout Wyoming, over most habitat types. They are very common at lower elevations but in Colorado have been reported as being locally common at elevations up nearly to timberline at 10,000 feet (3,048 m). They have a keen sense of smell but rather poor eyesight, which might account for the frequency of their being killed by moving vehicles. Although they can run at speeds of up to about ten miles per hour, they will rarely retreat from danger,

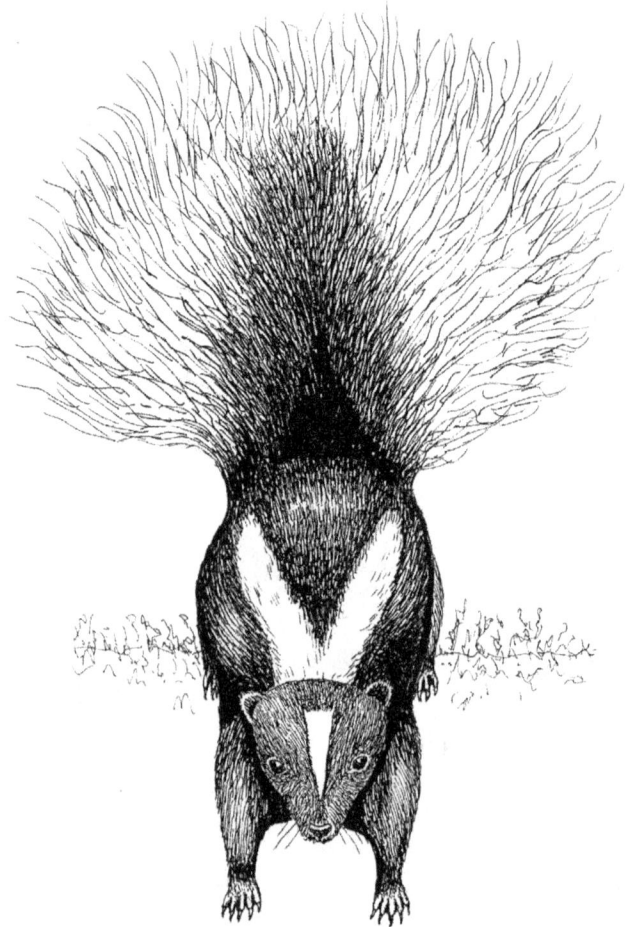

Striped skunk, adult threat posture

Western spotted skunk, adult threat posture (*left*), and Pacific marten, adult standing posture (*right*)

relying on their foul-odor mercaptan-based spray for protection. They can send this musk out for several meters ahead of them with good accuracy, but they do provide fair warning with their conspicuous tail-erection signaling and foot-stomping. Neither of these warnings was ever enough for the terrier I once owned as a youngster, which made both of us sometimes sincerely regretful.

Habitats and ecology: Striped skunks favor mixed woodlands, brushlands, and open fields with wooden ravines and rocky outcrops. They are abundant in cultivated fields and near farmsteads. They are opportunistic foragers, with vegetable materials composing up to 20 percent of their diet. Insects are important foods during the summer months, while during fall and winter carrion, small mammals, other vertebrates, and plant materials are of increased importance. Like bears, skunks move into dens during winter, but they do not hibernate. Male home ranges are larger than those of females, and in various studies those of both sexes have varied from 0.15 to 4.6 square miles (0.4–12 km²) (Armstrong, Fitzgerald, and Meaney, 2011).

Breeding biology: Breeding occurs in February or March, with ovulation occurring about 48 hours after copulation. The gestation period ranges from 59 to 77 days, with births occurring in May or early June. Litter sizes range from two to ten with an average of six; females have 12 mammae. The young are able to breed in the spring following their birth, at about ten months of age. In spite of their seemingly strong defense mechanism, most skunks do not live to become a year old because badgers, foxes, and owls prey on them.

Suggested reading: Clark and Stromberg, 1987; Armstrong, Fitzgerald, and Meaney, 2011; Buskirk, 2016.

Western Spotted Skunk. *Spilogale gracilis*. Localized.
Eastern Spotted Skunk. *Spilogale putorius*. Slightly
dispersed.

Identification: The two spotted skunk species are nearly identical in pelage patterns, having a complex pattern of white spots and stripes on their head and body. The forehead has a white spot, and a white stripe begins below the ear (broader in the western species) and extends back to the shoulders. Two other white stripes (broader in the western species) begin at the base of the forelegs and flanks and extend upward toward the middle of the back. Variably smaller spots are present near the tail. The tail of the eastern species is mostly black; that of the western is mostly white. *Eastern spotted skunk:* Length: 426–567 mm (16.7–21.9 in); tail: 140–235 mm (5.5–9.2 in). Weight: 425–661 g (0.9–1.5 lb). *Western spotted skunk:* Length: 425–575 mm (16.7–22.6 in); tail: 135–241 mm (5.3–9.5 in). Weight: 420–650 g (0.9–1.4 lb).

Voice: Growling, purring, and hissing sounds are used in threat or defense by both species. They also use foot-pattering as a warning signal when threatened. Unlike the striped skunk, the spotted skunks perform a unique handstand posture before emitting their powerfully effective musky spray. The spray is chemically much like that of the striped skunk (Clark and Stromberg, 1987).

Status: Spotted skunks are nocturnal and much less frequently encountered than striped skunks. They are also more arboreal, and the western species tends to be found among shrubby habitats in broken country, whereas the eastern species tends to occur in agricultural lands. Generally, the eastern species occurs east of a line from the eastern slope of the Bighorn Mountains southeast to the common border of Albany and Laramie Counties.

Habitats and ecology: Western spotted skunks are usually found in relatively arid habitats, such as shrublands where they can den in rock crevices, under cacti or shrubs, or in human-made structures or woodpiles. Similar sites are chosen by the eastern species. Both species consume a wide variety of plant and animal foods, much like those of the striped skunk.

Breeding biology: These two species differ most conspicuously in that the eastern spotted skunk breeds in the spring and the western in the fall. Both species have somewhat delayed implantation of the fertilized egg, of 14 days (eastern) or 50 to 65 days (western). The gestation period is 28 to 31 days in the western, with young born in the spring. In the eastern, the gestation period lasts 50 to 65 days, and the young are born in May or June. In both the litter size is from three to six young, averaging about four. The young are sexually mature and able to breed by about four or five months of age (Clark and Stromberg, 1987; Armstrong, Fitzgerald, and Meaney, 2011).

Suggested reading: Clark and Stromberg, 1987; Armstrong, Fitzgerald, and Meaney, 2011; Buskirk, 2016.

Family Felidae (Cats)

Bobcat. *Lynx rufus*. Ubiquitous.

Identification: This widely distributed but highly elusive cat occurs throughout Wyoming. It favors habitats with good hiding places such as rock caves, brush piles, fallen trees, and the like. It is easily distinguished from domestic cats by its bobbed tail (thus, bobcat), which is tipped with black. As with many other wild cats, the ears are slightly tufted and are contrastingly white behind. (Most cat species have similar white spots behind their ears; I have long wondered if they might serve as "taillights" for kittens trying to follow behind their mother at night.) Males average slightly larger than females. Length:

81–101 cm (32.0–39.7 in); tail: 100–165 mm (3.9–6.5 in). Weight: 9–12 kg (19.8–26.4 lb).

Voice: Bobcats have vocalizations much like domestic cats, including yowling during the breeding season, a cough-bark when threatened, and a loud scream.

Status: Bobcats almost certainly occur in every Wyoming county but are probably most common in broken, fairly open, and rocky country.

Habitats and ecology: Bobcats are largely nocturnal, but I once saw one during the early morning hours near a Platte River crane roost. Home ranges of adults are highly variable, from 150 to nearly 5,000 acres. Their foods are also highly diverse, but rabbits and rat-sized rodents predominate, and prey size ranges from mice to fawns. Fish, amphibians, reptiles, and birds are also eaten. Great horned owls have been reported to prey upon immatures, while coyotes and mountain lions often kill adult bobcats.

Breeding biology: Bobcats are solitary, forming only brief pairbonds during the breeding season, which usually occurs during winter and spring. The gestation period is about 62 days, and the litter size ranges from one to eight but averages three. As with other cats, the bobcat kittens' eyes are closed at birth but open at about ten days. By four weeks the kittens are able to eat solid foods. They are weaned by seven to eight weeks and begin to follow their mother on short trips. By about seven months they start to disperse from their natal range and become fairly independent. Females become sexually mature at one or two years, but males are usually mature only in their second year.

Suggested reading: Clark and Stromberg, 1987; Armstrong, Fitzgerald, and Meaney, 2011; Buskirk, 2016.

Canada Lynx. *Lynx canadensis*. Localized.

Identification: The Canada lynx differs from the bobcat in having a shorter black-tipped tail, more obvious ear tufts, and a paler and less clearly spotted and barred pelage that is gray in winter and pale tan in summer. Length: 670–820 mm (25.3–32.3 in); tail: 50–140 mm (2.0–5.5 in). Weight 5.1–18 kg (11–40 lb).

Voice: The voice of a lynx is distinctly catlike, with adult males uttering a deep, loud meow and females a whining purr. During the mating season, two males sometimes engage in a prolonged and almost frightening screaming contest (caterwauling) that is perhaps best described as banshee-like.

Status: This elusive cat is mostly limited to the Greater Yellowstone ecoregion, but it has also been reported from the Bighorn Mountains and the Medicine Bow range. It is rare everywhere and is mostly associated with dense coniferous forests. Over much of its range its population size is closely tied to that of the snowshoe hare, which across North America constitutes an average of 35 to 100 percent of its annual diet. In Colorado, preliminary data (548 kills) indicated that snowshoe hares composed 35 to 91 percent of their annual total prey, averaging 74 percent, with red squirrels of secondary importance (Armstrong, Fitzgerald, and Meaney, 2011).

Habitats and ecology: Lynx are limited to the dense coniferous forests of northwestern Wyoming and primarily inhabit the subalpine forests of Engelmann spruce–subalpine fir, where deep snow accumulates in winter. During summer and autumn it might also move into riparian areas. The presence of young aspen, where the cover is not too dense for snowshoe hares' mobility, is also possibly a factor that influences lynx usage. Lynx are highly mobile; some individuals of a group of several hundred lynx that were introduced into Colorado were later found as far away as Wyoming, Utah, Nebraska, Montana, Kansas, and Iowa. Home ranges in the southern Rockies have been found to vary from 15 to 105 square miles (39–506 km^2). Male home ranges are often about twice as large as those of females (Armstrong, Fitzgerald, and Meaney, 2011).

Breeding biology: Like Wyoming's other cats, male lynx are polygynous and females are seasonally polyestrous. Mating occurs during March and April, and the gestation period is about nine weeks. The young are born in late May and early June. Litter sizes range from one to six but average three young. The juveniles disperse during their first fall, and females probably breed as yearlings (Clark and Stromberg, 1987; Armstrong, Fitzgerald, and Meaney, 2011).

Suggested reading: Clark and Stromberg, 1987; Ruggiero, 1994; Armstrong, Fitzgerald, and Meaney, 2011; Buskirk, 2016.

Mountain Lion (Puma, Cougar). *Puma concolor*.

Ubiquitous.

Identification: The mountain lion is the largest North American cat (wolf-sized) and the only Wyoming species with a long tail that is tipped with black. Length: 200–250 cm (78–98 in); tail: 650–800 mm (25–31 in). Weight: 34–91 kg (75–200 lb).

Voice: Generally silent, but males are said to produce a wailing cry, and females utter a far-carrying, high-pitched yowl (caterwauling) when in estrus. Males are unable to roar like African lions, apparently because their hyoid bone is solid and can't vibrate, and thus they cannot produce a throaty roar. Kittens produce loud, rasping purring, and adults growl like overgrown house cats. Soft chirping sounds are used by mothers and young to locate one another, and typical cat-like meowing, hissing, and spitting sounds are also produced.

Status: Mountain lions are mostly found in western Wyoming, but they do extend into eastern grasslands along wooded riparian corridors. They generally prefer dense cover and rugged terrain where deer are abundant and are often found in shrublands and juniper woodlands. Perhaps they are most abundant in the foothills and low mountains of the Bighorn Basin (Clark and Stromberg, 1987).

Habitats and ecology: Mountain lions hunt by day or night but need sufficient cover for stalking prey, such as brushy, wooded vegetation or rough terrain. In a Utah telemetry study, the three most commonly used vegetation types in descending frequency were mixed ponderosa pine–Gambel oak forest, pinyon-juniper woodland, and spruce-fir forest.

Mountain lion, adult

They also need a source of free water, and in Utah mountain lions had a high incidence of sandstone ledge usage, presumably as resting or sleeping sites. It has been estimated that an adult mountain lion must kill a deer about every two weeks to survive winter. Females with young have higher food requirements, and kill rates also vary with age and social structure. Home ranges vary greatly, from as little as about 15 to more than 270 square miles (about 40–700 km^2) for females, and from 46 to 320 square miles (120–830 km^2) for males (Armstrong, Fitzgerald, and Meaney, 2011).

Breeding biology: Mountain lions breed at anytime of the year, depending on when females come into estrus. The very large home ranges of males would indicate that males are probably polygynous. Females are polyestrous, cycling every few weeks until they are bred. They use loud caterwauling calls to attract males. The gestation period is about 92 days, with young in Wyoming likely to be born in late summer or fall. Litter sizes range from 1 to 6 (females have six functional mammae), but average 2.6. Females continue to hunt for their young for as long as 22 months, meaning that their reproductive rates are low, averaging about four litters per decade. This long period of juvenile dependency also means that when a female with young is killed by humans for "sport," her entire family is likely to die too.

Suggested reading: Clark and Stromberg, 1987; Anderson and Lindzey, 2005; Bekoff and Lowe, 2007; Armstrong, Fitzgerald, and Meaney, 2011; Buskirk, 2016.

Overleaf: Mountain lion, adult female

Family Cervidae (Deer)

(American) Elk. *Cervus canadensis*. Ubiquitous.

Identification: Elk are easily distinguished from other North American deer by their large size and a creamy white rump area surrounding a very small tail of the same color. The antlers of males seasonally have a pair of single large, vertically oriented tines with up to five smaller ones projecting forward and upward. Length: 108–234 cm (3.5–7.6 ft); tail: 8–14 cm (3.1–5.5 in). Weight: 118–497 kg (260–1,095 lb).

Voice: The elk's most distinctive vocalization is the "bugling" of rutting males ("bulls") during September and October, a whistle that begins with a low grunt and shifts to a high-pitched scream that might last for several seconds. Sometimes the call is terminated with a chuckling sound. Bugling is most frequent in early morning and late afternoon. It may serve as a female-attractant, a threat to other bulls, or as a herd-spacing mechanism (Armstrong, Fitzgerald, and Meaney, 2011). Females ("cows") utter barking sounds when alarmed and sometimes also bugle. Cows often vocalize softly to their calves, and the calves to their mothers. Calves also squeal in alarm.

Status: Elk are distributed statewide but are most abundant in the Greater Yellowstone ecoregion, especially in Yellowstone and Grand Teton National Parks and the National Elk Refuge near Jackson. In the winter of 2016–17, the Jackson Hole elk population was nearly 10,500, or about half the maximum ever recorded, which was 21,200 in 1991. Another population, the northern Yellowstone National Park herd of elk, located within and north of the park, totaled 7,500 in the winter of 2017–18, the highest number since 2005. Historically, the park has provided summer range for as many as 10,000 to 20,000 elk, distributed among six to seven herds. In January 2018, the total state elk population was estimated at nearly 105,000 animals. Elk are a major food source for gray wolves, which have reduced the elk numbers in Yellowstone National Park significantly since they were introduced in the mid-1990s. This in turn has affected the abundance of willows and aspens, major elk browse sources, and influenced the status of other species that also depend on willows and aspens (Creel and Christianson, 2009; Ripple and Beschta, 2007).

Habitats and ecology: Mountain meadows on fairly steep slopes are ideal elk habitat, but the animals also locally extend into grasslands and sagebrush-dominated shrublands. Grasses and forbs (nongrass herbaceous plants) are primary winter foods, and grasses are the most important food source in spring, with forbs becoming increasingly important in summer. Browsing on willows, aspen, and other woody materials may be important during winter when grasses become unavailable. In Colorado, elk are usually found above 6,000 feet (1,800 m), and both sexes move to higher elevations in summer, returning to lower altitudes when snow depths impede foraging (Armstrong, Fitzgerald, and Meaney, 2011). Summer home ranges are fairly small, of about 3 to 96 square miles (8–250 km²), but seasonal migrations in the Yellowstone region may extend 60 miles or more. Perhaps the most famous of these passages might be the annual southward fall migration out of Yellowstone Park to the National Elk Refuge, a relatively short distance of about 50 miles, but one that is marked by a gauntlet of elk hunters who are lined up and ready to shoot elk as they leave the safety of the national park and strive to make it alive to the National Elk Refuge.

Breeding biology: Elk breed in the fall, stimulated by decreasing day length. Rutting bulls attempt to gather as many cows as possible under their control, by threats, sparring contests, and even intense fights that might become deadly. Dominant males may thereby gather as many as 30 cows, along with their calves, together for breeding. Females undergo several estrus cycles until they are bred. Single calves are born the following spring. Female calves become sexually mature within a year, but males that are at least three years old are likely to do the majority of the breeding. Although an elk might survive for up to 20 years, mortality in the wild comes much younger and is strongly influenced by hunting (Clark and Stromberg, 1987; Armstrong, Fitzgerald, and Meaney, 2011).

Elk, adult male

Elk, adult male

Suggested reading: Houston, 1968; Van Wormer, 1969; Cole, 1969; Boyce and Hayden-Wing, 1979; Houston, 1982; Clark and Stromberg, 1987; Royce, 1989; Armstrong, Fitzgerald, and Meaney, 2011; Buskirk, 2016.

White-tailed Deer. *Odocoileus virginianus*.

Ubiquitous.

Identification: Compared with mule deer, white-tailed deer have longer and wider tails, which are brown above and white below, and are raised in alarm, exposing their white undersides ("tail-flagging"). Mule deer have notably larger and slightly longer ears (about as long as the distance from the base of their ears to their nostrils), and a rather narrow white tail that is tipped with black. Adult male white-tailed deer have paired antlers with four or more tines that rise vertically from curved and forward-pointing main tines, whereas mule deer have antlers with both main and secondary tines that tilt upward and fork in a Y-like manner. The mule deer is further different from white-tails in its stiff-legged bounding style of running, unlike the white-tail's smooth gallop. Length: 134–215 cm (4.9–7.0 ft); tail: 15–36 cm (5.9–14 in). Weight: 40–215 kg (88–473 lb).

Voice: Adults snort in alarm and utter bawls when traumatized. Males utter grunts in low-level antagonistic situations, grunt-snorts at stronger intensities, and grunt-snort wheezes during high-level dominance interactions. Mothers and fawns utter several maternal-neonatal sounds, and females utter contact calls when separated from a group. Both sexes perform foot-stamping when alert for possible danger, and rutting males utter grunts and perform a "flehmen sniff" (a lip-curling and inhaling behavior common in hooved mammals during rutting) (Atkeson, Marchinton, and Miller, 1988).

Status: White-tailed deer occur statewide, occupying many habitats, but they favor dense deciduous riparian corridors. They avoid dense coniferous forests, open prairie, and very dry lowlands. They are attracted to agricultural lands for access to corn, wheat, fruits, and other crops, and they also choose wetland areas with dense cover. In 2015 Wyoming had an estimated deer population of 423,000, including 70,000 white-tailed and 353,000 mule deer. There are marked year-to-year changes in both species' populations, depending in large part on winter severity. The densest Wyoming white-tailed population is probably in the Black Hills (Clark and Stromberg, 1987).

Habitats and ecology: White-tailed deer have a very diverse diet that tends to be higher in grasses and forbs than is typical of mule deer, although local variations no doubt occur. Unlike mule deer, white-tailed deer do not make major seasonal migrations but often have permanent home ranges covering up to several square miles. Male home ranges are on average larger than those of females (Armstrong, Fitzgerald, and Meaney, 2011). White-tailed deer in Nebraska are slightly heavier on average than mule deer and over large areas of contact in central Nebraska have gradually displaced them.

Breeding biology: Male white-tailed deer do not try to assemble harems, but during the fall rutting season they mark parts of their home range with signposts. These include scrapes made on the earth by pawing an area and urinating on it, breaking nearby branches, and leaving scent marks from the metatarsal glands on adjacent shrubs or other vegetation. They also make polished areas on the trunks of nearby trees with their antlers by rubbing them and thus removing the bark. Males find and follow estrus females by smell, with dominant males being most likely to succeed in mating. The estrus period in females lasts about 24 hours and is repeated about a month later if mating does not occur. The gestation period is 201 days, with fawns being born the following spring. Yearling females produce single fawns, but among older females two young are usually born, and triplets are not rare. The fawns are highly precocial and can run soon after birth. They begin to eat solids by about two to three weeks after birth and are fully weaned by their fifth month. Males probably are also sexually mature as yearlings, but they do not reach full size until they are three or four years old, and only by then are they able to compete effectively for mating opportunities (Clark and Stromberg, 1987; Armstrong, Fitzgerald, and Meaney, 2011).

Suggested reading: Clark and Stromberg, 1987; Putnam, 1988; Atkeson, Marchinton, and Miller, 1988; Armstrong, Fitzgerald, and Meaney, 2011; Buskirk, 2016.

Mule Deer. *Odocoileus hemionus*. Ubiquitous.

Identification: See the white-tailed deer account for identifying these two species. Mule deer and white-tailed deer occasionally hybridize, with intermediate traits appearing. Perhaps the best means of identifying hybrids is to examine the metatarsal gland on the outside of the hind legs. On mule deer these glands are situated high on the lower leg (tarsus), are four to six inches long, and are surrounded by brown fur. The glands of whitetails are at or below the midpoint of the tarsus, are usually less than one inch long, and are surrounded by white hairs. Those of hybrids measure between two to four inches long and are sometimes encircled with white hair. Length: 116–200 cm (3.8–5.6 ft); tail: 10–23 mm (3.2–7.5 in). Weight: 50–200 kg (110–440 lb).

Voice: Vocalizations of the mule deer are essentially like those of the white-tailed deer.

Status: Mule deer are the most common deer in Wyoming, and in 2016 had an estimated population of 364,000 and a hunter kill of 23,379 deer taken by 49,859 hunters. The population has declined by about a third since the early 1990s, and in 2000 totaled about 550,000 animals.

Habitats and ecology: Mule deer are more arid-adapted than white-tailed deer, and are more mobile, seasonally moving

Mule deer, adult male

Mule deer, adult male

to lower altitudes to avoid heavy snow. They range from low elevations to alpine tundra but are most abundant in shrublands with broken topography. Winter diets in the Rocky Mountains consist mostly of browse (74 percent) and forbs (15 percent). Browse still composes 49 percent of the total in spring, with grasses and forbs adding 25 percent each. During summer, browse makes up half of the diet and forbs 46 percent, while in fall browse consumption increases to 60 percent and forbs decline to 30 percent. Unlike white-tailed deer, mule deer seem to be able to survive in the absence of free water, except in very arid habitats. Their annual home ranges are highly variable, from as small as 1.5 to 8.5 square miles (3.92–22 km²); in some areas they make seasonal migrations of up to 100 miles (160 km) (Sawyer, Lindzey, and McWhirter, 2005; Armstrong, Fitzgerald, and Meaney, 2011).

Breeding biology: Breeding occurs late in the fall and early winter, when the polygynous males detect and seek out estrous females. They also aggressively interact with other competing males, as described for the white-tailed deer. Gestation likewise lasts about 200 days, and two fawns are typically born to mature females. Captive does might live for up to 22 years, and bucks to 16 years. However, lifetimes in the wild

are much shorter, with average annual mortalities of 28 to 43 percent common in stable herds, and fawns make up about half of the mortality (Armstrong, Fitzgerald, and Meaney, 2011).

Suggested reading: Wyoming Game and Fish Department, 1978; Clark and Stromberg, 1987; Putnam, 1988; Armstrong, Fitzgerald, and Meaney, 2011; Buskirk, 2016.

Moose. *Alces alces shirasi.* Dispersed.

Identification: The enormous size of moose makes them almost unmistakable, but they also have a dark brown pelage, except for whitish legs, and a long bulbous nose. Adult males have a pendant dewlap, or beard, and seasonally very large palmate antlers with many short tines. Length: 240–290 cm (7.9–9.5 ft); tail: 60 cm (20 in). Weight: 400–500 kg (880–1,100 lb).

Voice: Moose are relatively silent, but cows utter low grunts when communicating with their calves or other moose. Cows in estrus also utter longer calls, essentially extended grunts of varied lengths, when trying to attract bulls. At this time cows also utter whining calls. Males utter low-pitched grunting sounds that serve to challenge other bulls, in conjunction with raking vegetation with their antlers, using sidewise thrashing movements.

Status: Moose in Wyoming are mostly limited to the Greater Yellowstone ecoregion of northwestern Wyoming but also are present in the Sierra Madre, Medicine Bow, and Laramie ranges as well as the Bighorn Mountains. Moose were introduced into the Bighorn Mountains in 1948, and the population peaked during the early 2000s. They appeared in southwestern Wyoming in 1978 as a result of immigration from Colorado. From there they spread into the Sierra Madre and Medicine Bow ranges and the Laramie Mountains. In 2015 the statewide population was estimated at 3,470 animals. During that year 352 moose, or about 10 percent of the population, were killed by hunters. By comparison, from 1989 to 1991 the population had grown to 12,500 to 13,000 animals, and about 1,500 were being killed annually by hunters. Declines since then in the Bighorns and elsewhere in Wyoming have been attributed to the overbrowsing of shrub and tree communities, resulting in poor health and reduced reproductive rates by cows—rather than the decline being a result of wolf predation.

Habitats and ecology: Moose are browsing animals, and their habitats extend from lowland riparian communities to the subalpine spruce-fir zone. They are especially common among abundant willows, which is their favorite food source, and they are also found in upland lodgepole pine forests, and to a lesser degree in aspen and spruce-fir communities and grass meadows. In Wyoming, the most important summer foods are willows, algae, pondweed, and occasionally sagebrush. As much as 91 percent of the moose summer foods analyzed in Rocky Mountain National Park consisted of six kinds of willows (Armstrong, Fitzgerald, and Meaney, 2011).

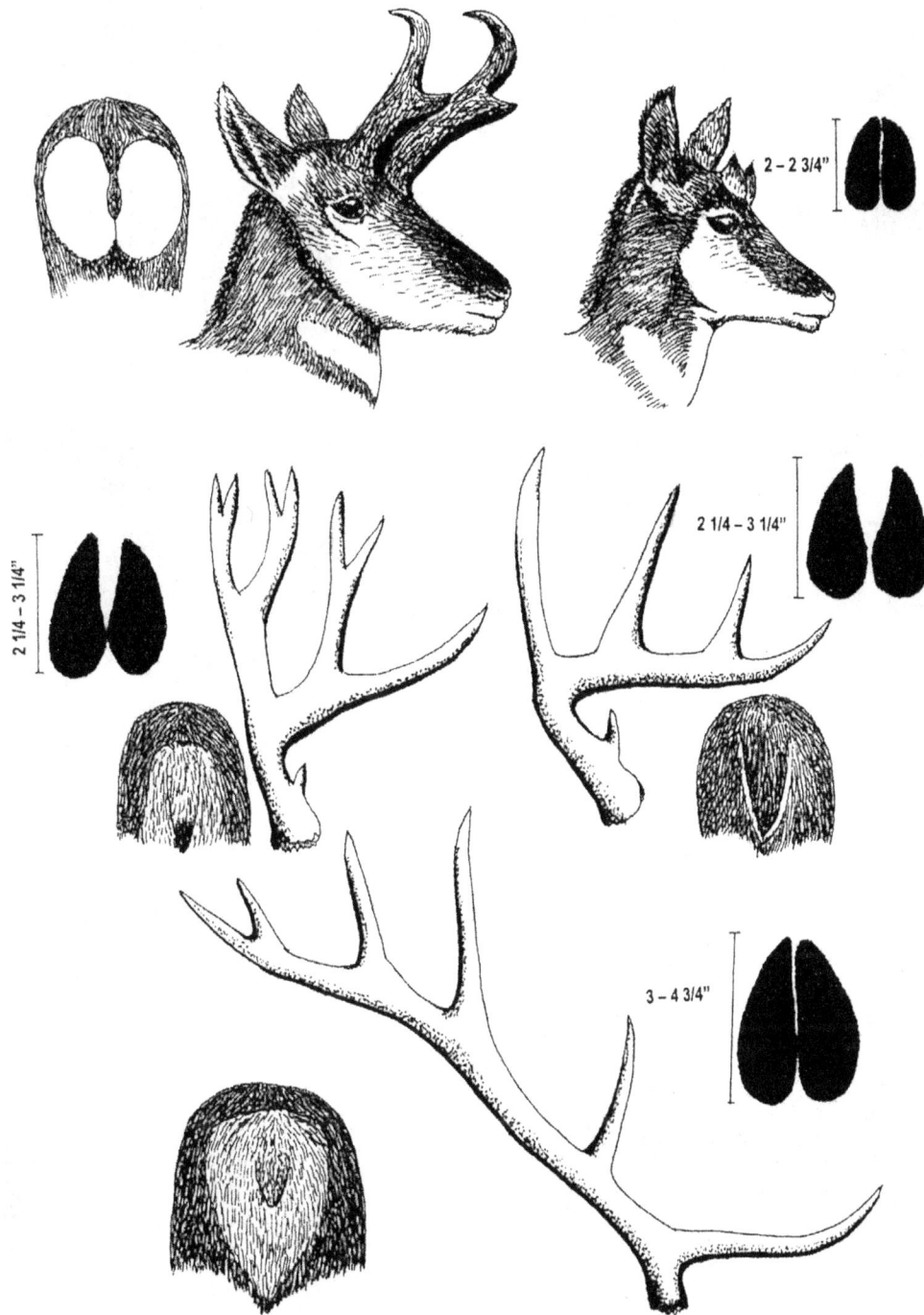

Heads of pronghorn male (*upper left*) and female (*upper right*) along with rump (*left*) and hoofprint (*right*); antlers, hoofprints, and rumps of male mule deer (*middle left*), white-tailed deer (*middle right*), and elk (*below*)

Breeding biology: Moose breed in early fall, from mid-September to early November, when both sexes become more vocal and males engage in challenges and fights with other bulls. Females undergo successive estrus cycles at 20- to 30-day intervals until they are bred. The gestation period is about 231 days, with most calves being born in late May and early June. Normally only a single calf is born, although twins occur in about 10 to 30 percent of births. Females usually breed when they are between 4 and 12 years of age (Clark and Stromberg, 1987; Armstrong, Fitzgerald, and Meaney, 2011).

Suggested reading: Altmann, 1959; Clark and Stromberg, 1987; Putnam, 1988; Franzman and Schwartz, 2007; Armstrong, Fitzgerald, and Meaney, 2011; Buskirk, 2016.

Overleaf: Moose, adult male

Family Antilocapridae (Pronghorns)

Pronghorn. *Antilocapra americana*. Ubiquitous.

Identification: Both sexes of pronghorns are mostly a sandy brown color, with white rumps, white on the underparts and lower legs, plus white patches on the neck and cheeks. Adult males have a broad black streak down the muzzle, black cheek patches, and a short black mane, Adult males also have short black horns (not antlers) that are slightly forked and curve backward toward their tips. The outer fibrous sheaths of the males' horns are shed annually. The smaller females lack black facial markings and have short, nublike horns, as do juvenile males. Length: 124–147 cm (4.1–4.8 ft); tail: 97–178 mm (3.1–5.8 in). Weight: 40–70 kg (88–154 lb).

Voice: Vocalizations include the male's advertisement or territorial snortlike roar that is uttered when a male enters another's territory and a similar short chuckle that is also used by males during antagonistic encounters. An alarm snort-wheeze is uttered by both sexes when they are threatened, and a high-pitched whine is emitted by males during courtship. Fawns when in need of attention use bleating.

Status: Wyoming is known to have the highest pronghorn population of any state; about half the nation's population is believed to be in Wyoming. In 2006 Wyoming had an estimated 575,000 pronghorns, the highest in nearly two decades. Since then it has been in decline, dropping to 360,000 by 2013. Habitat fragmentation, owing to energy development (petroleum, natural gas, and wind and solar power), have made serious inroads in the pronghorn's sagebrush habitat. Bluetongue and other diseases have caused die-offs in the thousands, and construction of fences, plus the presence of roads and railroads have interrupted traditional migration routes, which may be as long as 100 miles or more (Sawyer, Lindzey, and McWhirter, 2005).

Habitats and ecology: In northwestern Colorado, shrubs make up more than 90 percent of the fall and winter foods in the pronghorn's sagebrush-bitterbrush range, while in spring and summer forbs account for 60 to 84 percent of their diet. On a statewide and year-round study in Colorado, forbs and shrubs each accounted for 43 percent, with cactus providing 11 percent and grasses 3 percent. Although fairly drought tolerant, pronghorns prefer to drink water daily during warmer months, so water sources influence their local distribution. Home ranges vary greatly, from 160 to 5,700 acres (65–2,300 hectares), and territories range from 64 to 990 acres (25–400 hectares) on good ranges (Armstrong, Fitzgerald, and Meaney, 2011). Pronghorns are highly social, with herds in Wyoming ranging in size up to a thousand or more individuals. Their densities vary from 1.5 to 8.3 pronghorns per square mile (0.6–3.3 per 100 hectares) (Clark and Stromberg, 1987).

Pronghorn , adult male

Breeding biology: Pronghorn males are polygynous, and females become sexually mature at 16 months. They breed in the fall, from mid-September to mid-October, when males attempt to gather harems of females, or try to mate with any females entering their territories. The gestation period lasts 252 days, with young being born in the spring. Usually females give birth to twins, which soon become able to run swiftly. Within a few months the young can reach the top speed of adults, which is more than 60 miles per hour (Clark and Stromberg, 1987; Armstrong, Fitzgerald, and Meaney, 2011).

Suggested reading: Sundstrom, Hepworth, and Diem, 1973; Kitchen, 1974; Clark and Stromberg, 1987; Turbak, 1995; Byers, 1997; O'Gara and Yoakum, 2004; Armstrong, Fitzgerald, and Meaney, 2011; Buskirk, 2016.

Pronghorn, adult male

Bison, adult male head

Family Bovidae (Bison, Sheep, and Goats)

American Bison. *Bison bison*. Local.

Identification: Unmistakable. The bison is the largest North American mammal, unique with its black tapering horns that curve upward and outward, and, in males, massive shoulders and both a shaggy beard and a densely woolly forehead and shoulder pelage that extends back to the forelegs. Females are smaller and more cowlike in appearance. Length: 198–380 cm (6.7–12.5 ft); tail: 43–81 cm (14.1–26.7 in). Weight: 410–910 kg (900–2,000 lb).

Voice: The deep roaring bellow of a rutting male bison is distinctive. It lasts up to about six seconds, or sometimes more, with a bass-dominated frequency range of 1,500 to 7,500 Hz. Cows also roar but less loudly than males. Their usual vocalizations are soft, guttural grunts, and calves produce higher-pitched grunts as well as bleating (Gunderson and Mahan, 1980).

Status: Bison are now well established in the Yellowstone–Grand Teton parks, and large captive herds of bison occur elsewhere in Wyoming, such as at Terry Bison Ranch near Laramie (Johnsgard, 2005). During the 1980s, about 3,300 wild bison were in Wyoming (Clark and Stromberg, 1987), but as of 2016 there were about 5,000 bison in Yellowstone National Park alone. These are the descendants of a remnant population of 23 wild bison that escaped the mass slaughter of the late 1800s, plus 21 captives that were introduced into the park's Lamar Valley in 1902. There are also nearly 1,000 bison in Grand Teton National Park, descendants of a captive herd of 16 animals in a private park near Moran that escaped their exhibition area in 1968. By the time I first did field research in Grand Teton National Park during the mid-1970s, they had multiplied to a few dozen and were roaming freely about the park. In 1975, 18 of them began wintering on the National Elk Refuge. Since 1980 the Grand Teton–Jackson population has been increasing 10 to 14 percent annually.

Habitats and ecology: Bison were historically associated with shortgrass and mixed-grass prairies as well as meadows and, to a limited extent, shrub-grass and desert grasslands. Now the greatest numbers of bison are in private, state, and national parks, ranches, and preserves, and by 2018 the wild populations numbered more than 300,000 and the private herds about 500,000. In Yellowstone National Park, bison favor sedge- and grass-dominated meadows interspersed with lodgepole pine forests. During the fall and winter, sedges make up 37 to 50 percent of their diet, with grasses composing somewhat smaller percentages. Some shrubs and browse are also eaten, mostly in summer (Clark and Stromberg, 1987).

Breeding biology: Bison breed over a long period, mostly from July to October, when mature males often engage in pushing or fighting contests for access to females, bellow loudly, and closely follow (tend) females approaching or in estrus. Males over eight years old are the most sexually active, and in Yellowstone about half of the females first become pregnant at 3.5 years of age. Gestation lasts 270 to 285 days, and single (rarely twin) calves are born from April to June. At least in protected situations bison might live to 41 years of age (Clark and Stromberg, 1987).

Suggested reading: McHugh, 1958; Clark and Stromberg, 1987; Danz, 1997; Irby and Knight, 1998; Rinella, 2009; Armstrong, Fitzgerald, and Meaney, 2011; Buskirk, 2016.

Mountain Goat. *Oreamnos americanus*. Localized.

Identification: Mountain goats are uniquely white, except for a black nose, horns, hooves, and eyelids. Both sexes have conical horns that are slightly curved backward; those of males are longer and heavier than those of females. These alpine-adapted goats are limited to steep mountain slopes, usually above treeline, where the rubberlike lower surfaces of their hooves enable them to climb the steepest of rock inclines. Length: 124–179 cm (3.6–5.9 ft); tail: 84–203 mm (2.7–6.7 in). Weight: 46–136 kg (100–300 lb).

Bison, adult male

Voice: Adult mountain goats are very silent animals, but they do snort and make wheezing sounds that are related to maintaining personal space. Adult females also utter mewing-like calls when they are separated from each other, and calves make soft bleating calls.

Status: Mountain goats in Wyoming are limited to the northern Absaroka Mountains and the Snake River range south of Grand Teton National Park They were introduced into Yellowstone Park and are local in the park's northeastern corner on Baronette and Abiather peaks. In 2010 that park's population was 200 to 250 animals (Johnsgard, 2013).

Habitats and ecology: Mountain goats clip the short alpine vegetation, eating a variety of plants. In Montana, summer foods are about three-fourths sedges, grasses, and rushes, and a fourth component comes from forbs, whereas in winter the percentage of sedges, grasses, and rushes is similar, but about 12 percent consists of coniferous trees and 14 percent other vegetation (Clark and Stromberg, 1987). Mountain goats move seasonally downward in elevation between summer and winter over distances of 3 to 15 miles (5–24 km), as influenced by snow depth, and typically winter on south-facing slopes where vegetation is exposed. They also make shorter daily movements up and down slopes as the day progresses (Clark and Stromberg, 1987).

Breeding biology: Breeding occurs in fall, from mid-November through early December. Fighting among males is rare, but serous injuries sometimes occur. Males are polygynous, and seek out estrus females. The estrus period lasts 48 to 72 hours, and the gestation period is about 180 days. Kids are born in late May and June, and the incidence of twins ranges from 8 to 30 percent. The young nurse until early September, and females become sexually mature at 2.5 years of age (Clark and Stromberg, 1987; Armstrong, Fitzgerald, and Meaney, 2011).

Suggested reading: Clark and Stromberg, 1987; Lemke, 2004; Armstrong, Fitzgerald, and Meaney, 2011; Buskirk, 2016.

Bighorn (Mountain) Sheep. *Ovis canadensis*.
Widespread.

Identification: Bighorn sheep are the only North American sheep with massive recurved horns that in adult males may make almost a compete circle but in females and juvenile males are shorter, thinner, and only slightly recurved. Both sexes are medium brown in pelage color, except for a large white rump area and a short brown tail. Length: 149–195 cm (4.9–6.4 ft); tail: 80–127 mm (2.6–4.5 in). Weight: 75–168 kg (165–370 lb).

Voice: Adult sheep snort or make coughlike noises when alarmed but are otherwise quite silent. Adults bleat when searching for other group members, such as when females are searching for lambs. The crashing sound made by males when they head-butt is surprisingly loud and can be heard over substantial distances.

Status: Bighorn sheep are native to the mountains of northwestern Wyoming and the Bighorn Mountains. They were also once present in the Black Hills and elsewhere in eastern

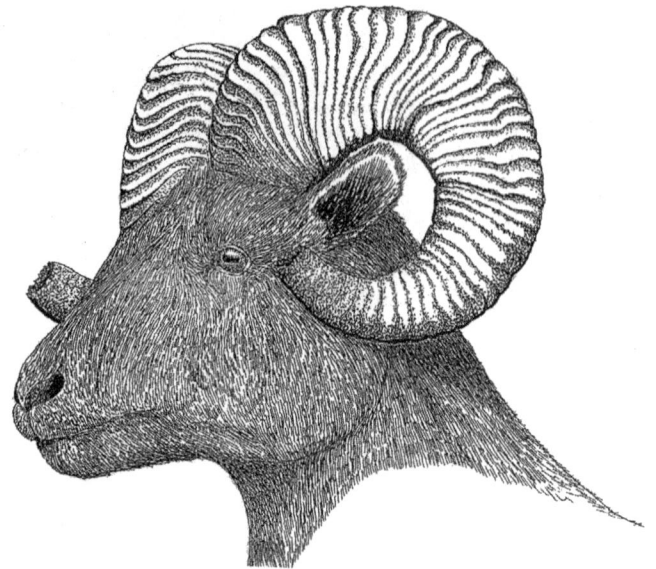

Bighorn sheep, adult male head

Wyoming (*O. c. audubonii*), but hunting and diseases extirpated that race. They have been reintroduced in various ranges in the state, including the Laramie and Medicine Bow (Snowy) ranges, the Seminoe and Ferris Mountains, the southern end of the Wind River Mountains, the west slope of the Bighorn Mountains, the Black Hills, and Wind River Canyon. In 2014 the Wyoming population was estimated at 6,450, with 85 percent in the northwestern corner of the state.

Habitats and ecology: Bighorn sheep are associated with high mountains, often above timberline. They are quite mobile and might move up to 3 miles (5 km) per day in Yellowstone National Park. They may also move in elevation as much as hundreds of feet lower and up to 30 miles (50 km) in distance in winter, to where snow is less deep and their home ranges are smaller. At higher elevations, grasses dominate their diets year round, but at lower elevations browse is the major winter food. During spring and summer the sexes separate, with older males associated in small bachelor bands and females, their lambs, and immature males grouped in larger assemblages (Armstrong, Fitzgerald, and Meaney, 2011).

Breeding biology: Male bighorn sheep are polygynous and compete strongly for mating rights. Estrus females also tend to seek out the largest rams. Males at least seven years old do most of the breeding, which occurs in November and December in Colorado. Gestation lasts about 170 to 180 days, with lambs being born in May or June. Usually only a single lamb is born per pregnancy. Lambs are highly precocial and are weaned by five to six months. Many diseases affect the bighorn population, and mortality is high during the first year of life. However, lifespans of up to 17 years have been reported (Armstrong, Fitzgerald, and Meaney, 2011).

Suggested reading: Geist, 1971; Clark and Stromberg, 1987; Armstrong, Fitzgerald, and Meaney, 2011; Buskirk, 2016.

Mountain goat, adults

The following 194 profile summaries include nearly half of Wyoming's total documented 422 bird species. They were chosen to represent nearly all the most common and widespread species, the more rare and threatened species, and some of the notable ("charismatic") species that tourists with a high interest in birds might like to see and learn about. At the end of each species profile is a reference ("*Suggested reading:* BNA") to the relevant species account in *The Birds of North America* monograph series, produced by the American Ornithologists' Union (now the American Ornithological Society) and the Academy of Natural Sciences of Philadelphia from 1992 to 2003 (cf. A. Poole, in bird references). Most major research libraries have this series, and the accounts (many of them since revised) are also now available online, through the Cornell Lab of Ornithology's Birds of North America Online (https://birdsna.org).

See the taxonomic synopsis at the beginning of chapter 5 for definitions of the code letters that follow each species' names. These code letters relate to each species' breeding or migratory status, Wyoming distribution and dispersion patterns, and seasonal occurrence as well as various categories of state or national conservation concern (CC) and whether the bird is a "charismatic" (Ch) species of special viewing or biological interest. Unlike the other sections, although vocal traits are mentioned in the following profiles, additional species identification information is not provided here because so many excellent bird field guides are now available, including online guides such as Cornell's All About Birds (https://allaboutbirds.org) and the National Audubon Society's Guide to North American Birds (https://www.audubon.org/bird-guide). For questions about the Wyoming Bird Records Committee or how to file a rare bird report, contact the Wyoming Game and Fish Department nongame bird biologist.

Family Anatidae (Ducks, Geese, and Swans)

Snow Goose. *Anser caerulescens*. Un; Mig.

Status: An uncommon to rare migrant throughout the state, especially eastwardly. Occasional in Grand Teton National Park. Increasingly common nationally in recent years.

Habitats and ecology: Generally associated with large marsh and wetland habitats; feeding in dry fields is less frequent than for Canada geese, and rootstalks and tubers of marshland plants are more regularly eaten.

Suggested viewing locations: More common during fall than spring; wetlands and agricultural fields east of the Rockies attract these birds during migration. In Wyoming, locations such as Yellowstone Lake, Table Mountain Wildlife Habitat Management Area (WHMA) (Goshen County), Bump Sullivan Reservoir (near Yoder), and Clifford F. Graham Reservoir (Uinta County) are good choices (Dorn and Dorn, 1990).

Ocean Lake and Lake DeSmet are also good viewing locations (Scott, 1993).

Suggested reading: BNA 514 (T. B. Mowbray, F. Cooke, and B. Ganter, 2000); Johnsgard, 2016d.

Cackling Goose. *Branta hutchinsii*. Co.

Status: A common migrant through the state, probably more abundant than is generally realized, owing to its similarity to the Canada goose.

Habitats and ecology: Cackling geese are most likely seen in wetlands east of the Rockies during migration in March–April and October–November. Because this species was recognized as specifically distinct from the Canada goose only a few decades ago, regional records for it are still quite limited. A few sight records are from Johnson and Natrona Counties.

Suggested viewing locations: This tundra breeder is most often seen on the eastern plains of the region in company with Canada geese, and it is difficult to visually distinguish from the smaller races of that species.

Suggested reading: BNA 2 (T. B. Mowbray et al., 2002); Johnsgard, 2016d.

Canada Goose. *Branta canadensis*. Res.

Status: The Canada goose is a common, virtually pandemic resident throughout the state, whereas the cackling goose is an Arctic-breeding spring and fall migrant, especially in the plains region.

Habitats and ecology: The extremely adaptable Canada goose sometimes nests within the city limits of large cities but also occurs on prairie marshes, beaver ponds, and forest-edged mountain lakes. Beaver lodges or muskrat houses provide safe and favored nest sites in many areas.

Breeding biology: Canada geese have strong, permanent pair-bonds, and most begin to breed when they are two or three years old. Males establish fairly large territories in marshes, usually including the same area and often the same nest site as in previous years. Pair-bonds are maintained by mutual displays, especially the triumph ceremony, and unless nest sites are limited or predator pressures are present, the nests tend to be well scattered. The nest is constructed primarily by the female, with the male standing guard and helping to some extent. Copulation occurs on the water, primarily during the egg-laying period, and incubation does not begin until the clutch of about five eggs is complete. Males remain close to the nest and take major responsibility for guarding it but do not help incubate. Incubation lasts 25 to 30 days. Both sexes tend the young, which soon begin to fend for themselves. During the fledging period of about 70 days, both parents undergo a flightless molt. Thereafter the family may

Bighorn sheep, adults

leave the area, with the family bonds persisting through the winter.

Suggested viewing locations: Canada geese occupy the majority of wetlands in the region, especially protected wetlands and larger lakes. Table Mountain WHMA and Springer Reservoir are especially favored Wyoming sites (Scott, 1993).

Suggested reading: BNA 682 (T. B. Mobray et al., 2002); Johnsgard, 2016d.

Trumpeter Swan. *Cygnus buccinator*. Lo; Ch, CC3, CC4; Res.

Status: The trumpeter swan is a common resident in the Yellowstone and Grand Teton parks, generally rare or accidental elsewhere. In the late 1970s, Yellowstone supported about 20 nesting pairs in or very near the park, while Red Rock Lakes Refuge to the west of Yellowstone contained about 27 nesting pairs. McEneaney (1988) judged that park's population to be less than 15 pairs during the late 1980s, with about 40 individuals typically present during summer months. Grand Teton National Park has long had nesting trumpeter swans. Johnsgard (1986) judged the Grand Teton population to be about six pairs during the early 1980s. The Greater Yellowstone ecosystem population has been in serious decline, and by 2009 fewer than 400 birds were present. Up to a half dozen pairs may still nest in Grand Teton National Park and the National Elk Refuge, usually on isolated lakes or beaver ponds. Only local nestings have occurred outside this region, including at Colony and Star Valley (Scott, 1993).

Habitats and ecology: In the Rocky Mountain region, this species is mostly limited to fairly large ponds (usually more than 30 acres) that have considerable aquatic vegetation and relative seclusion from disturbance by humans. Beaver ponds are most often used in the Jackson Hole area, and nests are sometimes built on their lodges.

Breeding biology: Trumpeter swans pair for life, and each pair returns to its nesting area in spring, as soon as the weather allows. Territories are established that average more than 30 acres (sometimes more than 100 acres), and they are vigorously defended; the adults even exclude their own offspring of previous years. The male performs such territorial defense, but the female participates in mutual "triumph ceremonies" after territorial disputes and also helps defend the nest site. Both sexes help construct the rather bulky nest, which may require a week or more. The eggs are laid at two-day intervals, and no incubation is performed until the clutch of about five eggs is complete. Thereafter the female performs all the incubation (a few records of males incubating exist but must be regarded as abnormal), while the male defends the nest. Incubation lasts about 33 days. Most of the cygnets hatch within a few hours of each other and are led from the nest within 24 hours of hatching. The nest may later be used for resting or brooding, but often the brood is led some distance from the nest for rearing on quiet and secluded ponds. The

Trumpeter swan, adult in flight

fledging period is approximately100 days in the Montana-Wyoming region, which occupies the entire summer and makes it impossible for birds to renest after nest failure.

Suggested viewing locations: The largest local population in the Rocky Mountains region is at Red Rock Lakes National Wildlife Refuge. The Madison River is an excellent location for finding these rare swans in Yellowstone National Park (McEneaney, 1988). Trumpeter swans are also usually easily visible at the National Elk Refuge near Jackson, and also on a pond at the north end of Jackson.

Suggested reading: BNA 105 (C. D. Mitchell and M. W. Eichholz, 2010); Johnsgard, 2016a.

Wood Duck. *Aix sponsa*. SD; Un, Ch; Su; Res.

Status: A local uncommon summer resident in eastern Wyoming; occasional (in Grand Teton) to rare (in Yellowstone) in the national parks.

Trumpeter swan, pair

Habitats and ecology: During the breeding season, these birds are found among woodlands having fairly large trees that offer nesting holes and also, frequently, those having acorns or similar nutlike foods in abundance. Even outside the breeding season the birds are usually associated with flooded woodlands rather than open marshes. In the interior Pacific Northwest, wood ducks begin breeding in mixed coniferous forests during the mature forest stage of succession (Sanderson, Bull, and Edgerton, 1980).

Breeding biology: Pair-bonds are established each year, after a prolonged period of courtship displays. No definite territorial behavior exists, but males assist females in seeking out suitable nest sites, which may take several days. Competition for nest sites is frequent and thus collective "dump nests" (having eggs of several females) are locally prevalent. The usual clutch size is 8 to 10 eggs. The female does the incubation, and males normally desert their mates before hatching. Incubation lasts 28 to 32 days. The female raises the brood, which fledges at about 60 days of age. Renesting after loss of the first clutch is fairly frequent, and a second brood may be raised on rare occasions.

Suggested viewing locations: Wooded streams such as the Laramie, North Platte, Belle Fourche, and Shoshone Rivers in Wyoming attract wood ducks (Dorn and Dorn, 1990).

Suggested reading: BNA 169 (G. R. Hepp and F. C. Bellrose, 2013); Johnsgard, 2017a.

Blue-winged Teal. *Spatula discors*. Ub; Co; Su; Res.

Status: A common summer resident throughout the region, but occasional (Grand Teton) to uncommon (Yellowstone) in the national parks. Blue-winged teal are common in the grasslands and foothills, especially in the prairie pothole country.

Habitats and ecology: This species favors relatively small, shallow marshes over those that are larger and deeper, especially locations that are surrounded by grass or sedge meadows. Migration in spring occurs fairly late, as does pair formation, but nonetheless renesting efforts are fairly common following nest failure.

Breeding biology: Pair-bonds are formed fairly late, mainly during the migration northward, but some displays may occur on the nesting grounds. Pairs are relatively tolerant of other pairs and often center their home ranges on very small ponds or even roadside ditches. The female chooses the nest site and builds the nest while the male waits nearby. The usual clutch size is of 8 to 10 eggs. After incubation begins the pair-bond is dissolved and males often fly elsewhere to complete their summer molt. Incubation lasts 23 to 24 days. Females take their broods to water within hours after hatching and usually raise them in rather heavy brooding cover. The fledging period is about six weeks, and females also begin to molt at about the time the young are fledged.

Suggested viewing locations: This species is widespread throughout the region during migration.

Suggested reading: BNA 625 (F. C. Rohwer, W. P. Johnson, and E. R. Loos, 2002); Johnsgard, 2017a.

Cinnamon Teal, *Spatula cyanoptera*. Ub; Un; Su; Res.

Status: An uncommon summer resident throughout the region, but common in both national parks (Grand Teton and Yellowstone). It is common in the grasslands and foothills, especially in western Wyoming and in somewhat saline wetlands.

Habitats and ecology: There is no obvious difference in habitat preferences between blue-winged teal and cinnamon teal, although the latter inhabits more alkaline marshes during the breeding season.

Breeding biology: The social behavior and breeding biology of the cinnamon teal are extremely similar to those of the blue-winged teal, and in a few areas they breed on the same marshes, nesting at the same time and using the same habitats. Nesting densities of cinnamon teal in the middle of their range are appreciably higher than those of blue-winged teal, however, and their home ranges tend to be very small.

Suggested viewing locations: Cinnamon teal are likely to be found in any shallow, somewhat alkaline marsh statewide.

Suggested reading: BNA 209 (G. H. Gammonley, 2012); Johnsgard, 2017a.

Northern Shoveler. *Spatula clypeata*. Ub; Co; Su; Res.

Status: A common summer resident essentially throughout the entire state; most common in wetlands of the plains and foothills; occasional (Grand Teton) to uncommon (Yellowstone) in the national parks.

Habitats and ecology: The specialized bill of the northern shoveler allows for filter-feeding of surface organisms, and submerged plants sometimes also provide a supply of organisms that can be reached from the surface.

Breeding biology: Shovelers begin pair formation on their wintering grounds and continue it through their arrival on the breeding grounds. Most of the displays are aquatic, but there are also "jump-flights" and aerial chases associated with courtship. The birds are seasonally monogamous (contrary to the early literature), and at least in captivity some birds remate with previous mates while others choose new ones. The pairs spread out over the breeding habitat and have been described as territorial by some ornithologists, while others have simply reported that they occupy overlapping home ranges of 15 to 90 acres in area. The usual clutch size is 8 to 12 eggs. The females do all the incubation, and the males abandon them during the incubation period. Incubation lasts 26 days. The fledging period is about six to seven weeks.

Suggested reading: BNA 217 (P. J. Dubowy, 1996); Johnsgard, 2017a.

Gadwall. *Mareca strepera*. Pan; Co; Su; Res.

Status: A common summer resident across the state, gadwalls commonly breed in both national parks (Grand Teton and Yellowstone). It is most common on the prairie marshes to the east.

Habitats and ecology: This prairie-adapted dabbling duck prefers shallow marshes with grassy or weedy nesting cover, especially where islands are present.

Breeding biology: Gadwalls form their pair-bonds relatively early, during a period of social courtship that involves aquatic display as well as aerial chases. Most birds are paired by the time they arrive on their nesting grounds, and pairs establish home ranges that may exceed 50 acres, often overlapping with the home ranges of other pairs. Territorial behavior as such is not significant, and nests are often close together, especially on islands. The female constructs the nest alone and is usually abandoned by her mate about a week or two after incubation has begun. The usual clutch size is 8 to 12 eggs. Incubation lasts 25 to 27 days. The hen raises her brood alone, usually on deep-water marshes that are unlikely to dry up before fledging, which requires seven to eight weeks.

Suggested viewing locations: This species is widespread throughout the region during migration.

Suggested reading: BNA 283 (C. R. Leschack, S. K. McKnight, and G. R. Hepp, 1997); Johnsgard, 2017a.

American Wigeon. *Mareca americana*. Ub; Co, CC2; Su; Res.

Status: A common summer resident throughout the state, breeding commonly in both national parks (Grand Teton and Yellowstone).

Habitats and ecology: Wigeons are associated with relatively open marshes and lakes that have abundant aquatic vegetation at or near the surface, and during the breeding season they favor areas with sedge meadows or shrubby or partially wooded habitats nearby. Wigeon are strongly vegetarian and spend more time grazing grassy vegetation than do most ducks.

Breeding biology: Wigeons form seasonally monogamous pair-bonds after a period of social courtship in winter and spring. Males perform fairly simple displays, mainly involving calling, chin-lifting, and raising the folded wings high above the back. After pair-formation, pairs establish a home range on marshes ranging from less than an acre to more than 20 acres in area. There is no territorial defense, although males evict other males from the vicinity of their mates. Nest sites are well hidden, and shortly after incubation begins males abandon their mates. The usual clutch size is 9 to 11 eggs, and the female incubates and rears the brood alone. Incubation lasts 24 or 25 days. Broods are reared on relatively open marshes, and fledging occurs at about 45 days of age.

Suggested viewing locations: This species is widespread throughout the region during migration.

Suggested reading: BNA 401 (A. E. Mini et al., 2014); Johnsgard, 2017a.

Mallard. *Anas platyrhynchos*. Ub; Co; Res.

Status: Mallards are an abundant resident throughout the region, breeding nearly everywhere ponds or marshes occur and in both national parks (Grand Teton and Yellowstone).

Habitats and ecology: This highly adaptable species nests in nearly all aquatic habitats but prefers nonforested areas over the forested, and shallow waters over those that are deeper. Mallards quickly locate and utilize protected areas, even when they are close to human activities, and thus they remain common in spite of intensive hunting pressures on them.

Breeding biology: Mallards begin social display early in the fall, with many adults probably forming new pair-bonds with earlier mates, and those hatched the previous summer beginning courtship for the first time. By spring, nearly all females have formed pair-bonds, and on arrival at their breeding grounds pairs spread out across the available habitat. Home ranges of such pairs vary greatly in size but at times may exceed 700 acres; spacing is enhanced by males evicting other males from the vicinity of their mates. Females choose their nest sites and are abandoned by their mates when incubation gets under way. The usual clutch size is 10 to 12 eggs. Incubation lasts 24 or 25 days. The newly hatched young are quickly led to water, and the fledging period is about 50 to 55 days. Mallards often try to renest if their first attempt fails; the clutch sizes of renesting efforts tend to be slightly smaller than the original clutches.

Suggested viewing locations: This species can be found on regional wetlands almost anywhere throughout the state.

Suggested reading: BNA 658 (N. Drilling, R. Titman, and F. McKinney, 2018); Johnsgard, 2017a.

Northern Pintail. *Anas acuta*. Pan; Co, CC2; Su; Res.

Status: A common year-round or summer resident throughout the region and an uncommon (Yellowstone) to occasional (Grand Teton) breeder in the national parks.

Habitats and ecology: This is a tundra- and prairie-adapted breeding species, and it is rarely found in heavily wooded wetlands. Pintails can breed on small and temporary ponds as well as permanent marshes, and they frequently nest on dry land in extremely exposed situations well away from water.

Breeding biology: Northern pintails form monogamous pair-bonds during a prolonged period of social courtship, which continues as the birds migrate north in spring. Most or all females are paired by the time the birds arrive on their nesting grounds, and the pairs tend to become well spaced as they establish large home ranges. Females begin nesting very early, shortly after hillsides are free of snow, and like most ducks they complete their clutches at the rate of one egg per day. The usual clutch size is seven to nine eggs. Incubation begins with the laying of the last egg, and by that time or shortly afterward the pair-bond is broken. Incubation lasts 25 or 26 days. When the brood hatches, the female leads them to water, sometimes shifting ponds and moving them nearly a mile from where they were hatched. The fledging period is 47 to 57 days in South Dakota and averages 41 and 46 days for females and males, respectively, in Manitoba.

Suggested viewing locations: This species is widespread throughout the region during migration.

Suggested reading: BNA 163 (R. G. Clark et al., 2014); Johnsgard, 2017a.

Green-winged Teal. *Anas crecca*. Ub; Co; Res.

Status: A common summer resident over nearly the entire state, breeding commonly in both national parks (Grand Teton and Yellowstone). It is very common in the prairie marshes and foothill areas during migration and often overwinters.

Habitats and ecology: Migrants are associated with almost all standing or slowly flowing aquatic habitats in Wyoming. Breeding normally occurs where ponds or sloughs are surrounded by a mixture of grassland, sedge meadows, and well-drained areas supporting shrubby or tall woody vegetation.

Breeding biology: Green-winged teal are highly social and display over a long period of late winter and spring while forming their pair-bonds, which are renewed annually. Pair-forming displays are numerous, elaborate, and highly animated. On reaching their breeding grounds, pairs spread out and establish home ranges that center on small ponds. Females select nest sites while accompanied by their mates, which usually remain attached to them until incubation is under way. The usual clutch size is 10 to 12 eggs. Incubation lasts 23 or 24 days. After the clutch has hatched, the female leads her young to shallow ponds, and they grow very rapidly. They fledge in no more than 44 days. Some Alaska estimates of fledging are as little as 35 days; fledging is unusually rapid at such high latitudes, where summer daylight allows for continuous feeding.

Suggested viewing locations: This species is widespread throughout the region during migration.

Suggested reading: BNA 193 (K. Johnson, 1995); Johnsgard, 2017a.

Canvasback. *Aythya valisineria*. Pan; Co, CC2; Mig.

Status: A local and uncommon summer resident over much of the state but rare at higher elevations and most abundant in prairie potholes and marshes. Reported as an uncommon breeder in Yellowstone National Park and a rare breeder in Grand Teton.

Habitats and ecology: During the breeding season, canvasbacks are found on shallow prairie marshes that have abundant growths of emergent vegetation along with open water areas that frequently are rich in aquatic plants, such as pondweeds.

Breeding biology: Canvasbacks renew their pair-bonds annually, and courtship is usually intense as the birds are returning to their nesting grounds. Several aquatic displays, including cooing calls and head-throw displays, are conspicuous then. As pairs form, they separate from the flocks and seek out nesting areas in smaller and shallower ponds than those used for courting. In densely populated areas, a substantial amount of nest parasitism occurs among canvasbacks and between canvasbacks and redheads. Although parasitic

redheads are prone to lay their eggs in canvasback nests, the latter usually lay eggs only in the nests of other canvasbacks. Thus mixed-species broods sometimes occur, but parasitized nests are less successful than those that are not parasitized. The usual unparasitized clutch size is seven to nine eggs. Incubation lasts 24 to 27 days. The fledging period is eight to nine weeks.

Suggested viewing locations: Deeper and larger marshes, lakes, and reservoirs attract this species on migration. Nesting in Wyoming is local although widespread in south-central and western areas, and probably includes Cokeville Meadows (Scott, 1993; Faulkner, 2010).

Suggested reading: BNA 659 (T. B. Mowbray, 2002); Johnsgard, 2017c.

Redhead. *Aythya americana*. Wi; Co; Res.

Status: Redheads are a common summer resident over most of the region and a locally uncommon to rare breeder. They are uncommon (Yellowstone) to occasional (Grand Teton) in the national parks and most common in prairie marshes.

Habitats and ecology: Breeding habitats consist of nonforested country with water areas sufficiently deep to provide permanent, fairly dense emergent vegetation as nesting cover. Water areas at least an acre in size are preferred for nesting, with substantial areas of open water for taking off and landing.

Breeding biology: Redheads have seasonal pair-bonds, established each winter and spring. Their displays and associated behavior are much like those of canvasbacks, and the two species often associate. On reaching their nesting grounds, pairs establish home ranges that typically include nesting site potholes and waiting site potholes, often shared with other pairs. Nest parasitism by redheads is high in most areas, and they drop eggs in the nests of a large variety of other marsh birds, although not all females are parasitic nesters. The usual unparasitized clutch size is 8 to 10 eggs, but many nests have 12 to 15 eggs. Males abandon their mates early during incubation and often fly elsewhere to molt. Incubation lasts 24 or 25 days. In Iowa, the young have been reported to fledge at 70 to 84 days of age, but shorter fledging periods have been reported for Canada.

Suggested viewing locations: This species is widespread on larger marshes and lakes throughout the Rocky Mountains region during migration. In Wyoming, Hutton Lake National Wildlife Refuge attracts migrating birds (Dorn and Dorn, 1990). Nesting in Wyoming is widespread, except for the eastern plains (Faulkner, 2010), and extends from Cokeville to Table Mountain (Scott, 1993).

Suggested reading: BNA 695 (M. C. Woodin and T. C. Michot, 2002); Johnsgard, 2017c.

Ring-necked Duck. *Aythya collaris*. Lo; Co; Res.

Status: The ring-necked duck is a common summer resident in montane woodland ponds in northwestern and southeastern Wyoming. It is a common breeder in both national parks

(Grand Teton and Yellowstone) but is rare or absent from prairie marshes during the breeding period.

Habitats and ecology: Unlike any of its near relatives, the ring-necked duck is strongly associated with beaver ponds and other forest wetlands, where it is often among the commonest of breeding ducks. Sedge-meadow marshes and boggy areas are preferred for nesting, and the presence of water lilies and associated heather cover seem to be an important part of breeding habitats.

Breeding biology: Pair-bonds in ring-necked ducks start to become established on the wintering grounds through social display that begins in January and February, but some displays persist until the birds arrive on their nesting grounds. Display patterns are much like those of redheads and canvasbacks, in spite of plumage differences. Pairs become spaced out over the breeding grounds but show little aggression when they come into contact, and nests are often close together on islands. The usual clutch size is 6 to 12 eggs. Incubation lasts 26 days. The pair-bond is usually broken near the end of the incubation period, and females raise their broods alone, on ponds often largely covered by water lilies. By the end of the fledging period of seven to eight weeks, the female will have begun her flightless period, and family bonds terminate.

Suggested viewing locations: This species is widespread on larger marshes and lakes throughout the region during migration. It is the most common nesting duck in the Jackson Hole area of northwestern Wyoming (Scott, 1993), and it breeds in the northwestern and south-central parts of the state (Faulkner, 2010).

Suggested reading: BNA 329 (C L. Roy et al., 2012); Johnsgard, 2017c.

Lesser Scaup. *Aythya affinis*. Di; Un, CC2; Su; Res.

Status: A regular migrant and uncommon summer resident in western and southern Wyoming, the lesser scaup is a breeder in both Yellowstone National Park (common) and Grand Teton National Park (occasional).

Habitats and ecology: This duck is largely a prairie-adapted breeder and is associated also with ponds in the foothill woodlands, especially those that support good populations of amphipods and other aquatic invertebrates.

Breeding biology: Lesser scaup form pair-bonds that persist for a single season during a prolonged period of winter and spring social display. Pairs establish relatively large but poorly defined home ranges, often centering on marshes two to five acres in area that include some deep water. Females build their nests alone and are abandoned by their mates shortly after they begin incubation. The usual clutch size is 8 to 12 eggs. Incubation lasts 26 or 27 days. Although scaup often nest in or near gull colonies, presumably for protection from nest predators, the gulls sometimes prey severely on ducklings, and much brood disruption is typical. Females often desert their ducklings early to begin their molt, and large broods consisting of ducklings from several families are frequent in some areas. The fledging period is about 47 to 50 days, a relatively short period for diving ducks.

Suggested viewing locations: This species is widespread on larger marshes and lakes throughout the region during migration. Nesting is widespread in Wyoming, except in the northeast (Faulkner, 2010) and at sites such as Cokeville Meadows National Wildlife Refuge, Seedskadee National Wildlife Refuge, and Table Mountain Wildlife Habitat Management Area (Scott 1993).

Suggested reading: BNA 338 (M. J. Anteau et al., 2014); Johnsgard, 2017c.

Harlequin Duck. *Histrionicus histrionicus*. Lo; Ra, Ch, CC3; Su; Res.

Status: The harlequin duck is a rare summer resident on mountain streams of northwestern Wyoming. It is an infrequent and apparently rare breeder in the Tetons, and local and rare on the Yellowstone River in Yellowstone National Park.

Habitats and ecology: Associated with clear, rapidly flowing streams where aquatic insects such as caddis larvae abound, it is often found where dippers also occur. McEneaney (1988) judged the Yellowstone National Park population of this beautiful diving duck at less than 20 pairs during the late 1980s.

Breeding biology: Harlequin ducks are noted for their extremely well-concealed nest sites chosen by the female, which are

Harlequin duck, adult male

often on rather inaccessible river islands, and are completely invisible from above. Tree-cavity nests are not common; instead the nest is on the ground and very close to water. The female may begin incubation before the clutch is complete, and may begin to line the nest as incubation gets underway. The average clutch is six eggs, and the egg-laying interval is rather long, of about three days. The incubation period is about 28 to 29 days, or rarely longer. Males leave their mates as incubation begins and gradually concentrate on favored foraging areas. Following hatching the female takes her brood to a secluded part of the river, where they usually move about very little. At times two broods may merge and be guarded by both females. Additional unsuccessful females may also participate in brood care. The fledging period is about 60 to 70 days.

Suggested viewing locations: The Yellowstone River provides some traditional locations for finding these torrent-loving birds in Yellowstone National Park, from its source at Yellowstone Lake downstream to Canyon Village. They still are most frequently seen a short distance below Fishing Bridge at LeHardy Rapids. Other reported Wyoming viewing sites include Snake River tributaries in Grand Teton National Park, the Middle Fork (Red Fork) of the Powder River, Shell Creek in the Bighorn Mountains, and Dinwoody Creek in the Wind River Range.

Suggested reading: BNA 466 (G. J. Robertson and R. I. Goudie, 1999); Johnsgard, 2016e.

Bufflehead. *Bucephala albeola*. Lo; Un; Res.

Status: An uncommon resident in northwestern Wyoming from Jackson Hole northward, the bufflehead can be found mainly in wooded wetlands where tree cavities (especially woodpecker holes) offer nesting sites. It is an uncommon breeder in both national parks (Grand Teton and Yellowstone).

Habitats and ecology: This species is so small that females can use the old nest holes of flickers (which are also used by bluebirds, starlings, and similar-sized hole-nesters) for nesting. In the interior Pacific Northwest, buffleheads begin breeding in mixed coniferous forests during the mature forest stage of succession (Sanderson, Bull, and Edgerton, 1980). At other times the birds are generally found on large and deep waters.

Breeding biology: Buffleheads form seasonal pair-bonds during a prolonged period of courtship that extends from winter through the spring migration. It is not known how often males remate with mates of the previous year, but females have a strong tendency to return to the place where they previously nested, and they often nest in the same cavity. Competition for nest sites among buffleheads and with other hole-nesting birds such as starlings and tree swallows makes nest-site availability an important facet of their biology. Males abandon their mates and often leave the area shortly after incubation gets underway; the usual clutch size is 8 to 12 eggs. The females leave their nests occasionally during the 29-day incubation period to feed. The young remain in the nest 24

to 36 hours after hatching; then, at their mother's signal, they jump down to the ground and leave as a group, usually during the morning. The fledging period is about 50 to 55 days, during most of which the female keeps the young within a guarded brood territory. However, some brood transfers and formations of multiple broods have been reported.

Suggested viewing locations: This species occurs commonly on deeper marshes and lakes throughout the region during migration and during winter. The densest regional breeding concentrations are probably in the forested wetlands of the Greater Yellowstone ecosystem (which include Park and Teton Counties of Wyoming and the national forests that surround Yellowstone and Grand Teton National Parks).

Suggested reading: BNA 67 (G. Gauthier, 2014); Johnsgard, 2016e.

Common Goldeneye. *Bucephala clangula*. Pan; Co; Win; Vis.

Status: The common goldeneye is a local nonbreeding summer visitor in Wyoming and a widespread migrant and wintering species throughout.

Habitats and ecology: During the breeding season, both the common goldeneye and Barrow's goldeneye are usually found in forested wetland habitats, where cavities in large trees offer nesting sites. At other times they occur on deeper and larger bodies of water, such as lakes. Because of confusion with Barrow's goldeneye, breeding reports of common goldeneye from Grand Teton and Yellowstone National Parks need confirmation, as do reports from the Wapiti and Dubois areas of Wyoming. Faulkner (2010) does not mention any proven Wyoming breeding records.

Suggested viewing locations: This species occurs commonly on deeper marshes and lakes throughout the region during migration and during winter. It is Wyoming's most common wintering duck but has not yet been proven to breed there.

Suggested reading: BNA 170 (J. M. Eadie, M. L. Mallory, and H. G. Lumsden, 1995); Johnsgard, 2016e.

Barrow's Goldeneye. *Bucephala islandica*. Pan; Co; Res.

Status: Barrow's goldeneye is a common breeding resident of the montane wetlands from west-central Wyoming northward. Elsewhere it is a common migrant or resident in most locations. This goldeneye is present in both national parks, ranging in abundance from common (Grand Teton) to abundant (Yellowstone).

Habitats and ecology: Breeding birds are associated with forested montane lakes, beaver ponds, and slowly flowing rivers in this region; elsewhere, nesting in cliff or rock crevices also is frequent.

Breeding biology: Courtship behavior in this species is prolonged and marked by loud, splashing head-throws by the male. Inciting by the females stimulates male display as well as threats and attacks among the males. The male displays preceding copulation are also prolonged and consist mostly of drinking, wing-stretching, and similar movements. After the paired birds have returned to their breeding areas, females begin to seek out nesting cavities. These are usually tree or stump cavities, but rock cavities may be used, or the nest may be well hidden under shrubs. Most tree cavities have entrances three to four inches in diameter. The average clutch is nine eggs, but competition for nest sites may result in two females laying in the same nest, resulting in larger clutches. Incubation requires 32 days. After hatching, the female establishes a brood territory from which other females and broods are excluded. Females typically abandon their brood when the ducklings are well grown but unfledged. Fledging requires about eight weeks, and thereafter the birds gradually move toward winter quarters.

Suggested viewing locations: This species occurs commonly on deeper marshes and lakes throughout the region during migration and during winter. Nesting is very common in the Jackson Hole area of Wyoming and occurs throughout the Greater Yellowstone ecosystem. Breeding is probable but unproven in the Bighorn Mountains (Canterbury, Johnsgard, and Downing, 2013).

Suggested reading: BNA 548 (J. M. Eadie, J-P. L. Savard, and M. L. Mallory, 2000); Johnsgard, 2016e.

Common Merganser. *Mergus merganser*. Wi; Co; Res.

Status: The common merganser is a common resident in most montane rivers during the breeding season and at other times can be found on nonforested rivers, lakes, and reservoirs. It is a common breeder in both national parks (Grand Teton and Yellowstone).

Habitats and ecology: This fish-eating species occurs in areas of clear water that support large fish populations, and it is much the commonest merganser of the region. Nesting occurs in tree cavities, rock crevices, and sometimes under boulders or dense shrubbery. Breeding occurs from the Rocky Mountains of western Colorado northward.

Breeding biology: During fall and winter, these mergansers usually stay in small flocks that sometimes feed cooperatively, but as spring approaches much time is spent in social display and in establishing pair-bonds, and flock sizes decrease. Females remain fairly gregarious while looking for nest sites, and they often nest close together. The clutch size varies from 7 to 14 eggs. Probably some dump-nesting by two or more females occurs in locations where nest cavities are limited. The males usually leave their mates before hatching, but on rare occasions have been seen with broods. Incubation lasts 28 to 32 days. The young are led to water a day or two after hatching, and the brood is usually raised in shallow rivers. At times the female carries part of her brood on her back, especially when they are frightened. The fledging period is 60 to 70 days.

Suggested viewing locations: This species occurs commonly on most deeper marshes, rivers, and lakes throughout the region during migration and winter. Many regional rivers and lakes are also used for nesting. The densest regional breeding concentrations are probably in the forested rivers and lakes of the Greater Yellowstone ecosystem.

Suggested reading: BNA 442 (J. Pearce, M. L. Mallory, and K. Metz, 2015); Johnsgard, 2016e.

Ruddy Duck. *Oxyura jamaicensis*. Wi; Co; Su; Res.

Status: An occasional to rare summer resident over much of Wyoming, the ruddy duck is mainly found in grassland marshy habitats. It is rarer in montane areas and a migrant more or less throughout the state. Breeding has been reported in both national parks (Grand Teton and Yellowstone).

Habitats and ecology: Nonbreeding birds are found on larger and generally deeper waters that have silty or muddy bottoms; breeding is on overgrown shallow marshes with abundant emergent vegetation and some open water.

Breeding biology: Ruddy ducks apparently mature in their first year, though not all females are thought to breed as yearlings. Pair-bonds are rather weak, and much display is related to territorial advertisement rather than courtship itself. Females show little or no pair-forming or pair-maintaining behavior, although males may remain in the vicinity of their mates after they have begun nesting. The usual clutch size is 6 to 10 eggs, but dump-nesting may produce larger clutches. Incubation lasts 24 days. Some males persist in remaining with females even after their broods have hatched, though they do not assist in rearing broods. The young are highly precocious and independent, so broods often become scattered long before the young birds fledge, which is about six to seven weeks after hatching.

Suggested viewing locations: This species occurs commonly on deeper marshes and lakes throughout the region during migration. In Wyoming, nesting occurs almost statewide on lower elevation (below 8,000 ft) wetlands having abundant emergent vegetation.

Suggested reading: BNA 696 (R. B. Brua, 2002); Johnsgard, 2017c.

Family Phasianidae (Pheasants, Grouse, and Turkeys)

Ruffed Grouse. *Bonasa umbellus*. SD; Co; Res.

Status: The ruffed grouse is widespread and relatively common in wooded areas of the Greater Yellowstone ecosystem, Bighorns, and Black Hills. It is most common west of the Continental Divide and is a common to uncommon breeder in both national parks (Grand Teton and Yellowstone).

Habitats and ecology: These grouse are especially associated with aspen woodlands because the buds and catkins provide a major food source. However, ruffed grouse also occur up to the spruce-fir zone of coniferous forest. Nesting is often in or near aspen clumps. Deciduous woods, especially aspen groves, are excellent places to find ruffed grouse, The highest population densities found by Hutto and Young (1999) in 566 transects of 11 unaltered forest, grassland, shrub, and wetland habitats of Idaho and Montana were in cottonwood-aspen communities.

Breeding biology: As the breeding areas become free of snow, male grouse establish territories that usually include a clump of aspens and one or more "drumming logs" from which they display daily. The drumming behavior of this species is a ritualized form of flight display; the bird does not leave the ground, and the sound generated by wing-beating attracts females to the log. When a female (or another male) appears, the male begins an elaborate strutting behavior that is a ritualized form of threat, leading to copulation if the intruder is female or to fighting if it is another male. After mating the female selects a nest site that is usually near a clump of aspens, whose catkins provide a nutritious food source during incubation. The usual clutch size is 9 to 12 eggs, and incubation lasts 23 to 24 days. After hatching, the young grow rapidly and can fly short distances after 10 to 12 days, but they remain with their mother until they are about four months old, and then they begin to disperse.

Suggested viewing locations: The Pacific Creek area of Grand Teton National Park is recommended for viewing ruffed grouse (Dorn and Dorn, 1990), as are the Valley Trail and Two Ocean Lake (Scott, 1993). The densest regional breeding concentrations are probably in aspen-poplar groves or mixed aspen-conifer forests of the Greater Yellowstone ecosystem.

Suggested reading: BNA 515 (D. Rausch et al., 2000); Johnsgard, 2016b.

Greater Sage-Grouse. *Centrocercus urophasianus*.

Ub; Co, Ch, CC1, CC2, CC3; Res.

Status: The greater sage-grouse is common but declining on sage habitats in the plains and foothills; it is rare or absent in the national parks except for the Jackson Hole area of Grand Teton National Park. In Wyoming, breeding has been reported from all counties.

Habitats and ecology: These birds are closely associated with sagebrush, which is their primary food and also is used for nesting cover. They occur locally in sage to 9,000 feet elevation.

Breeding biology: From early spring onward, large groups of male sage-grouse assemble on traditional "strutting grounds" in open sage country, where they compete for territories and where females later come for fertilization. The male displays are complex and highly stereotyped but include stepping, wing-brushing movements, and a series of rapid inflations and deflations of their esophageal "air sacs," with associated plopping sounds. Females are attracted to such groups of males just before their egg-laying period, and most mating occurs at about sunrise. Typically a single "master cock" dominates each display ground and accounts for most of the matings there. After copulation, the female leaves the strutting ground and probably does not return unless her clutch is destroyed. She usually nests some distance from the display

Greater sage-grouse, adult male strutting

ground, and about ten days are needed to complete a clutch of eight eggs. Incubation lasts 25 to 27 days. Males do not take part in incubation or nest defense, and the chicks hatch in a highly precocial condition. Their mother quickly moves them to moist areas where insect food is plentiful, and they fledge rapidly, in less than two weeks. However, the juveniles usually remain with their mother for most of the summer, gradually becoming more independent. Eventually the birds are forced to move to their wintering areas, which are usually at lower elevations and may be 50 miles or more from the nesting areas.

Suggested viewing locations: The range of this species is inexorably decreasing, as sagebrush areas are being cleared and converted to irrigated cultivation. However, Wyoming still supports the nation's largest sage-grouse population, probably one-third to one-half of the species' total. Sage-grouse display socially in spring, with as many as 50 or more males "strutting" on local display grounds, or leks. The densest regional breeding concentrations are probably in the sagebrush scrublands of south-central and southwestern Wyoming. Grand Teton National Park has long had a display ground near the Jackson Airport, but numbers have been low in recent years. The Farson area, along the road to Big Sandy Reservoir, is perhaps the best location in Wyoming for watching display. Hat Six Road over near Casper is an easily reached site from which birds can be seen from a parked vehicle. From Casper I-25 exit 182, go south on Wyoming Highway 253 for 8.2 miles to Natrona County Road 605, then go left to the lek, a few hundred yards farther on the south side (Scott, 1993).

Greater sage-grouse, adult male

Suggested reading: BNA 425 (M. A. Schroeder, J. R. Young, and C. E. Braun, 1999); Patterson, 1952; Braun, Oedekoven, and Aldridge, 2002; Bohne, Rinkes, and Kilpatrick, 2007; Johnsgard, 2016b; Paothong, 2017.

White-tailed Ptarmigan. *Lagopus leucura*. HL; Ra, CC1, Ch; Res.

Status: Limited to remote alpine areas, this species' current status in Wyoming is uncertain, and it might already be extirpated.

Habitats and ecology: White-tailed ptarmigans are confined to the alpine and timberline zones, moving slightly lower during winter, especially where willows remain exposed above the snow.

Breeding biology: Females build their nests in a variety of locations; their highly protective coloration makes it difficult for predators to find them. The clutch size is usually small, from four to seven eggs, which are laid at intervals of about 1.5 days, The incubation period is 22 or 23 days. Females typically defend their nests and broods very strongly, performing distraction displays during incubating or when brooding young. As the chicks become older, the female tends to place herself between an intruder and the brood, running back and forth while hissing. Males have at times been reported to defend the nest site but not the brood. Fledging occurs when the chicks are only about ten days old, when the females and broods soon move into areas that provide a combination of rocky habitat and an abundance of herbaceous vegetation. Hens remain with their well-grown broods through autumn, as the birds gradually move toward lower-altitude wintering areas.

Suggested viewing locations: In Wyoming, Quadrant Mountain in Yellowstone Park and Ishawooa Mesa in Park County have had older reported sightings (Dorn and Dorn, 1990). Documented sightings from the Medicine Bow range (or elsewhere in Wyoming) are no more recent than 2005 (Faulkner, 2010).

Suggested reading: BNA 68 (K. Martin et al., 2015); Johnsgard, 2016b.

Dusky Grouse. *Dendragapus obscurus*. Wi; Co; Res.

Status: The dusky grouse is a common and widespread resident in coniferous forests almost throughout the state. It is present in all the major ranges except the Black Hills, where it is extirpated. Breeding is common in both national parks (Grand Teton and Yellowstone). This species is recently separated taxonomically from a Pacific Slope population, which is now called the sooty grouse (*Dendragapus fuliginosus*) because of several plumage and behavioral differences.

Habitats and ecology: Closely associated with coniferous forests but also reaching alpine timberline during the breeding season, the dusky grouse can also be found as low as the ponderosa pine and sagebrush zones. In the interior Pacific Northwest, dusky grouse begin breeding in mixed coniferous forests during the shrub-seedling stage of succession

(Sanderson, Bull, and Edgerton, 1980). Nonbreeders descend in late spring and summer into the sagebrush zone. The highest population densities found by Hutto and Young (1999) in 566 transects of 11 unaltered forest, grassland, shrub, and wetland habitats of Idaho and Montana were in sagebrush communities, a common summer habitat for broods and nonbreeders.

Breeding biology: In early spring, males begin to establish breeding territories, usually where there is a combination of heavy cover for escape and fairly open vegetation for display sites. Their territorial proclamation is a series of low-pitched, owl-like hoots, during which the male lifts and partially spreads his tail feathers and opens the feathers of his neck to expose a yellowish skin patch that is surrounded by white-based dark feathers. This display attracts females for mating, after which the male plays no further role in reproduction. After mating, the female begins nesting. Nests are placed in varied locations but often are hidden among logs or under the roots of fallen trees. Eggs are laid at the rate of one every two or three days, and six to eight eggs constitute the usual clutch. The incubation period is 26 days. After hatching, the chicks grow rapidly and can fly short distances when only six or seven days old. Within two weeks they are able to fly up to 200 feet. The broods gradually move from herbaceous cover to woody thickets as the young develop. Gradually the broods break up, and the young disperse singly or in small groups as they slowly work their way downward toward wintering habitats.

Suggested viewing locations: Birding locations include coniferous woods throughout the state except the northeastern corner. Grand Teton and Yellowstone National Parks all support good populations of dusky grouse, and from spring to early summer "hooting" periods, the males' low-pitched display calls can often be heard and the birds thus easily located. They are hard to find when they are perched in trees, but actively displaying males are often quite fearless and can be easily approached.

Suggested reading: BNA (F. C. Zwickel and J. F. Bendell, 2018 [dusky grouse]; F. C. Zwickel and J. F. Bendell, 2018 [sooty grouse]); Harjer, 1974; Johnsgard, 2016b.

Sharp-tailed Grouse. *Tympanuchus phasianellus*. SD; Un, CC4; Res.

Status: A subspecies, *T. p. jamesi*, is an uncommon resident of plains and foothills, mainly in the northeastern and southeastern regions; additionally, a small local population of the race *T. p. columbianus* inhabits the Washakie Basin and Sierra Madre foothills of south-central Wyoming.

Habitats and ecology: The sharp-tailed grouse is associated with grasslands and grassy sagebrush areas, and sometimes also mountain meadows during the breeding season; it can also extend into cultivated fields during fall and winter. Brushy foothills and similar edge habitats are often used in Alberta.

Breeding biology: By early spring onward, male sharp-tailed

Sharp-tailed grouse, displaying male in alert posture

grouse begin to assemble and establish or reestablish territories in a communal male display area, or lek. Older, more experienced males tend to occupy the more central and desirable territories, sought out by females when they arrive for fertilization. Most of the males' elaborate "dancing" behavior is thus directed toward the other males and consists of ritualized hostile behavior, though a few displays and calls are reserved for females. Females visit the leks only long enough to be fertilized, and the males take no further part in reproduction. The usual clutch size is 10 to 12 eggs. Incubation lasts 23 or 24 days. The young hatch simultaneously and soon leave the nest to begin feeding on small insects. They can fly short distances by the time they are ten days old and may move up to a quarter-mile in a day even before fledging. They are nearly independent by the time they are six to eight weeks old and often disperse considerable distances at that time.

Suggested viewing locations: In spring, male sharp-tails "dance" on traditional lek display areas in groups of from a few to 20 or more males, during which dominance is determined and the relative access of males to females for fertilization is established. These activities begin in late winter and may continue until May. In Wyoming, grasslands around Sheridan, the Four Corners area of Weston County, and the rim of Goshen Hole in Laramie and Goshen Counties are good locations for locating leks (Dorn and Dorn, 1990). Other good viewing locations include Wagon Box Road south of Story, Bird Farm Road south of Big Horn, and along State Road 335 west of Big Horn (Scott, 1993).

Suggested reading: BNA 354 (J. W. Connelly, M. W. Gratson, and K. P. Reese, 1998); Lumsden, 1965; Johnsgard, 2016b.

Wild Turkey. *Meleagris gallopavo*. Di; Co; Res.

Status: A common resident in the state as a result of introduction efforts, wild turkeys can be found mainly in open forests of ponderosa pines or mixed woods, especially those with oaks or other mast-bearing trees. Turkeys are most common in the northern half of the state, from Park County southeast to Goshen County, and probably locally elsewhere as a result of introductions. Turkeys are virtually absent from the national parks (Grand Teton and Yellowstone).

Habitats and ecology: Breeding habitats vary greatly among the several subspecies. The southern Rio Grande race (introduced in 1996) is found in very arid habitats dominated by short grasses but including scattered trees needed for roosting, as well as a water supply and succulent vegetation. The long-established Merriam race of Wyoming is the most widespread race, and it is associated with red cedar and ponderosa pines, running water, and rugged topography.

Breeding biology: Turkeys spend the winter in small flocks that consist of adult males or larger groups of hens and family units. When the "gobbling" season begins, the males may establish individual gobbling or strutting areas, but groups of brothers typically associate, displaying in synchrony and allowing the most dominant of the brothers to fertilize any female that is attracted. Additionally, single highly aggressive males may dominate entire local populations, in a manner equivalent to the "master cock" situation in lekking grouse. Females have only brief contact with males until they are fertilized; then they establish their nests. The usual clutch size is 8 to 12 eggs, and incubation lasts 25 to 26 days. Within a week the chicks can make short flights, and they soon begin to roost in trees. When young males reach the age of six to seven months they may break away from their families and begin to establish brother unions that could persist their entire lives.

Suggested viewing locations: In Wyoming, the western Black Hills area of Crook County is a good birding location. There are also good populations in the Bighorns, the ranges near Casper and Laramie, and across all of northeastern Wyoming.

Suggested reading: BNA 22 (J. T. McRoberts, M. C. Wallace, and S. W. Eaton, 2014).

Family Podicipedidae (Grebes)

Eared Grebe. *Podiceps nigricollis*. Wi; Co; Su; Res.

Status: The eared grebe is a generally widespread summer resident in the state, except in the high montane lakes. The only national park for which there are breeding records is Yellowstone, where breeding has occurred on several lakes and ponds. Breeding in Grand Teton National Park is hypothetical.

Habitats and ecology: Associated in the breeding season with rather shallow marshes and lakes that have extensive reed beds and submerged aquatic plants, these grebes are generally found in larger and more open ponds than either pied-billed grebes or horned grebes, and unlike these other species, typically nest in large colonies.

Breeding biology: Pair-forming displays occur during spring migration while the birds are in flocks and continue after arrival on the breeding grounds. Courting occurs in the center of semicolonial breeding areas; no territorial behavior is evident. Displays are mutual and include an advertising call by unpaired or separated birds as well as head-shaking and "habit-preening." They also perform a "penguin-dance" by both members of a pair standing upright in the water facing each other, and a "cat-attitude" with withdrawn head and fluffed body feathers. The female builds the nest, and copulation occurs on the nest platform, without elaborate associated displays. The usual clutch size is three to four eggs. Incubation lasts 20 to 22 days. The nest is abandoned when the last egg hatches, and thereafter the young are tended by both parents, which can often be seen with their young riding on their back. The young birds are relatively independent by their third week, and the fledging period is about 45 days.

Suggested viewing locations: During migration, marshes and lakes throughout the region attract this species. Some good locations are Hutton Lake National Wildlife Refuge, Grayrocks Reservoir in Platte County, Goldeneye Wildlife Area in Natrona County, and Cliff Graham Reservoir in Uinta County (Dorn and Dorn, 1990).

Suggested reading: BNA 433 (S. A. Cullen, J. R. Jehl, Jr., and G. L. Nuechterlein, 1999).

Western Grebe. *Aechmophorus occidentalis*. Wi; Co; Su; Res.

Clark's Grebe. *Aechmophorus clarkii*. HL; Un, CC2; Su; Res.

Status: Both species are common (western grebe) and uncommon (Clark's grebe) summer residents on lowland marshes over much of the entire state, with possible breeding records in the national parks apparently confined to Yellowstone, where the western grebe is uncommon in summer (McEneaney, 1988), although there is no current evidence of breeding.

Habitats and ecology: Breeding typically occurs on permanent ponds and shallow lakes that are often slightly brackish and have large areas of open water as well as semiopen growths of emergent vegetation.

Breeding biology: Territorial activity by pairs is maintained only in the immediate vicinity of the nest, and most display activity occurs before the start of nesting, apparently serving primarily for pair-bond formation and maintenance. Most or all displays are performed by both sexes, often mutually. They include crest-raising while the birds swim together, with associated whistling notes and occasional withdrawal of the head and neck to the back. There is also a "high arch" posture with neck stretched and bill pointed downward and the tail raised, a similar but less extreme "low arch" posture, ritualized "habit-preening," and the "race." In this last-named display, two birds (sometimes only one and sometimes as many as six) call, then rise in the water and race side by side over the surface with arched necks, bills pointed diagonally upward, and wings partially raised. Behavior leading to the race display usually includes threat-pointing with the bill and mutual bill-dipping as the birds approach; diving often terminates the display. When more than two birds perform the race the additional birds are always males. Copulation normally occurs on the nest site, but it has been observed at the edge of a beach where the nests were on dry land. The usual clutch size is three to four eggs, and incubation lasts 23 days.

Suggested viewing locations: During migration, marshes and lakes throughout the region attract both of these two closely related species, such as Wyoming's Hutton Lake National Wildlife Refuge, Grayrocks Reservoir in Platte County, Flaming Gorge Reservoir in Sweetwater County, and Lovell Lakes in Big Horn County (Dorn and Dorn, 1990). In Wyoming, Clark's grebes have been found nesting in Sweetwater, Bighorn, and Fremont Counties, and westerns nest in these counties and several others, the largest colonies being in Big Horn, Carbon, Fremont, Park, and Uinta Counties (Faulkner, 2010).

Suggested reading: Western grebe: BNA 26a (N. LaPorte, R. W. Storer, and G. L. Nuechterlein, 2013. *Clark's grebe:* BNA 26b (R. W. Storer and G. L. Nuechterlein, 1992).

Family Columbidae (Pigeons and Doves)

Eurasian Collared-Dove. *Streptopelia decaocto*. Wi; Un; Res.

Status: The Eurasian collared-dove is an invasive and rapidly expanding resident species that is now common and occupies the entire state below 8,000 feet. First reported in Wyoming in 1998, 4,953 were counted during the 2017 Audubon Christmas Bird Count. Yellowstone National Park has at least one record (Faulkner, 2010), and this species has been increasingly seen in Jackson Hole since 2007.

Habitats and ecology: These birds are usually found in smaller towns and villages, especially around feedlots or granaries where waste grain is abundant.

Breeding biology: This species usually nests in urbanized or cultivated locations. The nest may be placed in a tree or, rarely, on the ledge of a building. Like those of other doves, the nest is frail, a thin structure of small twigs, and it is placed as high as 40 feet above the ground but often under 20 feet. Two white eggs similar to but larger than mourning dove's eggs constitute the clutch. The nesting period is prolonged, from at least March to August, and from three to as many as six broods might be raised in a single season. The incubation period is 14 days, and both sexes share in incubation. The young are able fly by 18 days of age and leave the nest vicinity by 21 days. Because of its repeated nesting behavior, this species has one of the highest rates of natural increase of any North American bird, which helps account for its explosive spread across the continent.

Suggested viewing locations: Look (and listen) for this bird in small towns and villages, especially near grain storage facilities.

Suggested reading: BNA 630 (C. M. Romagosa, 2012).

Mourning Dove. *Zenaida macroura*. Ub; Co; Su; Res.

Status: The mourning dove is a widespread and common breeder in the state, occupying nearly all vegetational zones up to the lower coniferous forest zone. It is present and breeding commonly in Grand Teton National Park and uncommonly in Yellowstone.

Habitats and ecology: Breeds from riparian woodlands and cultivated areas through grasslands and sagebrush to woodlands, aspen, and open coniferous forest habitats as well as in cities and farmsteads. It nests either on the ground or, preferentially, in shrubs or trees. The highest population densities found by Hutto and Young (1999) in 566 transects of 11 unaltered forest, grassland, shrub, and wetland habitats of Idaho and Montana were in sagebrush communities.

Breeding biology: Mourning doves begin to form pairs at the onset of the breeding season, when males that are dominant in winter flocks mate with high-ranking females; such pairs are the first to establish territories and appear to be the most successful in their reproductive efforts. The availability of choice nesting materials (usually twigs) is important in determining territorial boundaries in captive birds, whereas food and water sites are not defended. The two eggs are usually laid at about 24-hour intervals, and incubation is by both sexes, the male normally incubating during the day and the female at night. Typically the eggs hatch on successive days, so that one chick tends to be larger and more aggressive in food-begging than the other. By the time the young are 12 days old they are ready to leave the nest, and they normally fledge when they are 13 to 15 days old. By that time the adults have generally begun a second clutch in a new nest. In subsequent nesting, the two nests may be used alternately; in Texas as many as six nesting cycles have been reported by a single pair, using three different nest sites.

Suggested viewing locations: This is an extremely widespread species that is easily found throughout the region, both in towns and countryside.

Suggested reading: BNA 117 (D. L. Otis et al., 2008).

Family Cuculidae (Cuckoos)

Yellow-billed Cuckoo. *Coccyzus americanus*. HL; Ra, CC3, CC4; Su; Res.

Status: The yellow-billed cuckoo is a rare summer resident in the northern and eastern parts of the region (Bighorn and Black Hills ranges, North Platte valley). It is a vagrant in Grand Teton National Park. Because it is declining nationally at a substantial rate, it is a conservation Priority Species in Wyoming.

Habitats and ecology: This cuckoo is associated with thickety areas near water, second-growth woodlands, deserted farmlands, and brushy orchards. Dense woodlands are avoided.

Breeding biology: Cuckoos are relatively late spring migrants, arriving in Minnesota in May or even early June and inconspicuously taking up breeding territories. Their distinctive clucking and repeated hollow notes of *kaw* or *kowp* are frequently

uttered, especially on cloudy days or at night. The birds gather nesting materials from trees by breaking off small branches and carrying them back one at a time to the nest. The eggs are laid at irregular intervals, and incubation apparently begins during the egg-laying period, since hatching is staggered. The usual clutch size is three to four eggs. Apparently both sexes assist equally in incubation and also equally feed and brood the young. Incubation lasts 10 to 11 days. It has been suggested that when second broods are produced, one parent may remain to tend the first brood while the other looks after the second clutch and brood. The young birds remain in the nest about nine days but are still flightless when they leave the nest. At that time they are very agile in climbing about on branches. When they leave the nest the young birds are somewhat more than half the adult weight; the most recently hatched and smallest young may be left alone in the nest, often to be neglected and even to starve.

Suggested viewing locations: In Wyoming, the North Platte riverbottoms of Goshen County, the Powder River bottoms in Sheridan and Campbell Counties, and Ash Creek north of Sheridan provide good birding sites for this species (Dorn and Dorn, 1990).

Suggested reading: BNA 418 (J. M. Hughes, 2015); Bennett and Keinath, 2001.

Family Caprimulgidae (Nightjars)

Common Nighthawk. *Chordeiles minor*. Ub; Co, CC2; Su; Res.

Status: The nighthawk is a common summer resident throughout the region, mainly below the zone of coniferous forests but breeding commonly in both national parks (Grand Teton and Yellowstone).

Habitats and ecology: This species forages entirely in the air, on flying insects, and is especially common over grassland and urban areas, sometimes extending to shrub and desert scrub. Nesting occurs on the ground, usually in grasslands, or at the edges of woods, and sometimes on the asphalt rooftops of buildings. In Wyoming, the species is common statewide up to about 8,500 feet (Faulkner, 2010).

Breeding biology: Nighthawks are fairly late arrivals on northern nesting areas, and the males soon announce their presence by aerial displays. The most conspicuous of these is the *peent* call, uttered during a series of four to five wingbeats and serving to announce territorial ownership. Males also perform steep dives with down-flexed wings, each dive ending with a rush of air that produces a booming noise. Such dives are often almost directly over the nest site. Several other vocalizations are produced by males or both sexes. The females deposit their two eggs on almost any flat surface and often move them about in the course of incubation, sometimes as far as five to six feet from their original position. The eggs are often rolled in front of the bird as the female settles on them for incubation. Some investigators report that only the female incubates; other researchers have stated that one sex incubates at night and the other by day, which seems most likely given the foraging behavior of the species. Incubation lasts 19 days, and both sexes are known to help care for the young. As the adults bring food to their offspring they apparently place their bills inside the gaping mouths of the chicks and regurgitate food with a strong pumping of the head. Feeding is usually done at dusk, after sunset, and just before dawn but not at night. About three weeks are required for the fledging of the young, and they are independent at about 30 days.

Suggested viewing locations: This is a widespread species at lower elevations, and it is usually seen or heard coursing low over towns and cities near sundown across the region.

Suggested reading: BNA 213 (R. M. Brigham et al., 2011).

Common Poorwill. *Phalaenoptilus nuttallii*. Di; Co; Su; Res.

Status: A common summer resident in the region, the common poorwill is found mainly on drier habitats toward the south. The species is a vagrant at Grand Teton National Park.

Habitats and ecology: Generally this species is associated with rocky habitats and their arid-adapted shrubs or low trees, such as pinyon-juniper, saltbush, greasewood, sagebrush, and dry grasslands. Poorwills nest on the ground, often under scrub oaks, the leaves of which provide concealment for both adults and young. In Wyoming, the sage-dominated basins of Sweetwater, Carbon, Fremont, and Park Counties support good populations of this species (Dorn and Dorn, 1990), which is fairly common statewide except in Yellowstone National Park (Faulkner, 2010).

Breeding biology: Like other species in this family, poorwills are late-spring migrants, and soon after arrival they begin to utter their distinctive *poor-will* or *poor-will-low* notes during the evening. Virtually nothing is known of their courtship displays, which are presumably similar to those of the whip-poor-will. Nests are extremely difficult to locate, and the adult usually remains motionless on the nest until very closely approached. The clutch consists of two eggs. Although females perhaps do most of the incubating, males have been seen incubating at night and also have been observed brooding the young. When incubating or brooding, both adults and young keep their eyes almost completely shut, which adds to the effective camouflage of these already inconspicuous birds. When disturbed the adults often utter a loud hissing sound and maximally inflate themselves, which has a frightening effect on persons unfamiliar with the birds. Incubation lasts 20 or 21 days. The young are hatched in a downy coat, soon replaced with a juvenile plumage similar to that of adults. The fledging period is 20 to 23 days. Most poorwills migrate south by early fall, but in a few locations (such as California) individuals have been found torpid among rocks or vegetation during subfreezing temperatures. This discovery is the first known example of semihibernation among birds, although some others, such as hummingbirds, also enter a torpid state when exposed to cold overnight temperatures.

Suggested viewing locations: In Wyoming, the sage-dominated basins of Sweetwater, Carbon, Fremont, and Park Counties support good populations of this nocturnal species (Dorn and Dorn, 1990), which is fairly common statewide up to about 8,500 feet (Faulkner, 2010). Populations are not unusually dense anywhere in the region but are probably best developed in the southernmost and driest parts.

Suggested reading: BNA 32 (C. P. Woods, R. D. Csada, and R. M. Brigham, 2005).

Family Apodidae (Swifts)

White-throated Swift. *Aeronautes saxatalis*. Di; Co, CC1, CC2; Su; Res.

Status: The white-throated swift is a common summer resident in mountainous areas of the state, more common southwardly. It is uncommon in Yellowstone National Park and only a vagrant in Jackson Hole (Raynes and Wile, 1994).

Habitats and ecology: This swift is associated with steep cliffs, deep canyons, and generally mountainous terrain, sometimes observed as high as 13,000 feet. Nesting occurs in crevices of canyon walls, in completely inaccessible locations. Nesting occurs in Wyoming up to about 9,500 feet (Faulkner, 2010).

Breeding biology: Most white-throated swifts arrive at their nesting areas by mid-May. They soon begin to construct their unique nests, carrying individual feathers in their bills, sometimes apparently for miles. They also begin their aerial courtship, which may even include copulation while in flight. To initiate copulation, the birds fly toward each other from opposite directions, meet, and begin to tumble downward while clinging together, sometimes falling several hundred feet. But copulation apparently also takes place in the nesting areas, judging from some observations of egg collectors. The usual clutch size is four to five eggs. It may be presumed that both sexes incubate, but nothing specific is known about incubation behavior. Incubation probably lasts about 20 to 27 days, but it and the fledging period are still uncertain. During the nonbreeding period, the birds often roost in communal quarters, much like chimney swifts. Observations of one such roosting site in California indicated that the birds, numbering 100 to 200, returned to their roosting crevice shortly after sunset. Within five minutes the entire flock had entered the crevice, passing through an entry only two to three inches wide! In very cold weather, roosting birds may become torpid, although this is not known to be a regular adaptation of the species for coping with cold periods.

Suggested viewing locations: In Wyoming, these widespread swifts can be seen at the Yellowstone River canyon, in canyons of the Bighorn Mountains, and at Wind River Canyon (Washakie County). Hawk Springs Reservoir (Goshen County), Flaming Gorge Reservoir (Sweetwater County), along Beaver Rim (Fremont County), and the rim of Goshen Hole (Goshen County) also offer possibilities (Dorn and Dorn, 1990).

Suggested reading: BNA 526 (T. P. Ryan and C. T. Collins, 2000).

Family Trochilidae (Hummingbirds)

Broad-tailed Hummingbird. *Selasphorus platycercus*. Di; Co; Su; Res.

Status: The broad-tailed hummingbird is a common summer resident in the southern parts of the region, mainly west of the plains and south of Montana. This bird is a common breeding species in Grand Teton National Park but is rare in Yellowstone.

Habitats and ecology: Typically associated with ponderosa pine forests and aspen groves, this hummingbird also extends into mountain meadows, pinyon-juniper woodland, and riparian cottonwoods in this region. Willow-lined streams adjacent to meadows and parklands are favored foraging areas. In Colorado, breeding birds are abundant in foothills or mountains with aspens, pines, or Douglas-fir at about 6,500 to 7,500 feet elevation, but nests have been seen from 5,200 to 10,750 feet. During the summer the birds gradually move upward, finally reaching alpine meadows in late summer.

Breeding biology: In Jackson Hole, these hummingbirds arrive in early May, and the males soon become highly territorial, chasing other males from the vicinity. Their display consists of hovering in front of a female, orienting the brilliant red gorget toward her, then quickly climbing 30 to 40 feet and making a vertical dive downward, swooping directly past her. The female spends several days gathering cottonwood or willow down and spider webs to construct her nest, which is usually on a horizontal tree branch with another branch or crook directly overhead. Of ten Colorado nests studied, five were in aspens, four in spruces, and one in a subalpine fir. They were located from 3 to more than 30 feet above the ground, and the nest core was coated with moss, lichens, and fragments of aspen bark. All incubation is by the female; the promiscuous males play no role in parental care. Incubation lasts about 16 days. The young are initially fed on regurgitated food and then increasingly with tiny insects. The female feeds them by thrusting her bill into their throats and regurgitating with rapid pumping movements of the head. The young soon nearly outgrow their nest, which is well trampled down by the end of the 23-day nestling period. Females have been known to consume almost twice their own weight in sugar syrup during a single day, which provides some measure of the metabolic rate of these tiny birds.

Suggested viewing locations: In Wyoming, this hummingbird is present in the northwestern mountains, Bighorns, and Medicine Bow Mountains (Dorn and Dorn, 1990).

Suggested reading: BNA 16 (A. F. Camfield, W. A. Calder, and L. L. Calder, 2013); Calder, 1973; Johnsgard, 1997b.

Rufous Hummingbird. *Selasphorus rufus*. Co; Fa; Mig.

Status: The rufous hummingbird is a common migrant, especially during fall (late July to mid-August), and a possible summer resident in the northern parts of the state. It is reported to be a common summer resident in Jackson Hole by Raynes and Wile (1994), although this has been questioned (Faulkner, 2010).

Habitats and ecology: In general, this hummingbird uses coniferous forests for breeding, but the birds occupy a variety of forest-edge habitats, including mountain meadows and burned-over forest areas where nectar-bearing flowers such as gilia are abundant. In the Pacific Northwest, this species breeds in mixed coniferous forests during the shrub-seedling and pole-sapling stages of succession (Sanderson, Bull, and Edgerton, 1980). Brushy areas in the foothills are also used on migration, as are urban gardens and alpine tundra.

Breeding biology: Like other hummingbirds, this species does not form pair-bonds, but rather the males establish small territories that they defend and attempt to attract females to by conspicuous displays. The males' display flights typically trace a complete oval with a slanted axis. During the downward phase the bird produces a sound sequence that starts with a wing buzz, progresses to a staccato whining, and ends with a rattle. After mating, the female builds a nest in any of various locations, from near the ground to as high as about 40 feet, often among drooping branches of conifers. Nests are often placed along paths or in gullies, and are typically decorated with lichens and lined with willow down. Early nests are usually built low in trees, but those made late in the season tend to be near the tops. As with other hummingbird species, two eggs are laid, at intervals up to about three days. Incubation is by the female alone and lasts 17 days. The fledging period is 22 days.

Suggested viewing locations: In Wyoming, the Greater Yellowstone ecosystem and the Wind River range offer some possibilities for finding this generally uncommon species, which as of 2010 had not been a proven Wyoming breeding species (Faulkner, 2010).

Suggested reading: BNA 53 (S. Healy and W. A. Calder, 2006); Calder, 1973; Johnsgard, 1997b.

Calliope Hummingbird. *Selasphorus calliope*. Lo; Co, CC1; Su; Res.

Status: The calliope hummingbird is a common summer resident over most of the mountainous region west of the plains; it is probably the commonest breeding species in both national parks (Grand Teton and Yellowstone).

Habitats and ecology: Open meadow areas near coniferous forests, such as low willow or sage areas rich in plants like Indian paintbrush or gilia, are favored areas for this species in the Jackson Hole area. In the Pacific Northwest, this species breeds in mixed coniferous forests during the shrub-seedling and pole-sapling stages of succession (Sanderson, Bull, and Edgerton, 1980). Openings in woodlands, sometimes as high as timberline, are also frequented, and in late summer alpine meadows are commonly used by migrating birds. The highest population densities found by Hutto and Young (1999) in 566 transects of 11 unaltered forest, grassland, shrub, and wetland habitats of Idaho and Montana were in cottonwood-aspen communities.

Overleaf: Broad-tailed hummingbird, adult male

Breeding biology: Calliope hummingbirds are among the smallest of the North American bird species, weighing less than three grams. Yet, the territorial males are highly conspicuous in spring when they perch on exposed twigs and periodically launch into display flights. These consist of a series of swooping flights along a U-shaped course that may be 20 to 30 feet in length and begin from a height of about 30 to 60 feet. The sound produced during this flight is a mechanical *bzzzt*, apparently made by tail vibration, but the wings are able to generate a metallic *tzing* sound as the male hovers in front of a female. Like other hummingbirds, calliope males are promiscuous and abandon the female immediately after mating. Females usually build their nests on small conifer branches among a cluster of old pine cones, the nest becoming well concealed among the cones. When the nests are placed in aspens, they often mimic mistletoe knobs. The nests are placed at heights ranging from less than 2 feet to nearly 70 feet, but most are placed less than 10 feet above ground. The nest is made of mosses, small leaves, needles, bark, and other materials and is bound together with spider webbing. The two eggs are deposited about three days apart, and incubation requires 15 to 16 days. The young grow slowly, and the fledging period lasts from 18 to 23 days. Thus the total nesting period, from egg-laying to fledging, averages about 36 days. The short growing season in the mountains probably prevents any possibility of second nestings or renesting efforts.

Suggested viewing locations: In Wyoming, this species is easily found in the mountain meadows and willow flats of Grand Teton and Yellowstone National Parks.

Suggested reading: BNA 135 (W. A. Calder and L. L. Calder, 1994); Calder, 1973; Johnsgard, 1997b.

Family Rallidae (Rails and Coots)

American Coot. *Fulica americana*. Ub; Co; Su; Res.

Status: The coot is a widespread and common summer resident on wetlands throughout the state, especially at lower elevations. It is present and breeding in both national parks but is much more common in Grand Teton than Yellowstone.

Habitats and ecology: Coots are associated with ponds and marshes that have a combination of open water with emergent reed beds, which is where nesting occurs. Besides foraging on aquatic plants, the birds sometimes also graze on nearby shorelines and meadows.

Breeding biology: Coots are monogamous, having a potential lifelong pair-bond, and spend much of their time in advertising and defending territories. These are established soon after arrival on the breeding grounds, and although the male patrols the territory at first, later it is defended by both members of the pair. Pairs also construct display platforms for copulation and, as the egg-laying period approaches, construct one or more egg nests as well as brood nests later on. Both sexes participate in incubation, with the male most often incubating at night. Unlike gallinules, coots seem to have no

specific nest-relief ceremony. The usual clutch size is six to nine eggs. Incubation lasts 21 to 24 days. Hatching is typically staggered over several days. Apparently the male takes most of the responsibility for brooding the young birds, although the female may take the first-hatched chicks and leave the male to incubate and tend the later hatchlings. The young begin to beg shortly after hatching and soon begin to follow the adults during their foraging. After a month or so they are nearly independent, but they beg occasionally almost to the time they are independent, at about eight weeks of age. If the adults begin a second clutch they may expel the young of the first brood from the area while they are still fairly young.

Suggested viewing locations: Wetlands at lower elevations throughout the region provide easy viewing opportunities. Breeding probably occurs in all larger marshes of Wyoming (Scott, 1993).

Suggested reading: BNA 697 (I. L. Brisbin, Jr. and T. B. Mowbray, 2002).

Family Gruidae (Cranes)

Sandhill Crane. *Antigone canadensis*. Di; Un, Ch, CC3; Su; Res.

Status: Greater sandhill cranes (*A. c. tabida*) are locally uncommon summer residents in the more remote wetlands, pastures, and mountain meadows of the state. Lesser sandhill cranes (*A. c. canadensis*) are regular spring and fall migrants in eastern Wyoming.

Habitats and ecology: In the Rockies, sandhill cranes are especially associated with beaver impoundments, where the birds nest along shorelines or sometimes on beaver lodges, often in dense willow thickets. The birds are highly territorial and nests usually are well scattered. Their loud calls serve to advertise territories and to communicate over long distances.

Breeding biology: Cranes are monogamous, probably pairing for life after reaching reproductive maturity at about four years of age. Upon returning to their breeding areas, pairs establish territories as early as two to four weeks before nest building gets under way. Nest building is done by both sexes and may take from a day to a week or more. Two eggs are laid at a two-day interval. Both sexes participate in incubation, with the female apparently always doing the nighttime incubation. Incubation lasts 30 to 32 days. The eggs typically hatch 24 hours apart, and the chicks begin to feed immediately, with the first-hatched often taken away from the nest by one adult while the other remains to hatch the second chick. Perhaps because the young "colts" are very aggressive toward each other, they are often brooded separately. Fledging occurs at 67 to 75 days of age, and the family soon migrates as a unit.

Suggested viewing locations: In Grand Teton National Park, favored crane habitats include Willow Flats near the Jackson Lake dam, sedge meadows east of Rockefeller Lodge,

Greater sandhill crane, adult calling

and beaver ponds below Teton Point and along the Buffalo Fork River. In Yellowstone National Park, Lamar Valley and Hayden Valley are highly favored habitats; other good crane habitats exist at Willow Park, Swan Lake Flats, Blacktail Ponds, Antelope Creek, and near Fishing Bridge. Outside of these park areas, breeding occurs along the Snake River and Green River drainages, and locally east to the Bighorn Range (Faulkner, 2010).

Suggested reading: BNA 31 (B. D. Gerber et al., 2014); Johnsgard, 2011b.

Family Recurvirostridae (Stilts and Avocets)

Black-necked Stilt. *Himantopus mexicanus.* Lo; Un, Ch; Su; Res.

Status: The black-necked stilt is an uncommon and local summer resident in low-elevation wetlands throughout much of the state. Breeding has occurred in at least seven scattered locations. It is rare in Grand Teton National Park and a vagrant in Yellowstone.

Habitats and ecology: Breeding in this species usually occurs in the grassy shoreline areas of shallow freshwater or brackish pools of wetlands having extensive mudflats, or sometimes along the shorelines of salt lakes where vegetation is essentially lacking. Black-necked stilts are often found in company with American avocets, which use similar habitats.

Breeding biology: Like avocets, stilts form pair-bonds gradually and without associated elaborate displays through the persistent association of a female with a particular male, in spite of initial aggressiveness by the male. Stilts defend territories on their breeding grounds better than avocets do and advertise them by aerial displays. Copulation in stilts is preceded by slight ritualized breast-preening by both sexes, apparently identical to that of avocets. Nest building is probably done by both sexes, and materials are added to the nest throughout incubation. During periods of rising water, the nest may be raised considerably by such added materials, and both sexes apparently share incubation about equally. Incubation begins when the last or penultimate egg is laid, and lasts 25 days. The clutch size is three to five eggs, usually four. The eggs hatch relatively synchronously, and the young remain in the nest no more than 24 hours. They are probably brooded for at least a week and are independent at about four weeks.

Suggested viewing locations: In Wyoming, important migration stopping points include Lovell Lakes (Big Horn County), Hutton Lake National Wildlife Refuge (Albany County), Goldeneye Wildlife Area (Natrona County), and the Bridger Power Plant near Point of Rocks (Dorn and Dorn, 1990). Breedings have occurred periodically in Sweetwater County (Bridger Power Plant, Old Eden Reservoir) and at Lovell Lakes, Loch Katrine (Park County), Ocean Lake (Fremont County), and Table Mountain Wildlife Habitat Management Area (Goshen County) (Faulkner, 2010).

Suggested reading: BNA 449 (J. A. Robinson et al., 1999); Johnsgard, 1981.

American Avocet. *Recurvirostra americana.* Wi; Co; Su; Res.

Status: The American avocet is a locally common summer resident over much of the state, mainly on shallow marshes of the plains. It is uncommon in Yellowstone National Park and occasional in Grand Teton, with breeding reported only for Grand Teton National Park.

Habitats and ecology: During breeding this species favors ponds or shallow lakes with exposed and sparsely vegetated shorelines and somewhat saline waters that have large populations of aquatic invertebrates, which the birds gather by making scythe-like movements of their curved bill through the water.

Breeding biology: In Oregon, avocets arrive on their breeding areas 15 to 20 days before egg-laying to establish territories and begin courtship. They apparently form pairs during late winter, without associated elaborate posturing. Copulation is preceded by a rather simple breast-preening ceremony that may be initiated by either bird. Pairs form close bonds and forage together as well as defend their territory as a unit. Both sexes develop incubation patches and begin to incubate their clutch as soon as it is completed. The clutch size is three to five eggs, usually four. Early during incubation the male spends more time on the nest than the female, but the female is more attentive later on. Incubation lasts 22 to 24 days. The eggs hatch over a one- to two-day period, and the young soon become very active, feeding themselves almost from the outset. They fledge in four to five weeks, and thereafter the families begin to form flocks, which remain intact until the following breeding season.

Suggested viewing locations: In Wyoming, the sites listed for the black-necked stilt also attract avocets, as does Dave Johnson Power Plant east of Glenrock and Carmody Lake (about three miles northwest of Sweetwater Station in Fremont County) (Dorn and Dorn, 1990). The alkaline lakes in the Red Desert region also often have high numbers (Scott, 1993; Faulkner 2010).

Suggested reading: BNA 275 (J. T. Ackerman et al., 2013); Johnsgard, 1981.

Family Charadriidae (Plovers)

Snowy Plover. *Charadrius nivosus*. Lo; VR; Su; Res.

Status: The snowy plover is a local and very rare summer resident, largely limited to saline flats and sandy riverbeds in southwestern Wyoming. It is a species of regional concern by the US Forest Service.

Habitats and ecology: Barren salt plains represent prime breeding habitat for this arid-adapted species, and sandy riverbeds or barren shorelines of reservoirs are used secondarily.

Breeding biology: After arriving on their breeding areas and establishing territories, males begin to advertise with various calls and displays including "scraping," a ritualized nest-building behavior. One of the other male displays is a slow "butterfly flight" accompanied by a trilling call. Although the birds commonly breed around salt water, they can drink no more saline water than other shorebirds and must obtain liquid by eating insects or other succulent foods. Thermal extremes are also common in their often vegetation-free and highly reflective environment. Thus, during hot weather parental activity increases, with the birds spending most of their time standing over the eggs or chicks rather than sitting on them. The clutch size is two to four eggs, usually three. The eggs are laid about three days apart, but hatching is synchronous. Both sexes incubate, and incubation lasts 24 days. Both sexes also defend the eggs and young, performing effective "broken-wing" behavior when threatened. The young fledge in 27 to 31 days.

Suggested reading: BNA 154 (G. W. Page et al., 2009); Johnsgard, 1981.

Killdeer. *Charadrius vociferus*. Ub; Co; Su; Res.

Status: The killdeer is a common summer or permanent resident throughout the state, both on the plains and in montane areas, but they are more common at lower elevations. Killdeers are a common breeder in both national parks (Grand Teton and Yellowstone).

Habitats and ecology: Widely distributed in open-land habitats, including pastures, roadsides, gravel pits, golf courses, airports, and sometimes suburban lawns. Gravelly areas are favored, and graveled rooftops are sometimes used for nesting in urban areas. Migrating and wintering birds are more closely associated with water, but the birds also use mud flats and open fields.

Breeding biology: Although some birds are paired at the time they arrive on their nesting areas in southern Canada, most arrive unpaired. Males advertise their territories in a variety of ways, such as uttering the familiar *killdeer* calls while flying with slow, deep wingbeats, and by sham-nesting or "scraping" displays that resemble nest-building behavior. Such scraping displays are performed not only by unmated males but also before copulation, during hostile encounters, and during actual nest construction. Once pair-bonds are formed, the pair remains together and both sexes defend their territory, although they may do some foraging outside the defended area. The clutch size is three to five eggs, usually four. Incubation lasts 24 to 26 days. Both sexes also incubate the eggs and care for the young, but males tend to be more aggressive toward humans, while females vigorously evict other killdeers from the nest vicinity. The familiar injury-feigning display, or "broken-wing act," is primarily directed toward potential mammalian predators; large grazing mammals such as horses and cattle are more likely to be threatened or even attacked. Evidently the male undertakes most of the brooding duties, which last about three weeks. Fledging occurs by the time the young are 40 days old.

Suggested viewing locations: This abundant plover can be seen on grassy meadows or shorelines almost anywhere in the region at lower elevations. Dense breeding populations occur in all such habitats of nonmontane areas.

Suggested reading: BNA 517 (B. J. Jackson and J. A. Jackson, 2000); Johnsgard, 1981.

Mountain Plover. *Charadrius montanus*. Wi; Un, CC3; Su; Res.

Status: The mountain plover is a local uncommon summer resident on the drier shortgrass and desert scrub plains and a vagrant in Grand Teton National Park.

Habitats and ecology: This species breeds exclusively in early spring on arid grasslands where the grasses are usually no more than three inches in height, and sometimes in semidesert areas far from water with cacti and scattered shrubs. During the nonbreeding seasons the birds are also found in relatively dry habitats.

Breeding biology: In northeastern Colorado, mountain plovers arrive in late March and soon disperse over their breeding grounds. Males commonly reestablish their old territories, whereas females also return to the same general area but may visit several territories before choosing mates. Territorial males advertise with calls and an aerial "falling-leaf" display, and occasionally with a slow "butterfly flight." As with other plovers, "scraping" is the most frequent courtship display of the male, a behavior that produces several potential nest sites throughout his territory. The clutch size is usually three eggs. Incubation lasts 28 to 31 days. At least some females begin a second clutch with new mates within about two weeks of completing their first clutches, leaving their first mates to attend to the original clutches. Evidently the female often incubates the second clutch herself, but current evidence indicates that only one sex is involved in incubation and brooding duties for each clutch and brood.

Suggested viewing locations: In Wyoming, breeding birds are most likely to be seen on the Laramie Plains, especially near Bamforth Lake, and along the Carbon-Albany county line (Dorn and Dorn, 1990). Shirley Rim between Wyoming Highways 77 and 487 is also a good location (Scott, 1993). Most Wyoming nesting occurs in the Bighorn, Great Divide, Laramie, Shirley, and Washakie Basins (Faulkner, 2010).

Suggested reading: BNA 211 (F. L. Knopf and M. B. Wunder, 2006); Graul, 1975; Dinsmore, 2001; Johnsgard, 1981.

Family Scolopacidae (Sandpipers, Snipes, and Phalaropes)

Upland Sandpiper. *Bartramia longicauda*. Lo; Un, CC3, CC4, Ch; Su; Res.

Status: The upland sandpiper is an uncommon summer resident in native grassland areas east of the Bighorn, Laramie, and Medicine Bow ranges; it is a conservation priority species in Wyoming.

Habitats and ecology: This species is generally associated with wet meadows, hayfields, mowed prairies, or midlength prairies. It avoids both shortgrass steppe areas and extremely tall grasses; in addition, it is often found far from water and rarely, if ever, wading for its food.

Breeding biology: In North Dakota, the first spring arrivals appear about two weeks before the start of nesting, and they are usually paired. Territorial birds perform a flight display consisting of circling with quivering wing beats while uttering a musical purring or chattering call, and then finally diving abruptly back to earth. In North Dakota, nesting begins almost simultaneously, and the eggs are laid at approximately daily intervals. The clutch size is usually four eggs. Incubation lasts 21 days. Both sexes incubate, and adults typically feign injury when discovered on the nest. A fairly long interval occurs between the first "pipping" cracks and the hatching of the last egg, which may vary from less than 24 hours to about three days. The chicks are brooded by both parents, and by the time they are 30 days old they appear to be full grown and presumably are fledged.

Upland sandpiper, adult incubating

Suggested viewing locations: Most Wyoming breeding occurs east of a line extending from Sheridan to Laramie, and locally farther west in Carbon, Natrona, and Big Horn Counties (Faulkner, 2010). Favored nesting sites include the North Platte River valley northwest of Douglas, areas south of Lusk along Highway 85, and meadows just west of LaGrange (Dorn and Dorn, 1990). Breeding also occurs on the high plains north of Shirley Basin, near the headwaters of Bate's Creek (Scott, 1993).

Suggested reading: BNA 580 (C. S. Houston, C. Jackson, and D. E. Bowen, Jr., 2011); Higgins and Kirsch, 1975; Johnsgard, 1981.

Long-billed Curlew. *Numenius americanus*. Wi; Un, CC3, CC4, Ch; Su; Res.

Status: An uncommon summer resident in grassland areas over much of the state, the long-billed curlew is mostly absent at higher elevations. It is a common summer resident in Grand Teton National Park but rare in Yellowstone. This curlew is a conservation priority species in Wyoming.

Habitats and ecology: On the breeding grounds this species occurs in shortgrass areas, grazed taller grasslands, and overgrazed grasslands with scattered shrubs or cacti. Hilly or rolling areas seem favored over flatlands, and the birds often nest rather far from standing water. However, migrating birds are usually found on beaches or other shoreline habitats.

Breeding biology: In the Nebraska Sandhills, long-billed curlews arrive by early April, usually in flocks of fewer than 12 birds. The rest of the month is spent in prenesting activities, including establishing core areas and foraging areas. Core areas typically consist of rolling sands, and they are advertised by extended flight displays and calling above the ultimate nest site. Meadows adjacent to nesting locations are used for foraging and are advertised by similar flight displays. The foraging area is part of the defended territory, and other curlews are forcibly excluded. The clutch size is usually four eggs. Both sexes incubate, and both sexes care for the brood. Incubation lasts 27 or 28 days. The fledging period is about 30 days. When a nest or brood is disturbed, the alarm call of the resident pair quickly attracts nearby pairs to help in distraction behavior.

Suggested viewing locations: In Wyoming, breeding occurs in the meadows west of Pinedale and Daniel Junction, and in the Bear River marshes of Cokeville Meadows National Wildlife Refuge (Scott, 1993). Curlews also breed near Lusk (Niobrara County), at Chapman Bench (Park County), and in northern Sublette County (Faulkner, 2010). They may also be seen at the National Elk Refuge (Dorn and Dorn, 1990).

Suggested reading: BNA 628 (B. D. Dugger and K. M. Dugger, 2002); Forsythe, 1972; Johnsgard, 1981.

Baird's Sandpiper. *Calidris bairdii*. Co; Mig.

Status: Baird's sandpiper is a common migrant throughout the state, more common on the plains and rare in both national parks (Grand Teton and Yellowstone).

Habitats and ecology: Migrants are associated with wet meadows and shallow ponds, often feeding in grassy areas somewhat away from water, but also along muddy shorelines, where they tend to peck at food sources rather than probe for them.

Suggested viewing locations: In Wyoming, Hutton Lake National Wildlife Refuge (Albany County), Table Mountain Wildlife Unit (Goshen County), the south end of Boysen Reservoir (Fremont County), and Yellowstone Lake area attract migrating birds (Dorn and Dorn, 1990). Bigger lakes, such as Keyhole Reservoir, and Loch Katrine (Park County) are favored by migrants in Wyoming (Scott, 1993).

Suggested reading: BNA 661 (W. Moskoff and R. Montgomerie, 2002); Johnsgard, 1981.

Least Sandpiper. *Calidris minutilla*. Co; Mig.

Status: The least sandpiper is a common migrant throughout the state. It is more common on the plains, where it is generally among the commonest of the "peeps," but it is rare in both national parks (Grand Teton and Yellowstone).

Habitats and ecology: While on migration these sandpipers are found on a variety of moist habitats, often in company with semipalmated, Baird's, or western sandpipers, probably feeding on much the same invertebrate foods as these species.

Suggested viewing locations: In Wyoming, Hutton Lake National Wildlife Refuge (Albany County), Table Mountain Wildlife Habitat Management Area (Goshen County), and the south end of Boysen Reservoir (Fremont County) attract migrating birds (Dorn and Dorn, 1990).

Suggested reading: BNA 115 (S. Nebel and J. M. Cooper, 2008); Johnsgard, 1981.

Long-billed Dowitcher. *Limnodromus scolopaceus*. Co; Mig.

Status: The long-billed dowitcher is a common migrant throughout the state. It is rarer in montane areas and occasional to rare in the national parks (Grand Teton and Yellowstone).

Habitats and ecology: Migrating birds use marshy habitats that provide foraging at "knee-deep" depths.

Suggested viewing locations: In Wyoming, Hutton Lake National Wildlife Refuge (Albany County), Table Mountain Wildlife Habitat Management Area (Goshen County), and Lowell Lakes (Bighorn County) attract migrating birds (Dorn and Dorn, 1990), as do Keyhole Reservoir (Crook County), the Bridger Power Plant ponds (Sweetwater County), and Loch Katrine (Park County) (Scott, 1993). Prairie marshes east of the mountains are the best regional birding sites.

Suggested reading: BNA 493 (J. Y. Takekawa and N. D. Warnock, 2000); Johnsgard, 1981.

Spotted Sandpiper. *Actitis macularia*. Ub; Co; Su; Res.

Status: The spotted sandpiper is a common summer resident throughout the state, breeding commonly in both national parks (Grand Teton and Yellowstone).

Habitats and ecology: This sandpiper species is associated with forest streams, pools, and rivers, usually at lower elevations but extending locally to alpine timberline. It utilizes a wide array of open terrains with water present, and rarely even some in the absence of nearby water. Shaded watercourses are favored, and sometimes the birds are found along rapidly flowing mountain torrents.

Breeding biology: Male and female spotted sandpipers arrive on their breeding grounds at about the same time, and pair-bonds are rapidly formed during a period of intense aggression, especially among females, which are larger and more aggressive than the males. Females establish territories, and pairs are formed by males entering such territories and being either accepted or expelled by unmated females. When a male leaves the shoreline area and enters nesting cover with a female, a bond has been formed, and the female may lay her first egg within five days of the male's arrival. The clutch size is usually four eggs. The eggs are laid at approximately daily intervals, and by the time she lays the third egg the female begins to show a resurgence of sexual activity, with increased singing and territoriality. Although some females remain monogamous and assist with incubation, others allow their first mates to undertake incubation duties and accept a second mate. Successive mating with as many as four mates in a single season has been found, and typically the female helps incubate the final clutch. The young birds leave the nest as soon as their feathers dry and reportedly are able to fly as early as 13 to 16 days after hatching.

Suggested viewing locations: In Wyoming, good observation areas in summer include Lovell Lakes (Big Horn County), Hutton Lake National Wildlife Refuge (Albany County), Goldeneye Wildlife Area (Natrona County), and Yellowstone Lake (Dorn and Dorn, 1990).

Suggested reading: BNA 289 (J. M. Reed, L. W. Oring, and E. M. Gray, 1997); Johnsgard, 1981.

Lesser Yellowlegs. *Tringa flavipes*. Pan; Co; Mig.

Status: A common migrant nearly throughout the state, the lesser yellowlegs is occasional in Grand Teton National Park and rare in Yellowstone.

Habitats and ecology: Breeding typically occurs in habitats that have a combination of rather open and tall woodlands, with low and sparse brushy undergrowth, and are fairly close to grassy or marshy ponds. Broken hills, covered with burned or fallen timber, and low poplar second growth, are favored Alberta nesting habitats. Outside the breeding season the birds occur along mud flats and shallow ponds, often with vegetated shorelines, and sometimes flooded fields.

Suggested viewing locations: This species migrates across the plains mostly to the east of the Rockies, stopping in Wyoming at sites such as Hutton Lake National Wildlife Refuge (Albany County) (Dorn and Dorn, 1990). Wyoming's Table Mountain Wildlife Habitat Management Area (Goshen County), Loch Katrine (Park County), and the Bridger Power Plant ponds (Sweetwater County) also attract migrating yellowlegs (Scott, 1993).

Suggested reading: BNA 427 (T. L. Tibbitts and W Moskoff, 2014); Johnsgard, 1981.

Greater Yellowlegs. *Tringa melanoleuca*. Pan; Co; Mig.

Status: The greater yellowlegs is a common migrant throughout the state. It is occasional in Grand Teton National Park) and rare in Yellowstone.

Habitats and ecology: During migration these birds occupy the edges of marshes and slow-moving rivers, foraging along the shorelines and sometimes wading out belly-deep to probe in the mud or skim the surface for invertebrates. On the breeding grounds the birds favor muskeg areas, with a mix of ponds, trees, and clearings, and sometimes extend into subalpine scrub near timberline. In Alberta, a favored nesting habitat consists of muskeg with spruce and tamarack.

Suggested viewing locations: This species migrates across the plains mostly to the east of the Rockies, stopping in Wyoming at sites such as Hutton Lake National Wildlife Refuge (Albany County).

Suggested reading: BNA 355 (C. E. Elphick and T. L. Tibbitts, 1998); Johnsgard, 1981.

Wilson's Phalarope. *Phalaropus tricolor*. Ub; Co; Su; Res.

Status: Wilson's phalarope is a common summer resident over most of the state's lowlands, becoming rarer in the mountains. Breeding has been reported from both national parks, common in Grand Teton and uncommon in Yellowstone.

Habitats and ecology: Breeding habitats are typically wet meadows that adjoin shallow marshes, which range from fresh to highly saline. Ditches, river edges, and shallow lakes are sometimes also used for breeding. Migrating birds use similar areas.

Breeding biology: Although female phalaropes are appreciably larger and more brightly colored than males, recent studies have cast doubt on the idea that they are regularly polyandrous. Pair-bonds apparently are formed after the birds arrive on the breeding areas, during a period of behavior that is intensely aggressive but scarcely indicative of typical territoriality. The female probably makes the nest scrape after the pair is formed, but the male adds the nest lining. The clutch size is usually four eggs. Eggs are laid about 48 hours apart, and presumably the female plays no further role in parental care. Incubation lasts 20 to 21 days. The male incubates, and he leads his brood from the nest to foraging areas only a few hours after they hatch. The fledging period has not been reported, but in the closely related northern phalarope it is less than three weeks.

Suggested viewing locations: This species is a fairly common migrant on lower altitude wetlands of Wyoming (such as Ocean Lake in Fremont County and most shallow lakes and marshes), and a local breeder on freshwater and especially saline wetlands wherever crustaceans such as brine shrimp are often abundant.

Suggested reading: BNA 83 (M. A. Colwell and J. R. Jehl, Jr., 1994); Johnsgard, 1981.

Red-necked Phalarope. *Phalaropus lobatus*. Pan; Co; Mig.

Status: The red-necked phalarope is a common migrant throughout the state, mainly in the plains and rare in both national parks (Grand Teton and Yellowstone).

Habitats and ecology: Breeding habitats of this species are subarctic ponds, marshes, and lagoons that have adjacent grassy or sedge vegetation, which is where nesting occurs. Proximity to lakes or other fairly permanent bodies of water may also be a part of the habitat characteristics. On migration the birds are found in the same areas as Wilson's phalaropes and are often seen in company with them.

Suggested viewing locations: This species is too infrequent in Wyoming to suggest reliable birding sites, but migrants favor saline wetlands that are also used by the Wilson's phalarope. Migrants are often seen around larger lakes such as Table Mountain Wildlife Habitat Management Area (Goshen County), Ocean Lake, Hutton Lake, and wetlands.

Suggested reading: BNA 538 (M. A. Rubega, D. Schamel, and D. M. Tracy, 2000); Kangarise, 1979; Johnsgard, 1981.

Family Laridae (Gulls and Terns)

California Gull. *Larus californicus*. HL; Co; Su; Res.

Status: The California gull is a common summer resident and local breeder over much of the state, mainly on the plains. The only breeding in the national parks is in Yellowstone, where 200 to 300 pairs breed yearly on the Molly Islands of Yellowstone Lake. The breeding range of the species in the general region is increasing, but nationally the population trend has been downward.

Habitats and ecology: Like the ring-billed gull, this species usually nests on gravelly islands of large lakes or reservoirs or along their shorelines, and in many areas the two species nest in close proximity. In Alberta, the California gulls tend to nest on more elevated and boulder-strewn sites, while ring-bills occupy more level terrain. Ring-billed gulls also tend to cluster their nests, whereas California gulls space their nests more randomly.

Breeding biology: California gulls arrive from their wintering grounds along the Pacific coast some weeks before the onset of nesting. Territorial establishment and courtship activities begin as soon as they arrive, even if the nesting areas are still covered by snow. The clutch size is usually three eggs. They are laid at an average interval of two days, so that most clutches are completed in four to five days. Egg-laying within colonies is highly synchronized, and in a sample of 100 nests nearly all the eggs were laid within two weeks. Incubation is performed by both sexes and averages about 26 days, with a range of 23 to 28 days. Although these gulls are serious egg predators for other species, relatively few eggs are eaten or disappear within the nesting colony, and hatching success is often high. The chicks are relatively precocial, and though they are usually raised in close vicinity of their nest, they are also well able to run and elude danger from an early age. They fledge at ages of 36 to 44 days, averaging 40 days.

Suggested viewing locations: Wyoming locations for seeing this species include the lakes on the Laramie Plains, Ocean Lake (Fremont County), Yellowstone Lake, Flaming Gorge Reservoir, and wetlands in the Casper area (Dorn and Dorn, 1990). Breeding colonies have been regular at Bamforth and Ocean Lakes, Pathfinder Reservoir, and Yellowstone Lake (Scott, 1993).

Suggested reading: BNA 259 (D. W. Winkler, 1996).

Black Tern. *Chlidonias niger.* Lo; Ra, CC3, CC4; Su; Res.

Status: The black tern is a rare summer resident over much of the state, mainly in plains marshlands, rare in the national parks (Grand Teton and Yellowstone). It is declining nationally and is a Wyoming Species of Greatest Conservation Need.

Habitats and ecology: Typical nesting habitat consists of small to large marshes with extensive stands of emergent vegetation and some areas of open water. Fish populations are not necessary, as the birds feed mostly on insects while on the nesting grounds. Nests are more often placed among emergent vegetation than on muskrat houses, although the latter are sometimes used.

Breeding biology: Prenesting behavior in black terns is marked by two types of display flights, including "fish flights" (the birds usually carry insects rather than fish), normally performed by two birds, and "flock flights," involving most or all of the birds of an entire nesting area. In the courtship phase, one bird (probably the male) postures and calls while standing on a potential nest site. Also, the two birds make aerial glides downward from several hundred feet while maintaining a fixed position relative to each other. Nesting sites of the previous year apparently are not reused. The nests seem to be built from materials gathered in the immediate vicinity of the nest, rather than carried in. The clutch size is usually three eggs. Incubation lasts 20 to 22 days. Both sexes assist in incubation, and both brood the young for at least eight days after hatching. Little brooding is done thereafter, though the chicks are unable to fly until they are more than 20 days old. Young birds are fed almost exclusively with insects and continue to feed on them for a time after they fledge.

Suggested viewing locations: In Wyoming, migrants or summering birds may be found at the lakes on the Laramie Plains, the Table Mountain Wildlife Habitat Management Area (Goshen County), and the Cokeville/Bear River area (Dorn and Dorn, 1990). They breed regularly at sites on the Laramie Plains, such as at Hutton Lake National Wildlife Refuge and Cokeville National Wildlife Refuge, and periodically probably nest elsewhere (Faulkner, 2010).

Suggested reading: BNA 147 (S. R. Heath, E. H. Dunn, and D. J. Agro, 2009).

Forster's Tern. *Sterna forsteri.* HL; Un; Su; Res.

Status: Forster's tern is an uncommon summer resident, mainly on plains marshes. It is occasional in Grand Teton National Park to rare in Yellowstone.

Habitats and ecology: Large marshes that have extensive reed beds or muskrat houses for nest sites are the typical breeding habitats of this species, which breeds colonially in such locations, with as many as five nests sometimes situated on a single muskrat house. Such sites that are close to open water areas for foraging are especially favored nesting locations.

Breeding biology: Shortly after they arrive on their nesting marshes, Forster's tern pairs begin to seek out nest sites. They are relatively colonial, and as many as five nests may be placed on a favorable site, such as a large muskrat house. The floating rootstalks of cattails may also serve as a nest site, but such locations are more often used by black terns. Nest building is initiated almost simultaneously by all members of a colony The clutch size is usually two or three eggs. Incubation lasts 23 to 25 days, and both sexes incubate. Wind and wave action, house-building activities by muskrats, and possibly intraspecific hostility are probably major causes of egg loss, which seems to be relatively high in this species. Little information is available on the growth of the young, but presumably they fledge in less than a month, as is typical of the common tern.

Suggested viewing locations: In Wyoming, migrants and summering birds may be found at the lakes on the Laramie Plains, Ocean Lake (Fremont County), Lovell Lakes (Big Horn County), Cliff Graham Reservoir (Uinta County), and Goldeneye Wildlife Area (Natrona County) (Dorn and Dorn, 1990). Breeding has been documented at only a few Wyoming sites, such as on the Laramie Plains, at Ocean Lake, and at Cokeville National Wildlife Refuge (Faulkner, 2010).

Suggested reading: BNA 595 (M. K. McNicholl, P. E. Lowther, and J. A. Hall, 2001).

Family Gaviidae (Loons)

Common Loon. *Gavia immer*. Un, Ra, CC4, Ch; Mig.

Status: The common loon is an uncommon migrant and local breeder on a few lakes, especially Yellowstone Lake. It is an occasional migrant and rare breeder in Grand Teton National Park, nesting on Jackson Lake or other lakes such as Grassy Lake.

Habitats and ecology: Breeding typically occurs on clear and sometimes deep mountain lakes where fish are abundant, human disturbance is at a minimum, and small islands provide nest sites. In some areas, muskrat houses or similar artificial islands may also be used.

Breeding biology: Loons are highly territorial, and shortly after arriving on their breeding grounds they establish a territory that may be up to about 60 acres in area, which they advertise by their familiar "yodeling" call. Most of the elaborate displays include bill-dipping, raising the head and breast, a "circle dance" between territorial opponents, "splash-diving," rearing upright in the water with folded or spread wings, and a low flying rush over the water. Copulation occurs on shore and is not marked by elaborate display behavior. The clutch size is usually two eggs, and incubation lasts 29 or 30 days. Both parents care for the young, which often ride on their backs during their first few weeks of life. The fledging period is about 10 to 11 weeks.

Suggested viewing locations: McEneaney (1988) judged the Yellowstone National Park population to be less than 15 pairs during the late 1980s. Yellowstone Lake is probably the best location for finding breeding loons in that park, especially its southwestern arm (McEneaney, 1988; Scott, 1993). All but 7 of Wyoming's 27 known breeding sites are from Yellowstone Park (Faulkner, 2010). Migrating loons may often be seen at some of the larger and deeper reservoirs across the eastern plains.

Suggested reading: BNA 313 (D. C. Evers et al., 2010).

Family Phalacrocoracidae (Cormorants)

Double-crested Cormorant. *Phalacrocorax auritus*. Lo; Un; Su; Res.

Status: The double-crested cormorant is a common migrant and local summer resident statewide. Yellowstone National Park has a few breeding birds, on the Molly Islands of Yellowstone Lake. They are common at various locations and reportedly breeders in Grand Teton National Park.

Habitats and ecology: These cormorants are associated with lakes and rivers with good fish populations. They often nest on islands or on cliffs and sometimes in trees.

Breeding biology: Cormorants are at least seasonally monogamous, usually breeding initially when three years old. Courtship occurs on water and includes much chasing and diving. Males choose the territory, which includes the nest and adjacent perching spot. Copulation occurs on the nest, mainly during the nest-building period. The clutch size is usually three or four eggs. Both sexes assist in incubation, which begins before the clutch is complete; thus hatching is staggered over several days. Incubation lasts 25 to 29 days. The young leave the nest by about six weeks but continue to be fed by their parents until nine weeks of age, when family bonds disintegrate.

Suggested viewing locations: During migration, marshes, reservoirs, and lakes throughout the region attract this species, especially those with good fish populations. Nesting in Wyoming occurs at Pathfinder Reservoir, Ocean Lake, and elsewhere across the state except in the arid southwest.

Suggested reading: BNA 441 (B. S. Dorr, J. J. Hatch, and D. V. Weseloh, 2014).

Family Pelecanidae (Pelicans)

American White Pelican. *Pelecanus erythroryhnchos*. HL; Co, Ch; Su; Res.

Status: The American white pelican is a common summer resident and a very local breeder (Natrona, Albany, and Park Counties). Yellowstone National Park supports a breeding colony of this species on the Molly Islands of Yellowstone Lake.

Habitats and ecology: Pelicans are associated with lakes and rivers that have large fish populations within reach by surface-feeding. They are gregarious, typically foraging and nesting in groups, and sometimes foraging well away from nesting grounds, which are typically low islands. The Molly Islands in the southern part of Yellowstone Lake are small, low islands with an extremely limited nesting area that is often subject to high wave effects.

Breeding biology: Pelicans are at least seasonally monogamous, and little display activity occurs on the nesting areas. Territorial defense is limited to the area immediately around the nest site, and most described displays occur at or near the nest. These include a "head-up" display with inflated or expanded gular pouch, which may serve as a greeting display to the mate and threat toward others; a "bow," with the bill pointed toward the feet and waved from side to side; and a "strutting walk" with the male following the female. Copulation occurs on land and is preceded by wing-quivering and squatting by the female. The clutch size is usually two eggs. Incubation lasts 29 to 36 days. Both sexes incubate and share in feeding the young, but they feed only their own chicks. At the age of 50 to 60 days the young of the colony form a large "pod," and they fledge at about 10 to 11 weeks.

Suggested viewing locations: During migration, marshes, reservoirs, and lakes throughout the region attract this species, especially those with good fish populations. During summer in Wyoming, nonbreeders may appear at almost any large reservoir (Dorn and Dorn, 1990), such as Ocean Lake, Wheatland Reservoir, Keyhole Reservoir, and Jackson Lake (Scott, 1993). Nesting occurs on Yellowstone Lake, Pathfinder Reservoir (Natrona County), and Bamforth National Wildlife Refuge (Albany County). Pelicans are regularly seen in the Teton area, especially on Jackson Lake and the adjoining Snake River.

Suggested reading: BNA 57 (F. L. Knopf and R. M. Evans, 2004); Schaller, 1964.

Family Ardeidae (Herons and Egrets)

American Bittern. *Botaurus lentiginosus*. Lo; Un, CC3, CC4; Su; Res.

Status: The American bittern is a widespread but uncommon and inconspicuous summer resident. It breeds locally, especially along overgrown edges of beaver ponds or in dense reedy marshes. Few nesting records have been made for Grand Teton National Park, but the bittern is rare in Yellowstone.

Habitats and ecology: This bittern is associated with reed beds and other emergent marsh vegetation, and it is rarely observed feeding in open water in the manner of other herons or egrets. Foods include frogs, snakes, and other animal life in addition to fish, and thus the species is not limited to areas where fish occur.

Breeding biology: Relatively little is known of the social behavior of this elusive bird, but males evidently establish and advertise territories with their distinctive "pumping" call, especially at dawn and dusk. However, the male starts no nest during this period. Females are attracted to such territories and form possibly polygamous pair-bonds. Copulation has been observed on open ground, after the male raised a pair of normally hidden white and airy "shoulder" plumes. He then persistently advanced toward the female while repeatedly lowering and swaying his head from side to side (Johnsgard, 1982, 2016c). After overtaking the retreating female, he simply climbed on her back, grasped her nape, and copulated. No specific postcopulatory behavior was noted (personal observations). The female evidently chooses the nest location (about 50 yards from the area of copulation in the case personally noted) and apparently does all the nest building and incubation. The clutch size is usually four eggs. The male takes no part in defending the nest, but the female defends it fiercely. Incubation lasts 24 to 29 days. The young remain in the nest for about two weeks. The fledging period is still uncertain, but it is 50 to 55 days in a closely related European species.

Suggested viewing locations: Nesting records are scattered, such as on the Laramie Plains and at Goshen Hole and Cokeville Meadows National Wildlife Refuge (Faulkner, 2010). I (Johnsgard, 1982) found a nest at Christian Pond, Grand Teton National Park. Nesting may also occur at Hutton Lake National Wildlife Refuge and Table Mountain Wildlife Habitat Management Area (Scott, 1993).

Suggested reading: BNA 18 (P. E. Lowther et al., 2009); Hancock and Elliott, 1978; Johnsgard, 2016c.

Great Blue Heron. *Ardea herodias*. Ub; Co; Su; Res.

Status: The great blue heron is a common summer resident that breeds locally throughout the state. It nests locally wherever conditions permit but is absent from high montane lakes. It nests commonly in Grand Teton and Yellowstone National Parks.

Habitats and ecology: This species occurs in a variety of habitats that support fish life, but it usually breeds where there are trees. However, rarely it will nest on the ground, on rock ledges, or among bulrushes. Large cottonwoods are a favored location for nesting colonies in the Tetons (such as in the Oxbow area).

Breeding biology: Great blue herons are seasonally monogamous, and both sexes arrive at the nesting ground at about the same time. These birds probably breed initially when two years old. The male selects the breeding territory, which usually centers on an old tree nest. Several obviously hostile displays are associated with territorial defense. Additionally, numerous highly ritualized territorial advertising displays occur, including the "stretch," "snap," and others. These are predominantly male displays, performed at the nest site, and serve to attract females and aid pair formation. Mutual behavior between members of a pair includes twig-passing, feather-nibbling, bill-stroking, and similar activities. Copulation is sometimes preceded by displays, such as feather-nibbling. When building or improving the nest, the male gathers materials and the female works them into the nest. The clutch size is usually four eggs. Both sexes incubate, and nest-relief ceremonies are performed. Incubation lasts 25 to 29 days. The eggs typically hatch over an interval of five to eight days, and adults feed the young by regurgitating food into the bottom of the nest. Although the young can make short flights in the nest vicinity when seven weeks old, they usually continue to use the nest and are fed by the adults until they are about 10 to 11 weeks old.

Suggested viewing locations: In Wyoming, Hutton Lake National Wildlife Refuge (Albany County), Table Mountain Wildlife Habitat Management Area (Goshen County), Oxbow Lake (Grand Teton National Park), and Seedskadee National Wildlife Refuge (Sweetwater County) are good places to look for this species (Dorn and Dorn, 1990). The densest regional breeding concentrations in Wyoming are probably along the Bighorn, Green, North Platte, Powder, and Snake Rivers (Faulkner, 2010).

Suggested reading: BNA 25 (R. G. Vennesland and R. W. Butler, 2011); Hancock and Elliott, 1978; Findholt, 1984.

Snowy Egret. *Egretta thula*. Lo; Ra, Ch; Su; Res.

Status: The snowy egret is a regular but rare and local summer resident. There are no national park breeding records, although vagrant summering birds have been seen in Yellowstone and Grand Teton National Parks. Breeding has occurred in several Wyoming counties (Albany, Fremont, Hot Springs, Lincoln, and Natrona).

Habitats and ecology: These birds occur in a wide range of aquatic habitats but seem to prefer somewhat sheltered locations for breeding and often occur in company with other larger heron species. When foraging the birds are fairly active and sometimes rush about in shallow water in an apparent attempt to flush out their prey.

Breeding biology: After returning to their breeding grounds, males establish a territory that centers on a potential nest site but need not include an old nest. Besides hostile displays, the male performs several sexual displays that include both a stationary and an aerial "stretch" as major advertisements. A single "circle flight" around the potential mate is also common, and a yet more impressive display is a towering circular flight from 50 to 150 yards above the female, followed by a spectacular tumbling downward to land beside her. A mutual interaction called the "jumping over" display, in which one bird makes a short jump flight over the back of the other, is a probable indication that a pair-bond has been formed. The male gathers material, and the female constructs the nest. Copulation occurs on the nest site or on a limb close to it. The first egg may be laid before the nest is completed, and eggs are laid about two days apart. The clutch size is usually three to four eggs. Incubation lasts 20 to 24 days. Since incubation (by both sexes) begins before the clutch is complete, the first young hatches about 18 days after the last egg is laid. After 20 to 25 days the young are ready to leave the nest.

Suggested viewing locations: In Wyoming, Hutton Lake National Wildlife Refuge (Albany County) and Table Mountain Wildlife Unit (Goshen County) are good places to look for this species (Dorn and Dorn, 1990). Nesting occurred at Hutton Lake National Wildlife Refuge in 2005.

Suggested reading: BNA 489 (K. C. Parsons and T. L. Master, 2000); Hancock and Elliott, 1978; Findholt, 1984.

Family Threskiornithidae (Ibises and Spoonbills)

White-faced Ibis. *Plegadis chihi.* Lo; Un, CC4, Ch; Su; Res.

Status: The white-faced ibis is an uncommon migrant and local summer resident at lower elevations. Nesting has occurred along the Bear River drainage and at Hutton Lake National Wildlife Refuge (the state's largest colony) as well as a few other sites on the Laramie plains and probably as far north as Ocean Lake (Faulkner, 2010). This species is increasing nationally at a substantial rate, and additional breeding locations might be expected.

Habitats and ecology: These ibises are generally associated with freshwater or brackish marshes that have an abundance of cattails, bulrushes, or phragmites.

Breeding biology: Remarkably little is known of the social behavior of this species. Monogamous pair-bonds are formed, and both sexes help construct the nest, which takes about two days. Incubation begins with the laying of the last egg. The clutch size is usually three or four eggs. Both sexes also incubate, and during nest relief they do mutual billing and preening and utter guttural cooing notes. The adults continue to add material to the nest during incubation and the fledging period, for about six weeks. The adults feed the young by regurgitation, with the young inserting their bills into that of the parent, or at times they disgorge food into the nest

to be picked up by the young. By the time the young are about seven weeks old they fly with their parents to foraging grounds, returning with them at night for roosting.

Suggested viewing locations: In Wyoming, Hutton Lake National Wildlife Refuge (Albany County), Hawk Springs Reservoir near La Grange, and Table Mountain Wildlife Habitat Management Area (Goshen County) are good places to look for this species during migration (Dorn and Dorn, 1990).

Suggested reading: BNA 130 (R. A. Ryder and D. E. Manry, 1994); Findholt, 1984.

Family Cathartidae (New World Vultures)

Turkey Vulture. *Cathartes aura.* Wi; Co; Su; Res.

Status: The turkey vulture is a common summer resident nearly throughout the state, especially at lower elevations, and a rare visitor in the national parks (Grand Teton and Yellowstone).

Habitats and ecology: Turkey vultures are scavengers that consume only dead remains of large animals, such as livestock, which they find visually or by using their fine olfaction. They are generally found below 8,000 feet.

Breeding biology: Turkey vultures are monogamous, but little is known of their pairing behavior or their age of sexual maturity. However, the nests are well scattered even where nest sites are restricted, and a pair often uses a cave or other possible nest site as a roost for some time before laying eggs there. The clutch size is usually two eggs. Incubation lasts 31 to 37 days. Both sexes participate in incubation, and the incubating bird usually takes morning and afternoon breaks to preen and sit in the sunshine. Injury-feigning at the nest has been reported, and young birds will often disgorge their food or bite when approached. The young are relatively precocial and soon move to the mouth of the nesting cavity to sun themselves. The fledging period is surprisingly long, 70 to 80 days.

Suggested viewing locations: Generally, hilly areas near reservoirs are good places to look for this species, or in range country where road-killed carcasses are likely to be found. Nests in Wyoming are usually located on cliffs, such as at Casper Mountain or the Mendicino Hills near Guernsey (Scott, 1993).

Suggested reading: BNA 339 (D. A. Kirk and M. J. Mossman, 1998).

Family Pandionidae (Ospreys)

Osprey. *Pandion haliaetus.* SD; Un, CC3, Ch; Su; Res.

Status: The osprey is an uncommon summer resident in montane areas near lakes or streams, breeding commonly in both national parks (Grand Teton and Yellowstone). It is mostly a migrant on the plains, except around reservoirs, rivers, and lakes that support good fish populations.

Habitats and ecology: Commonly seen along clear rivers and lakes, these birds sometimes nest on rock pinnacles (such as in Yellowstone Canyon), but more often they nest in tall trees near water, in large nests resembling those of eagles. In the interior Pacific Northwest, ospreys begin breeding in mixed coniferous forests during the mature forest stage of succession (Sanderson, Bull, and Edgerton, 1980). The erection of artificial nesting platforms in areas lacking good natural sites has helped expand the breeding range.

Breeding biology: Ospreys arrive as the ice is melting from their nesting grounds, and males soon begin courtship flights. These swooping and soaring flights may serve to attract females; they also continue for a time after pair-bonds are established or reestablished. Nest building or repair of the old nest starts very soon, the male bringing most of the larger sticks and the female bringing in the lining materials as well as doing the final shaping of the nest. From the time she arrives until the young are nearly fledged, the female catches few, if any, fish, and thus relies on the male for virtually all her food. Mating occurs on the nest site or a nearby branch and continues during the egg-laying period. The clutch size is usually three or four eggs. Both sexes incubate, but the female undertakes most of the responsibility and does all the nighttime incubation. Incubation lasts 32 to 33 days. The eggs hatch at intervals of up to five days, which results in considerable differences in the sizes of the young. For the first month of brooding the female rarely leaves the nest, and the male does all the hunting. As the young approach fledging at about 55 days of age, the female may also help in hunting. After fledging the young continue to use the nest for roosting and as a feeding platform, but they soon attempt

Osprey, adults at nest

to catch fish on their own. They do not mature sexually until they are three years old.

Suggested viewing locations: Artist and Lookout Points in Yellowstone National Park are excellent locations for observing nesting by these birds (McEneaney, 1988). McEneaney judged the Yellowstone population to be at 50 to 60 pairs during the late 1980s. Torrey Lake near Dubois is another good viewing location (Dorn and Dorn, 1990). Wyoming populations have increased considerably since 1978 (Faulkner, 2010), and national populations increased nearly 3 percent annually between 1966 and 2015 (Sauer et al., 2017).

Suggested reading: BNA 683 (R. O. Bierregaard et al., 2003); Johnsgard, 1990.

Family Accipitridae (Hawks and Eagles)

Golden Eagle. *Aquila chrysaetos.* Ub; Un, Ch; Res.

Status: The golden eagle is an uncommon resident throughout the state and most common in montane or rimrock country that is relatively open. It breeds in both national parks: uncommon in Yellowstone and occasional in Grand Teton.

Habitats and ecology: This eagle is a mountain- and plains-adapted species that often occurs in grasslands, semidesert areas, pinyon-juniper woodlands, the ponderosa pine zone of coniferous forests, and sometimes (for foraging) mountain meadows or alpine tundra. It nests over a broad altitudinal range, usually on cliffs or in trees, rarely on the ground. In the interior Pacific Northwest, golden eagles begin breeding in mixed coniferous forests during the mature forest stage of succession (Sanderson, Bull, and Edgerton, 1980).

Breeding biology: Golden eagles are monogamous, and pairs occupy large home ranges (averaging about 35 square miles in California). Aerial displays are most common before the nesting season but may occur at other times too. They consist of soaring and swooping by one or both members of the pair. Both work on the massive nests, and several alternate nests may be maintained. The clutch size is usually two eggs. They are laid at intervals of three to four days, and incubation begins almost immediately with the laying of the first egg. Incubation lasts 43 to 45 days. The female does most of the incubation, but the male begins to assist in brooding soon after the young have hatched. By about 50 days of age the young are feathered, and they fledge at about 65 to 70 days. However, they remain dependent upon their parents for at least some food for as long as three months after fledging.

Suggested viewing locations: In Wyoming, cliff and canyon country across the state provides viewing possibilities year-round, and the Sarasota and Monet areas are favored in winter (Dorn and Dorn, 1990). Nesting occurs virtually throughout Wyoming, which may have the largest population of golden eagles in the United States. The densest regional breeding concentration of golden eagles is in the Great Divide Basin of south-central Wyoming. A famous winter eagle roost (of both golden and bald eagles) is located

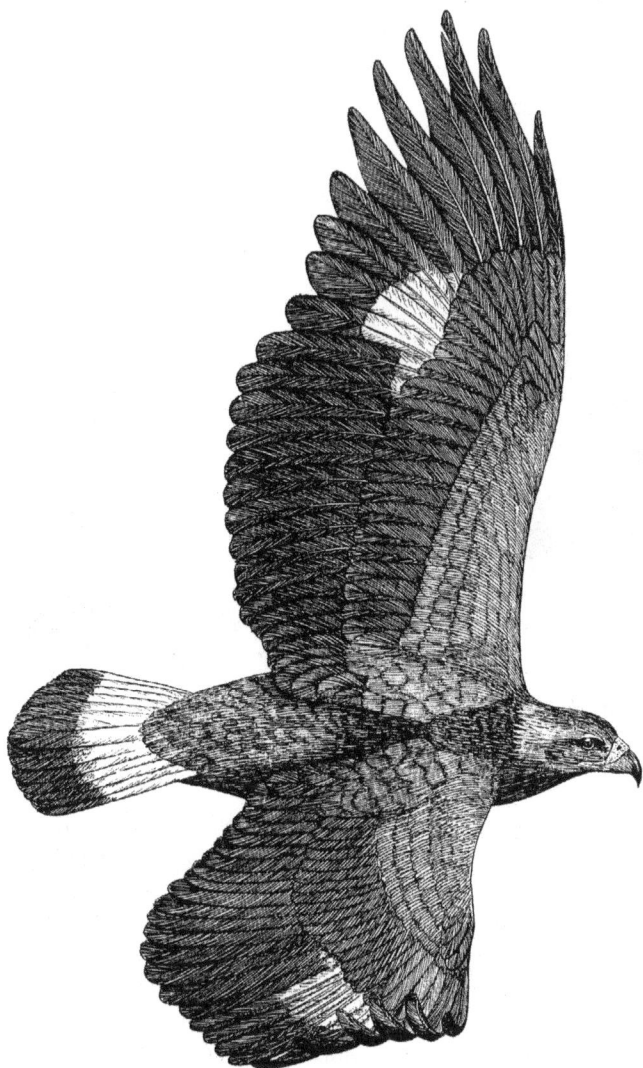

Golden eagle, adult in flight

in Jackson Canyon, about five miles southwest of Casper on Wyoming Highway 220.

Suggested reading: BNA 684 (M. N. Kochert et al., 2002); Snow, 1973a; Ohlendorf, 1975; Johnsgard, 1990.

Bald Eagle. *Haliaeetus leucocephalus*. All, Di; Un, CC3; Res.

Status: The bald eagle is an uncommon summer resident and common winter migrant in Wyoming. Nationally, bald eagles increased in population at an estimated rate of 5.18 percent annually between 1966 and 2015. The species was removed from the status of federally threatened and endangered species in 2007. By 2004 there were at least 95 breeding pairs in Wyoming, and by 2006 there were more than 10,000 pairs in North America, excluding Alaska and Canada. Given the recent rate of increase, there were possibly 20,000 breeding pairs of bald eagles south of Canada by 2018.

Habitats and ecology: This species feeds locally almost exclusively on carrion, especially road-killed deer, but in many regions it is primarily a fish-eating species. Immature eagles tend to be scavengers at large carcasses or other easy sources of meat until they become proficient predators. During summer, breeding pairs are widely dispersed along rivers and around lakes. Bald eagles are likely to be seen along any of the fairly clear rivers, lakes, and reservoirs in Wyoming, especially where fish populations are high or where local fish kills provide a sudden source of abundant food. Few eagles remain year-round, but some northern migrants overwinter in Wyoming and add to local resident populations (Faulkner, 2010).

Breeding biology: After maturing and acquiring adult plumage at four to five years of age, eagles pair monogamously and remain paired permanently. They perform aerial displays, one of which involves locking talons and tumbling downward through the sky for several hundred feet. These flights occur during the nest-building period. Copulation occurs at the same time, and egg-laying soon follows. The usual clutch size is two eggs, but up to three are sometimes laid. Both sexes assist in incubation, and the young hatch at intervals of several days. Incubation lasts 35 to 45 days. The female and young are brought food by the male. As the birds grow, both parents gather food for them, but rarely do more than two eaglets survive to fledging. This occurs at about 70 days of age, but the young birds follow their parents for some time afterward, until they are evicted from the area by the adults.

Suggested reading: BNA 506 (D. A. Buehler, 2000); Snow, 1973b; Johnsgard, 1990.

Northern Harrier. *Circus hudsonicus*. Pan; Un, CC3; Res.

Status: The northern harrier is an uncommon resident throughout the state, nesting locally, especially in nonforested habitats such as grasslands, croplands, and meadows. It is an uncommon breeder in Grand Teton National Park and occasional in Yellowstone. Declining nationally, the harrier is a Sensitive Species I in the US Forest Service Region 2 (Rocky Mountain region).

Habitats and ecology: Grassy areas, especially those near water, are favored by these birds, which nest on the ground rather than in trees, as do most hawks.

Breeding biology: Males migrate separately from females and arrive on the nesting grounds first. They display aerially by performing a series of spectacular dives and swoops, especially in the presence of females. Later the pair may display in this way and also by locking talons in flight. The nest is constructed mainly by the female, though the male may help gather materials. Frequently the birds are semicolonial, with up to six nests reported within a square mile. The eggs are laid at intervals of several days, and the female may begin to incubate at almost any time during the egg-laying period. The clutch size is usually four to six eggs. Incubation lasts 29 to 39 days. Males feed their incubating mates, and, on the basis of a group of six nests studied in Manitoba,

Bald eagle, adult in flight

sometimes provide food for two females. The young hatch at staggered intervals and while they are very small are brooded continuously by the female while the male brings in food. Later the female also hunts, but she usually receives the food the male brings in by aerial transfer. She is the only parent to feed the young directly. Where males are tending two nests, the females must do more hunting by themselves, and starvation of young nestlings is frequent. The young fledge at about five weeks, the smaller males a few days sooner than females.

Suggested viewing locations: In Wyoming, Hutton Lake National Wildlife Refuge (Albany County) and Table Mountain Wildlife Habitat Management Area (Goshen County) are good places to look for this species (Dorn and Dorn, 1990), which breeds almost statewide on lower-elevation grasslands and marshes.

Suggested reading: BNA 210 (K. G. Smith et al., 2011); Johnsgard, 1990.

Sharp-shinned Hawk. *Accipiter striatus*. Wi; Un; Res.

Status: The sharp-shinned hawk is an uncommon summer resident or year-round resident of montane woodlands of the state. It breeds uncommonly in Yellowstone National Park to occasionally in Grand Teton. This hawk is rarely found far from woodlands, even on migration.

Habitats and ecology: Fairly dense forests, either mixed or coniferous, are the preferred habitats of this species, which is swift and elusive, and usually nests in dense groves of trees. Aspens, riparian woodlands, and coniferous forests are all used for breeding. In the interior Pacific Northwest, sharp-shinned hawks begin breeding in mixed coniferous forests during the pole-sapling stage of succession (Sanderson, Bull, and Edgerton, 1980). In Wyoming, they breed in aspen and mid-elevation coniferous forests, probably favoring young, dense, and even-aged stands (Faulkner, 2010).

Breeding biology: Like other hawks, these are monogamous, and the female is appreciably larger than her mate, foraging on somewhat larger prey, primarily birds. In Utah, the birds appear at their nest sites as much as a month before egg-laying. They probably spend much of that time constructing new nests, since old ones are rarely used, even if they are still intact. However, a crow or squirrel nest is sometimes modified for use. The clutch size is usually four to five eggs, and incubation lasts 30 to 35 days. After the clutch is complete, both sexes assist in incubation, and the young eventually hatch almost simultaneously. They grow rapidly, with the males fledging at 24 days of age and the somewhat larger females at 27 days.

Suggested viewing locations: This widespread species may occur anywhere in wooded areas. In Wyoming, a migration corridor occurs from the Bighorns south to Pine Mountain, then to Casper Mountain, the Laramie ranges, and on south into Colorado (Scott, 1993).

Suggested reading: BNA 482 (K. L. Bildstein and K. D. Meyer, 2000); Johnsgard, 1990.

Cooper's Hawk. *Accipiter cooperii*. Di; Un; Res.

Status: The Cooper's hawk is an uncommon year-round resident of montane woodlands that breeds uncommonly in

Yellowstone National Park to occasionally in Grand Teton National Park.

Habitats and ecology: This hawk is associated with mature forests, especially deciduous or mixed forests and less often in pure coniferous stands. Aspen groves are favored breeding locations; nonbreeders use riparian woodlands, scrub oaks, and mountain meadows.

Breeding biology: In New York, Cooper's hawks arrive in their nesting areas in March, and the male establishes a territory about 100 yards in diameter. From this area he calls and feeds any female that might appear. As a pair is being formed, they perform courtship flights, either alone or together. Such flights may be seen for a month or more. During that time the male selects a nest site; rarely he may use an old nest but more frequently he chooses a new location. The male gathers most of the nest material and does most of the actual nest building, and he also continues to feed his mate during this period. The clutch size is usually four eggs. Incubation lasts 35 to 36 days. The female incubates while the male provides food for her, and he briefly guards the nest while she is eating. At the time of hatching the female carries the eggshells away from the nest and may even help the young birds out of the shell. For the first three weeks after hatching the female rarely leaves the nest, and thus all foraging is done by the male. The young birds fledge at slightly more than a month of age, the females about four days later than males, but they all remain dependent on their parents for food until they are about two months old.

Suggested viewing locations: This is a widely distributed species that can occur anywhere in wooded areas, especially in montane forests. In the interior Pacific Northwest, Cooper's hawks begin breeding in mixed coniferous forests during the pole-sapling stage of succession (Sanderson, Bull, and Edgerton, 1980). In Wyoming, they are most common in mid-elevation aspen and coniferous forests, especially in riparian habitats (Faulkner, 2010). They also visit suburban bird feeders in winter, as do the sharp-shinned hawks.

Suggested reading: BNA 75 (O. E. Curtis, R. N. Rosenfield, and J. Bielfeldt, 2006); Johnsgard, 1990.

Northern Goshawk. *Accipiter gentilis*. Wi; Un, CC3; Res.

Status: The northern goshawk is an uncommon resident in woodland and montane forests, breeding uncommonly in Yellowstone National Park to commonly in Grand Teton.

Habitats and ecology: This species is found in many habitats from aspen groves to timberline, but it favors dense conifers or aspens near water for breeding and ranges into low woodlands, riparian woods, and sage areas at other times. In the interior Pacific Northwest, goshawks begin breeding in mixed coniferous forests during the mature forest stage of succession (Sanderson, Bull, and Edgerton, 1980). In south-central Wyoming, the birds favor trees of large diameter in mixed lodgepole–quaking aspen forests (Faulkner, 2010).

Breeding biology: Goshawks are believed to pair for life, even though they may spend the winter period in somewhat different areas. As the nesting season approaches, the female returns to her old nest site and begins calling to attract her mate. She may also attract the male by performing aerial displays. The male is likewise known to display in flight, by flying in an undulating fashion with alternating dives and swoops, perhaps as a territorial advertisement. The pair occupies a large home range from 6 to 15 miles in diameter, encompassing the nesting tree. Often the birds use an old nest, with the female simply helping the male refurbish it, but if a new nest is built the male constructs it entirely alone, while the female watches from a nearby perch. Copulation occurs throughout the nest-building and egg-laying period, which may take two months. The clutch size is usually two or three eggs. Incubation lasts 36 to 41 days. The female does most of the incubation, with the male periodically bringing her food. By the time the young are 35 days old, they move out of the nest onto nearby branches, and they fledge when they are about 45 days of age. However, they are not completely independent of their parents until they are about 70 days old.

Suggested viewing locations: An elusive species, this hawk is most likely to be found in mature montane coniferous forests. Most Wyoming records are from the northwestern part of the state.

Suggested reading: BNA 298 (J. R. Squires and R. T. Reynolds, 1997); Smith and Keinath, 2004b.

Swainson's Hawk. *Buteo swainsoni*. Ub; Co, CC1; Su; Res.

Status: The Swainson's hawk is a common summer resident over most of the state. It breeds commonly in both national parks (Grand Teton and Yellowstone).

Habitats and ecology: Associated with open grasslands, sagebrush, agricultural lands, and rarely with riparian areas, this hawk typically nests in isolated trees but sometimes in bushes, on human-made structures, or on cliffs.

Breeding biology: Swainson's hawks are monogamous and arrive on their breeding ground in eastern Wyoming about a month before egg-laying begins. They soon begin nest building; sometimes use old magpie nests are used for a base, and infrequently they reuse their own old nests. The clutch size is usually two or three eggs. Although males rarely assist in incubation, they do bring prey to the incubating female. Incubation lasts 28 days. The female broods the young during the first 20 days after hatching but thereafter spends considerable time hunting. The young fledge in 28 to 35 days.

Suggested viewing locations: Grasslands east of the Rockies offer the best chances of finding Swainson's hawks. In Wyoming, such areas in Albany and Laramie Counties provide such possibilities (Dorn and Dorn, 1990). In Wyoming, nesting occurs statewide at elevations under 9,000 feet (Faulkner, 2010).

Suggested reading: BNA 265 (M. J. Bechard et al., 2010); Dunkle, 1977; Johnsgard, 1990.

Red-tailed Hawk. *Buteo jamaicensis*. Ub; Co; Res.

Status: The red-tailed hawk is a common resident nearly throughout the state, breeding commonly in both national parks (Grand Teton and Yellowstone). It is less frequent in more open plains, where it is replaced by the Swainson's hawk, as well as in nearly treeless areas, where the ferruginous hawk is found.

Habitats and ecology: This typically tree-nesting buteo does also extend to open woodlands and even treeless areas, where nesting may occur on cliffs. However, trees, especially large cottonwoods and pines, are favored nest sites. The highest population densities found by Hutto and Young (1999) in 566 transects of 11 unaltered forest, grassland, shrub, and wetland habitats of Idaho and Montana were in ponderosa pine communities.

Breeding biology: Red-tailed hawks pair monogamously and arrive at their nesting areas already mated. Nonetheless, courtship flights are common in early nesting phases, with the birds dramatically soaring and swooping together and occasionally locking talons in flight. Copulation often follows such flights. The nest is built by both birds well before egg-laying, and after it is completed the female stays near it while the male feeds her and brings nest-lining materials. The clutch size is usually two to three eggs. Incubation lasts 28 to 32 days. Both sexes help incubate, but the female assumes most of the responsibility and is fed by her mate during this period. The young are hatched at intervals of several days and grow rapidly. By the time they are a month old they may climb out onto adjoining branches, and they can fly at about 45 days. After leaving the nest they are fed progressively less by their parents and become relatively independent in about a month.

Suggested viewing locations: This nearly ubiquitous buteo is likely to be found anywhere in open country from sage or greasewood scrub to subalpine areas. Telephone poles or scattered trees offer attractive perching and lookout sites.

Suggested reading: BNA 52 (C. R. Preston and R. D. Beane, 2009); Johnsgard, 1990.

Ferruginous Hawk. *Buteo regalis*. Ub; Un, CC3, CC4; Res.

Status: The ferruginous hawk is an uncommon resident in Wyoming, primarily in open-country habitats. It is rare in both national parks, with only Grand Teton reporting breeding.

Habitats and ecology: Ferruginous hawks can be found during the breeding season in grasslands, sagebrush, and sometimes also mountain meadows. It nests in pygmy conifers and on cliff ledges, rock outcrops, and sometimes on human-made structures such as windmills.

Breeding biology: Pairs return to their breeding territory each year and usually use the same nest, so it gradually increases in size over time. Both sexes bring nesting material in the form of sticks and nest lining, which the female molds to fit her body. The clutch size is usually three to four eggs. Incubation lasts 32 to 36 days. Evidently the female does most of the incubation. After hatching, the male did most of the brooding in a nest observed in Washington. The young may leave the nest when only about a month old, but they do not fledge until they are about 44 to 48 days of age. They start catching live prey only a few days after fledging.

Suggested viewing locations: In Wyoming, favorable viewing possibilities exist in open country around Laramie, Pine Tree Junction, east of McFadden, and north of Baggs (Dorn and Dorn, 1990). Wyoming's breeding population of about 800 pairs is considered the continent's second largest, but it may be declining because of habitat loss and alteration (Faulkner, 2010).

Suggested reading: BNA 172 (J. Ng et al., 2017); Snow, 1974a; Johnsgard, 1990.

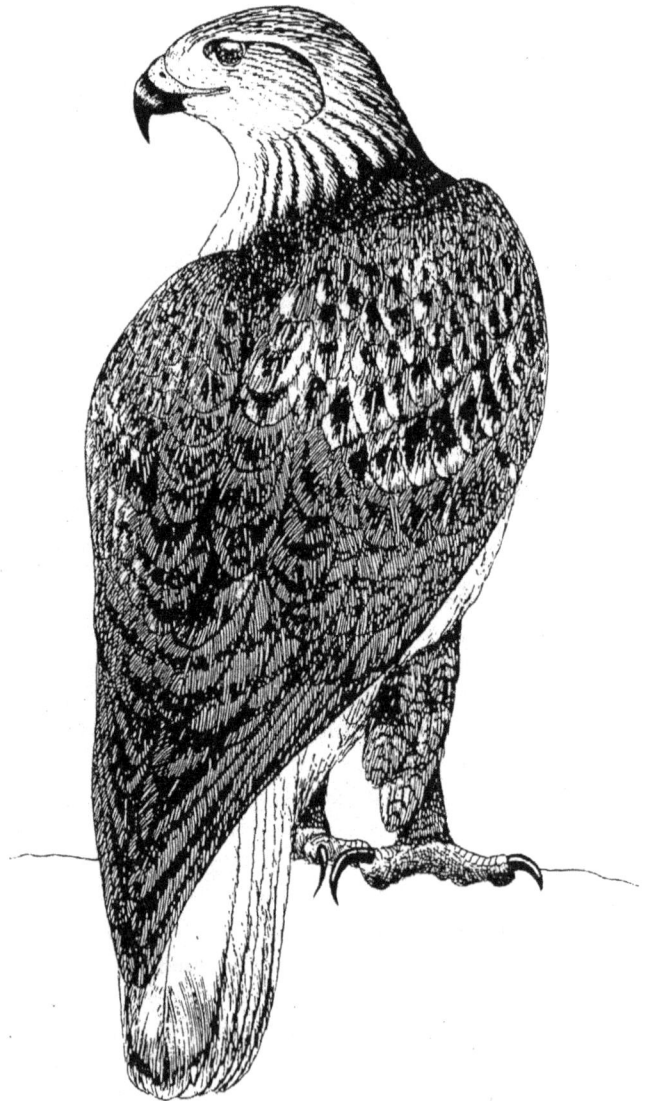

Ferruginous hawk, adult

Rough-legged Hawk. *Buteo lagopus*. Pan; Co; Win; Mig.

Status: The rough-legged hawk is a common winter visitor throughout the state, especially in open habitats. It is common in Grand Teton National Park to to uncommon in Yellowstone.

Habitats and ecology: Rough-legged hawks are usually found hunting in grasslands, sagebrush, or sometimes over marshes or mountain meadows.

Suggested viewing locations: During winter this species might be found in open country across the region. In Wyoming, favorable viewing locations include Goshen Hole (5–10 miles southwest of LaGrange), Campstool Road east of Cheyenne, Sweetwater Valley below Sweetwater Station, the Bridger Valley (Uinta County), south of Buffalo, and northeast of Sundance (Dorn and Dorn, 1990).

Suggested reading: BNA 641 (M. J. Bechard and T. R. Swem, 2002); Johnsgard, 1990.

Family Strigidae (Typical Owls)

Flammulated Owl. *Psiloscops flammeolus*. HL; Ra, CC1, CC3, Ch; Su; Res.

Status: The flammulated owl is an apparently rare summer resident, but it is probably more widespread than currently known. There is at least one breeding record for the mountains of southeastern Wyoming. It is a vagrant in Grand Teton National Park.

Habitats and ecology: Flammulated owls are associated with aspen and ponderosa pine forests during both breeding and nonbreeding periods, particularly ponderosa pines. Old-growth pinyon-juniper woodlands are also sometimes used. In the interior Pacific Northwest, this species begins breeding in mixed coniferous forests during the young forest stage of succession (Sanderson, Bull, and Edgerton, 1980). The only summer reports for Wyoming are from Carbon County, with the first actual nesting record not obtained until 2005 (Faulkner, 2010).

Breeding biology: Flammulated owls are strongly migratory, with males arriving in nesting areas before females and occupying their previous year's territories. Females often join their previous year's mate in the same territory, or they may mate with a male on an adjacent territory. Territories are rather small, in one study averaging only about 300 yards in diameter. Male flammulated owls announce their territorial presence by uttering a series of very low, mellow hoots at about two-second intervals. Paired males continue to sing during the incubation and brooding periods, and unpaired males sing all summer. Duetting is apparently rare in this species. Nests are placed in preexisting cavities, usually those made by flickers or similar-sized woodpeckers. In Colorado, nesting at elevations from 6,000 to 10,000 feet has been noted. The nests usually are in pines or aspens that have woodpecker holes about 10 to 20 feet above ground. Clutches are typically small, only two to four eggs, usually three. Incubation requires 21 to 22 days, with the female doing all the incubating and brooding, while the male provides her with food. By ten days after hatching the young have squired their juvenile plumages, and fledging occurs by 21 to 25 days. After 34 to 40 days following fledging, the young are no longer fed by their parents and begin to disperse.

Suggested viewing locations: Sightings have recently been made in Medicine Bow National Forest, especially around Battle Creek campground.

Suggested reading: BNA 93 (B. D. Linkhart and D. A. McCallum, 2013); Johnsgard, 2002b; Rashid, 2010.

Eastern Screech-Owl. *Megascops asio*. HL; Un; Res.

Western Screech-Owl. *Megascops kennicottii*. HL; Ra; Res.

Status: Screech-owls are uncommon residents statewide in Wyoming. Range limits of these two owls are still uncertain, but at least in most Western Slope montane areas of the region the western form is present. From the eastern slope of Montana's Continental Divide and the eastern slopes of Wyoming's Wind River range eastward the eastern species is apparently the resident form. Western screech-owls are rare but have bred in both national parks (Grand Teton and Yellowstone).

Habitats and ecology: Screech-owls are associated with a variety of wooded habitats (including farmyards, cities, and orchards) and from riparian edges through pinyon-juniper and oak-mahogany woodlands to aspens and ponderosa pine forests. Breeding density data are not available. In Wyoming, the eastern species is found almost exclusively in large riparian cottonwood trees, while the western species has been reliably reported from Park, Teton, and Sublette Counties (Faulkner, 2010).

Breeding biology: Screech-owls are so small and inconspicuous that they may well nest in an urban backyard without the owner's ever being aware of their presence. Most often the eastern species can be detected by its distinctive wailing call, or a series of short whistled notes that often speed up and become a trill, similar to the noise of a ball bouncing to a standstill. The western species utters an accelerating series of short whistles and a short trill followed by a longer one. From eight to nine days are needed to complete a clutch of four eggs. From the time incubation begins, the male probably hunts for both members of the pair, but he does not incubate the eggs. Incubation lasts 26 days. When the young have hatched, over a period of about three days, they are fed about equally by both parents. Early studies by A. A. Allen indicated that a surprising variety of prey is brought to the nestlings, including numerous adult songbirds such as sparrows, warblers, phoebes, and tanagers. In a 45-day period, 77 birds of 18 species were brought to the young, as well as numerous insects, mammals, salamanders, crayfish, and other prey. The young begin to fly when they are about 28 to 30 days of age, but they continue to be fed by the parents for some time.

Suggested viewing locations: Screech-owls are often seen or heard in city parks, woodlots, or other areas that are fairly close to humans. They are most easily seen by imitating their calls, or by using playbacks of recordings to lure them closer. In Wyoming, eastern screech-owls have been reported around Casper, Wheatland, Sybille Canyon (between Wheatland and Bosler), and Sheridan (Scott, 1993).

Suggested reading: BNA (G. Ritchison et al., 2017 [eastern screech-owl]; R. J. Cannings et al., 2017 [western screech-owl]); Johnsgard, 2002b.

Great Horned Owl. *Bubo virginianus.* Ub; Co; Res.

Status: The great horned owl is a common resident in wooded habitats throughout the state. It breeds in both national parks, uncommon in Yellowstone and common in Grand Teton.

Habitats and ecology: Powerful and adaptable, this species occurs everywhere from riparian woodlands through the coniferous forest zones and also extends into city parks, farm woodlots, and rocky canyons well away from trees. Nesting is thus highly variable but often occurs in abandoned bird or squirrel nests or in tree crotches, on rock ledges, or even on the ground.

Breeding biology: Great horned owls are strongly monogamous, and pairs keep in contact by using their familiar hooting calls, *who who who! whoo-whooo.* The male's call is appreciably lower in pitch than the female's. They begin their nesting season amazingly early, usually nesting in the same area and sometimes in the same nest they used the previous year. Incubation begins as soon as the first egg is laid, perhaps partly to keep the eggs from freezing but also to ensure staggered hatching of the young. The clutch size is usually two to three eggs. Incubation lasts 33 to 35 days. Both sexes reportedly incubate, but the female probably does most of it while the smaller male hunts for the pair. The young are hatched in a scanty down coating and do not open their eyes for a week or more. They are brooded by their parents for nearly a month and cannot fly until they are about nine to ten weeks old. Even after they fledge they continue to beg for food until they are driven away from the area by their parents.

Suggested viewing locations: Like screech-owls, great horned owls sometimes live quite close to humans, such as in well-wooded city parks, and can be detected by using playbacks or imitations of their calls. Breeding densities are quite uniform throughout the entire region.

Suggested reading: BNA 372 (C. Artuso et al., 2013); Johnsgard, 2002b.

Burrowing Owl. *Athene cunicularia.* Ub; Co, CC3, CC4, Ch; Su; Res.

Status: The burrowing owl is a locally common summer resident on the plains over much of the state; it is rare in both national parks (Grand Teton and Yellowstone). This species is declining nationally and is a conservation priority species in Wyoming.

Habitats and ecology: This is the only North American owl closely associated with plains rodents such as prairie dogs, and as the range and abundance of these mammals have decreased, so too has the status of the burrowing owl. It is largely an insectivorous species, often eating large beetles but also taking many small mice.

Breeding biology: Based on studies in New Mexico, burrowing owls arrive on their nesting areas either singly or paired, with males returning to the same burrows they occupied previously. Unpaired males display from their burrow locations by bowing and uttering a dovelike, double-noted *coo-coooo* "song" through the night. Pair formation may occur in a single evening. The clutch size is usually five to six eggs, and incubation lasts 27 to 30 days. Evidently only females incubate, and males feed their mates during pair formation, incubation, and brooding. When the young are three to four weeks old, the female begins to forage for herself and her brood, and at about this time the young birds are capable of flight.

Suggested viewing locations: In Wyoming, prairie dog towns near Cheyenne, north of Pine Mountain in Sweetwater County, and east of Continental Peak along the Sweetwater-Fremont county line are good places to look for this owl (Dorn and Dorn, 1990). They are most abundant in the eastern third of Wyoming (Faulkner, 2010), although they are no doubt declining, as both statewide and nationally prairie dog populations are also declining.

Suggested reading: BNA 61 (R. G. Poulin et al., 2011); Lincer and Steenhof, 1997; Johnsgard, 2002b.

Burrowing owl, adult and owlet

Great Gray Owl. *Strix nebulosa*. HL; Un, CC1, Ch; Res.

Status: The great gray owl is an uncommon to rare resident in the state's montane forest areas. It is uncommon in both national parks (Grand Teton and Yellowstone).

Habitats and ecology: In Alberta, these birds usually nest in poplar woodlands, often near muskeg areas. Nests are usually in old hawk nests of various large species from 10 to 80 feet above ground in conifers or hardwood trees. In the interior Pacific Northwest, this species begins breeding in mixed coniferous forests during the young forest stage of succession (Sanderson, Bull, and Edgerton, 1980).

Breeding biology: Studies in Alberta indicate that these owls usually nest in poplar woodlands, preferably near muskeg areas and well secluded from human activities. During the breeding season the male utters a long, drawn-out four-noted call that lacks the depth and throatiness of the great horned owl's, while the female's response is shorter and more screechy. Nesting in Alberta begins in late March or early April; the birds usually take over an old raptor nest with little or no attempt to recondition it. The clutch size is usually three to five eggs. Incubation lasts 28 to 30 days. Evidently the female does all the incubating, and the male provides food for his mate and later for the brood as well, chiefly small rodents such as meadow voles and red-backed mice. Very few larger mammals such as squirrels are taken, and almost no birds have been reported among the prey. The young are helpless and covered with white down when first hatched, and the male must hunt all day and presumably during the night to keep the brood and the female supplied with food. The young leave the nest when about 24 days old and by then are able to climb trees effectively, although they are unable to fly well until they are nearly six weeks old. They continue to follow their parents for at least another month and probably are then still fed to some degree.

Suggested viewing locations: In Wyoming, Yellowstone National Park is probably the best place to search for this owl. McEneaney (1988) judged the Yellowstone population to be less than 100 birds during the late 1980s. He noted that Canyon Junction and the Tower-Roosevelt area are excellent locations for finding these birds in Yellowstone National Park. The Jackson Hole area also is known to support breeding birds, but elsewhere in the state there are only suggestive indications of breeding (Faulkner, 2010).

Suggested reading: BNA 41 (E. L. Bull and J. R. Duncan, 1993); Johnsgard, 2002b.

Long-eared Owl. *Asio otus*. Di; Un, Ch; Res.

Status: The long-eared owl is an uncommon resident over much of the state. It is occasional to rare in both Grand Teton and Yellowstone National Parks, but it has bred in Grand Teton.

Habitats and ecology: A widespread species, this owl is often associated with coniferous or deciduous forests but also woodlots;

Great gray owl, adult and owlet

orchards; large, wooded parks; and even sagebrush or pinyon-juniper woodlands during the breeding season. Trees surrounded by open country seem to be favored for nesting. In the interior Pacific Northwest, this species begins breeding in mixed coniferous forests during the young forest stage of succession (Sanderson, Bull, and Edgerton, 1980). In Colorado, they have been found at elevations as high as 10,000 feet.

Breeding biology: A few weeks before egg-laying, courtship calling begins, marked by a series of short three-noted calls similar to mourning dove calls, uttered at intervals of about three seconds. Aerial display flights include wing-clapping noises as well as acrobatic flying maneuvers. The clutch size is usually four to five eggs. The eggs are laid at irregular intervals of one to five days, and a clutch of seven eggs may be completed in 10 to 11 days. Incubation lasts 25 to 30 days. Only the female incubates, and because of the early onset of incubation the young are hatched over a period of about 7 to 12 days. For the first 15 days of brooding, the female does not leave the nest area and is fed by the male. By the time they are 25 or 26 days old, the young are sufficiently developed to leave the nest and float to the ground, but they are not capable of full flight until they are about 30 to 32 days of age.

Suggested viewing locations: In Wyoming, these elusive owls might be found almost anywhere in woodlands, ranging from riparian shrubs to mature and dense montane forests.

Overleaf: Great gray owl, adult

They are sparsely distributed across the state, usually being found below 9,000 feet in areas of mixed woodlands and more open landscapes (Faulkner, 2010). They especially favor isolated stands of conifers during winter (Dorn and Dorn, 1990; Kingery, 1998).

Suggested reading: BNA 133 (J. S. Marks, D. L. Evans, and D. W. Holt, 1994); Johnsgard, 2002b.

Short-eared Owl. *Asio flammeus*. Di; Un, CC1; Res.

Status: The short-eared owl is an uncommon resident nearly statewide at lower elevations, associated with open meadows and marshes. It is rare in both national parks (Grand Teton and Yellowstone).

Habitats and ecology: This bird is a prairie-adapted species, usually breeding in areas of grassland, marshes, Arctic tundra, and low brushland. Nests are usually on the ground but sometimes in burrows. More diurnal than most owls, these owls are often seen hunting during daylight.

Breeding biology: The short-eared owl is one of the most diurnal of the grassland owls, and during spring it can sometimes be seen performing acrobatic courtship flights high above the prairies, marked by strong wing-clapping, swooping, diving, and somersaulting maneuvers and by a quavering, chattering cry as the bird plummets toward the ground. Copulation sometimes follows such aerial displays or may occur in their absence. Eggs are laid over a considerable period, at intervals of two to seven days. The clutch size is usually four to eight eggs. Incubation lasts 24 to 28 days. The female incubates alone, but her mate brings food to her during this period. The eggs usually hatch at intervals of about three days, and about two weeks after hatching the young begin to move some distance away from the nest. When they are about six weeks old, the young birds begin to catch some of their own food, such as insects and amphibians, but even after they are flying well at the age of two months the adults continue to care for them. About 90 percent of this owl's food consists of rodents, which makes the species extremely valuable from the human standpoint.

Suggested viewing locations: In Wyoming, this species occurs widely across the state, mainly in meadows, grasslands, and marshes; there it is usually found in a patchy and irregular distribution (Faulkner, 2010). Birders might search for short-eared owls in the grasslands south of Van Tassel or the National Elk Refuge (Scott, 1993).

Suggested reading: BNA 62 (D. A. Wiggins, D. W. Holt, and S. M. Leasure, 2006); Johnsgard, 2002b.

Boreal Owl. *Aegolius funereus*. HL; Un, CC3, Ch; Res.

Status: The boreal owl is an uncommon resident in the montane forests of northwestern, north-central, and southeastern Wyoming; it is an occasional breeder in Grand Teton National Park. This owl is a conservation priority species in Wyoming and a US Forest Service regional Sensitive Species.

Habitats and ecology: Boreal owls are associated with old-growth forest, usually of mixed coniferous species (often of Douglas-fir, ponderosa pine, and lodgepole pine) and subalpine spruce-fir forests. Nocturnal and much more often heard than seen, they are most common in places where red-backed voles (*Clethrionomys gapperi*), their primary prey, are abundant.

Breeding biology: The male boreal owl has a territorial "staccato" song that is uttered from late winter through spring during nighttime hours, mostly between dusk and midnight. The song consists of a series of trills of fairly constant pitch that last only about two seconds (or less), and each trill contains pulsed notes of about 12 per second. The vocalization is reportedly loud enough to carry up to about two miles under very favorable conditions and is readily audible at a mile's distance. It is produced persistently by unpaired males for 20 minutes or more for as long as the male remains unpaired. Early in the courting season it is uttered only for a few hours near a potential nest, but later it may last most of the night. A more prolonged version occurs when a female enters the male's territory. Although permanent pair-bonding appears to be typical, there have been several known cases of polygyny with the male taking a second mate, and also cases of a female attempting to raise two broods that had been fathered by different males. Nest sites are typically tree cavities, although nest boxes have often been used in Scandinavia. In North America, cavities made by northern flickers are often used. The clutch size is usually four eggs. Clutch sizes may exceed six in years when vole populations are high, or as low as two or three when voles are few. The incubation period ranges from 25 to 32 days, and the fledging period varies from 28 to 36 days, averaging about 32 days, The young are independent of their parents by five to six weeks of age and become sexually mature in less than a year.

Suggested viewing locations: This species has been seen or heard at Teton Pass and Togwotee Pass, and at Lake Marie in the Snowy Range (Scott, 1993). Breeding is known to occur in the Bighorn Mountains and probably also occurs in the Greater Yellowstone ecosystem and Sierra Madre–Medicine Bow region (Faulkner, 2010).

Suggested reading: BNA 63 (G. D. Hayward and P. H. Hayward, 1993); Johnsgard, 2002b; Rashid, 2010.

Northern Saw-whet Owl. *Aegolius acadicus*. SD; Co, Ch; Res.

Status: The northern saw-whet owl is a widespread and common (but very elusive) resident throughout the montane forests of the state including both national parks, where it is occasional at Grand Teton to rare at Yellowstone.

Habitats and ecology: Saw-whet owls occur widely, from riparian woodlands through aspen groves to the coniferous forest zones but not to timberline. The foothills and ponderosa pine zones are probably their favored habitats, where they nest in old woodpecker holes, but old-growth pinyon-juniper woodlands are also used. In the Pacific Northwest, this species begins breeding in mixed coniferous forests during the young forest stage of succession (Sanderson, Bull, and Edgerton, 1980). Forests and woodlands with open understories are favored for foraging.

Breeding biology: The weak voice and relatively quiet nature of this species make its nesting easily overlooked. The courtship call consists of a series of monotonous, spaced notes that is primarily heard during the early parts of the nesting period in March and April. Males court females by flying around them and landing nearby, often presenting small prey. As soon as egg-laying begins the female becomes very reluctant to leave the nest, and the combination of a large clutch size and an egg-laying interval of 24 to 74 hours results in a highly staggered period of hatching. The clutch size is usually five to six eggs. During the 26- to 28-day incubation period and early brooding, the male is occupied with getting food, which often consists of small mice, frogs, and occasionally birds. The young remain in the nest about four weeks and by the end of this period are able to fly moderately well. However, parental care continues until late summer when the distinctive juvenile plumage is lost and the first adultlike plumage is assumed.

Suggested viewing locations: In Wyoming, the saw-whet owl has been found in Grand Teton National Park and seen at Teton Pass and below both sides of Togwotee Pass (Scott, 1993). Few specific nesting areas are known in Wyoming, but the species is apparently most abundant in the northwestern mountains (Faulkner, 2010).

Suggested reading: BNA 42 (J. L. Rasmussen, S. G. Sealy, and R. J. Cannings, 2008); Johnsgard, 2002b; Rashid, 2010.

Family Alcedinidae (Kingfishers)

Belted Kingfisher. *Megaceryle alcyon*. Ub; Co; Su; Res.

Status: The belted kingfisher is a common summer resident across the state, overwintering in some years or where ice-free waters are present. It breeds in both national parks, common in Grand Teton and uncommon in Yellowstone. This kingfisher is a nationally declining species.

Habitats and ecology: Belted kingfishers are found near water rich in small fish populations, usually where nearby road cuts, eroded banks, gravel pits, or other exposed earthen surfaces provide opportunities for nesting, and usually also where nearby trees provide convenient perching and observation sites between flights.

Breeding biology: Belted kingfishers take up residence in suitable habitats that allow for large home ranges. At times they may forage up to five miles from the nest site, and a population density of about one pair per 1.8 square miles of habitat has been estimated in Minnesota. Small fish averaging about three to four inches in length compose more than half their diet. Both sexes participate in nest excavation, which may require up to three weeks. The clutch size is usually six to eight eggs, and the incubation period lasts 23 to 24 days. Both sexes incubate, apparently beginning after the last egg is laid. After hatching the male also assists in getting food. At

night he usually roosts away from the nest, sometimes in a separate burrow or in a forested area. The young are relatively helpless and spend much time clinging to one another, apparently to maintain body warmth. They remain in the nest for at least a month, which is when they are first able to fly, and then they stay near it for the next few days while their parents teach them to catch fish. An adult captures a fish, beats it until it is nearly senseless, and drops it back into the water. The young are encouraged to capture such easy prey and gradually learn to catch normal fish. Within ten days after fledging they are relatively independent and soon leave the vicinity of the adult pair.

Suggested viewing locations: This widespread species can be seen along most Rocky Mountain rivers or lakes that have good fish populations, nearby perching sites, and steep clay banks for nesting.

Suggested reading: BNA 84 (J. F. Kelly, E. S. Bridge, and M. J. Hamas, 2009).

Family Picidae (Woodpeckers)

Lewis's Woodpecker. *Melanerpes lewis*. Di; Un, CC1, CC2; Su; Res.

Status: The Lewis's woodpecker is a local and uncommon summer resident in forested areas of the state, especially at lower elevations. It is occasional at Grand Teton National Park to rare at Yellowstone, but it has bred at Grand Teton. This woodpecker is a nationally declining species and a US Forest Service regional Sensitive Species.

Habitats and ecology: This unusual woodpecker is especially associated with pine forests that are rather open and with burned-over or otherwise dead-tree areas that have abundant snags or stumps—which means that its distribution tends to be labile and adapted to local conditions. Streamside cottonwood groves in the ponderosa pine or pinyon-juniper zones are also used, and old cottonwoods are favorite nesting trees. In the interior Pacific Northwest, this species begins breeding in mixed coniferous forests during the young forest stage of succession (Sanderson, Bull, and Edgerton, 1980). They are more likely to be found at lowland and foothill sites than in montane forests; juniper and oak–mountain mahogany woodlands are often used for foraging. The birds are mainly adapted to catching free-living insects rather than excavating for insects in wood.

Breeding biology: Unlike other North American woodpeckers, this species is adapted to feeding on free-living insects and is remarkably adept at aerial fly-catching. As the breeding season approaches, the male begins to utter his harsh *churr* breeding call, which serves to attract mates and defend or announce nest sites. Males also drum, but mutual tapping and female drumming have not been reported. Copulation is typically preceded by reverse mounting, as in other woodpeckers. Males take the predominant role in selecting the nest site and defending the nest. Since old nest cavities are

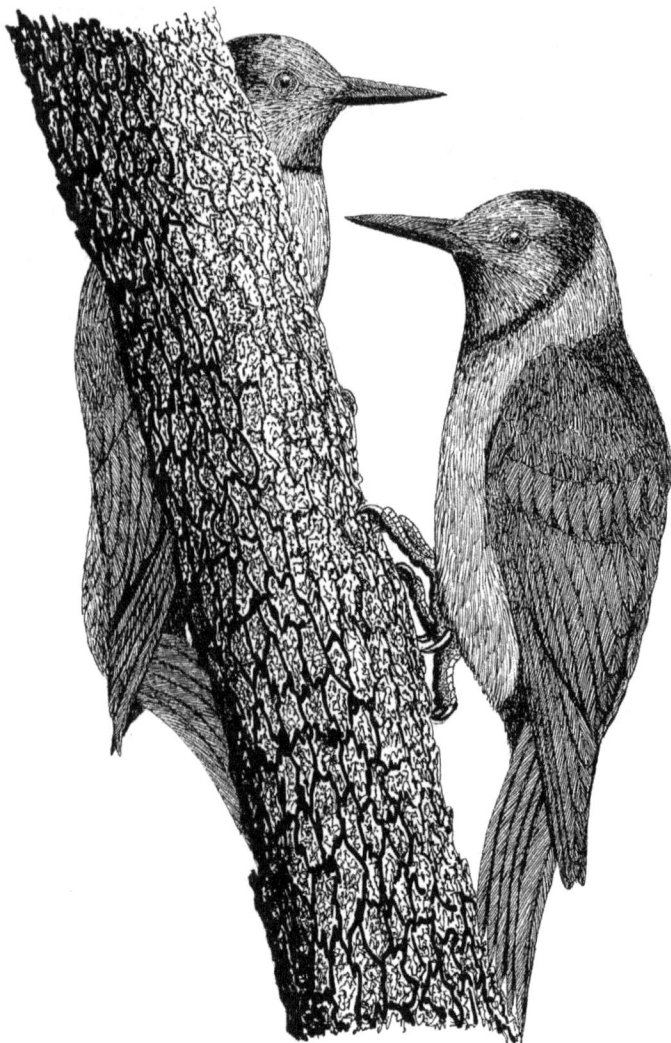

Lewis's woodpecker, adult pair

usually used, little excavation is needed. The clutch size is usually six to seven eggs, and the incubation period lasts 12 to 14 days. Males incubate and brood at night, and both sexes share these responsibilities during the day. The fledging period is probably 31 days. A few days before the young can fly, they move out of the nest cavity and begin to climb about. As the pair leaves the nest vicinity, each parent takes part of the brood and continues to feed them occasionally until they are able to catch insects on their own.

Suggested viewing locations: The densest regional breeding populations of this species are in northeastern Wyoming, but local distributions depend greatly on local forest conditions. Most known breeding records are from the Bear Lodge, Bighorn, Laramie, and Medicine Bow Mountains (Faulkner, 2010). Potential Wyoming viewing sites include the Black Hills National Forest and the entire Greater Yellowstone ecosystem, both areas having been greatly affected by bark beetle damage. Some suggested local viewing sites include Stockade Beaver Creek

Road (Weston County), Cottonwood Creek (Platte County), and Ash Creek (Sheridan County) (Dorn and Dorn, 1990), although specific viewing "hot spots" often last only four to five years before the birds move on to more newly disturbed sites.

Suggested reading: BNA 284 (K. T. Vierling, V. A. Saab, and B. W. Tobalske, 2013); Bock, 1970.

Red-headed Woodpecker. *Melanerpes erythrocephalus.* SD; Un, CC1; Su; Res.

Status: The red-headed woodpecker is a local and uncommon summer resident at lower elevations east of the Bighorn Mountains. It is a vagrant in Grand Teton National Park.

Habitats and ecology: Associated with open deciduous forests, woodlots, and riparian areas, this woodpecker species sometimes extends into the ponderosa pine zone too, but aspens and riparian cottonwood forests are the species' major habitats in this region. Like the Lewis's woodpecker, this species tends to nest in dead trees or the dead portions of living trees; it does less excavating for insects in wood than do most woodpeckers but often forages in open habitats at ground level.

Breeding biology: After males of this species return to their nesting areas in spring, they call and drum from their roosting and prospective nesting holes, apparently to attract mates to their excavations. When a female approaches, the male begins tapping from within the cavity, then typically flies away to allow the female to inspect the hole. He may also solicit copulation by inviting reverse mounting by the female while he is perched near the nest cavity. Mutual tapping at the nest hole seems to indicate that a pair-bond is formed and that the female accepts the nesting cavity. The clutch size is usually five eggs, and the incubation period lasts 12 to 14 days. Both sexes assist in incubation and brooding, with the male performing these activities at night. In one Illinois study, 3 of 15 pairs nested a second time in one season, sometimes while still feeding their first fledglings. The young birds tend to follow their parents for some time after leaving the nest, until they are chased away after about 25 days.

Suggested viewing locations: The densest Wyoming breeding populations are in river valleys along the state's eastern boundaries, in riparian cottonwood woodlands of the eastern plains and foothills, such as the North Platte River (Goshen County), the Powder River and Clear Creek (Sheridan County), and Lance Creek (Niobrara County) (Dorn and Dorn, 1990).

Suggested reading: BNA 518 (K. G. Smith et al., 2000).

Williamson's Sapsucker. *Sphyrapicus thyroideus.* SD; Un; Su; Res.

Status: The Williamson's sapsucker is a local and generally uncommon summer resident over much of the state, mainly in the northwestern, north-central, and southeastern mountains. It is uncommon in Yellowstone National Park to occasional in Grand Teton.

Lewis's woodpecker, adult

Habitats and ecology: Sapsucker breeding in this region usually occurs in the aspen or coniferous zones, especially in ponderosa pine forests with mixed aspens. In Colorado, these are mainly between about 7,000 and 8,500 feet but sometimes as high as 10,700 feet. High ridges in the Douglas-fir zone are also used for foraging, but nesting is usually done in aspens. In the interior Pacific Northwest, this species begins breeding in mixed coniferous forests during the mature forest stage of succession (Sanderson, Bull, and Edgerton, 1980).

Breeding biology: Males of this migratory species are the first to arrive in breeding areas in spring, and they soon establish territories that they advertise by drumming and territorial conflicts. Most males begin new nest-hole excavations each spring. Males also perform tapping at a potential nest site, to which the female may respond similarly if she accepts the location. Males roost inside such excavations until the eggs are laid. The clutch size is usually five to six eggs. Both sexes incubate, with the females gradually taking on a greater share. Incubation lasts 13 days. The period of brooding lasts for eight to ten days after hatching, and adults bring both sap and insects to nestlings. Fledging occurs 31 days after hatching.

Suggested viewing locations: In Wyoming, good viewing areas include the Sierra Madre Mountains in Carbon County, the Brooks Lake area of Fremont County, and the trail from Rendezvous Peak to Granite Canyon in Grand Teton National Park (Dorn and Dorn, 1990). McEneaney (1988) judged the Yellowstone Park population to be less than 400 pairs during the late 1980s, with the best observation opportunities being around Tower Falls and the Blacktail Plateau. Wyoming's densest populations are in the Greater Yellowstone ecosystem, with fewer records in the Laramie and Bighorn ranges, and none from the Black Hills National Forest (Faulkner, 2010).

Suggested reading: BNA 285 (L. W. Gyug et al., 2012).

Downy Woodpecker. *Dryobates pubescens*. Ub; Co; Res.

Status: The downy woodpecker is a common resident in wooded habitats throughout the state, including the national parks (reportedly uncommon in Yellowstone National Park).

Habitats and ecology: A wide variety of wooded habitats are used by downy woodpeckers, including farm lots, orchards, city parks, and natural habitats ranging from riparian forests to pinyon-juniper woodlands, oak–mountain mahogany scrub, and aspen or coniferous forests.

Breeding biology: Like hairy woodpeckers, this species is nonmigratory, and the birds spend the entire year near their breeding areas. Toward late winter territorial drumming and associated conflicts begin, and mates often drum at dawn to locate each other when their roosting holes are widely separated. In spring, both sexes begin to seek out suitable nest sites, and when one has located a potential site it begins drumming and tapping to attract the other. Short courtship flights occur near the nest and may strengthen site attachment or stimulate copulation, which takes place near the nest. Both sexes

help excavate the nest, but usually the female is most active. The clutch size is typically four to five eggs. The incubation period lasts 12 days. The male assists with incubation and spends the night in the nest. For the first week after hatching, one or the other adult remains at the nest at all times, but as the young birds develop both adults spend much time foraging. By the time the young are two weeks old, they can climb to the nest entrance to be fed, and they are ready to fly less than four weeks after hatching. Pair-bonds break down after the breeding season, and each sex excavates fresh roosting holes for use in winter, when each forages independently of its mate.

Suggested viewing locations: These woodpeckers are nearly ubiquitous throughout the region and are easy to find around bird feeders in winter or in riverbottom woodlands with cottonwood trees (Dorn and Dorn, 1990).

Suggested reading: BNA 613 (J. A. Jackson and H. R. Ouellet, 2018).

Hairy Woodpecker. *Dryobates villosus*. Ub; Co; Res.

Status: The hairy woodpecker is a common resident in wooded areas throughout the state, including both national parks (Grand Teton and Yellowstone).

Habitats and ecology: Optimum breeding habitat consists of fairly extensive areas of woodlands of conifers (especially lodgepole pines) or hardwoods, but nesting occurs in riparian forests, aspen groves, and various coniferous forests nearly to timberline. Generally, aspens and other hardwoods are preferred over conifers for breeding. In the interior Pacific Northwest, this species begins breeding in mixed coniferous forests during the young forest stage of succession (Sanderson, Bull, and Edgerton, 1980).

Breeding biology: Hairy woodpeckers are largely nonmigratory and begin to form pairs in midwinter, about three months before the start of nesting. Typically, males are attracted at this time to territories that the females establish the previous fall. During this time the pair performs drumming duets, and both sexes use drumming to locate a mate when they are visually separated. The male also uses drumming as a territorial display. When searching for a suitable nest site, the birds perform a slow tapping, which tends to attract the mate. At about this time females begin to solicit copulation, but copulation reaches a peak during actual excavation. The males do most of the excavation, excavating throughout the day and sometimes sleeping in the cavity at night. The clutch size is usually four eggs, and the incubation period lasts 11 to 15 days. During incubation the male continues to be most attentive to the eggs, even during daylight. Both sexes brood the nestlings for more than two weeks. After about 17 days the adults begin to feed the young from outside the nest rather than going inside with food. By the time the young are 28 to 30 days old, they emerge from the nesting hole and are able to fly strongly.

Suggested viewing locations: This widespread species is often found in mature aspen stands in summer and among cottonwoods in winter (Dorn and Dorn, 1990). Relatively dense regional breeding populations probably occur in northwestern Wyoming.

Suggested reading: BNA 702 (J. A. Jackson, H. R. Ouellet, and B. J. Jackson, 2003).

American Three-toed Woodpecker. *Picoides dorsalis.* SD; Un, CC3, Ch; Res.

Status: The American three-toed woodpecker is a local and variably uncommon resident in montane areas of the region; it is uncommon in Yellowstone National Park to occasional in Grand Teton. It is a conservation priority species in Wyoming and a US Forest Service regionally Sensitive Species.

Habitats and ecology: Like the black-backed woodpecker, this species is a fire-adapted form, typically moving into a burned forest area immediately after the fire, breeding, and dispersing four to five years later. Bluebirds, nuthatches, and many other cavity nesters then use the nest cavities until the snags eventually topple. Partly open, high-elevation coniferous or aspen forests are typical habitats. Spruce-fir forests are favored habitats in the northern Rockies.

Breeding biology: This relatively little-studied species, like the other three-toed woodpecker, rapidly colonizes stands of recently burned trees that are being attacked by bark-boring beetles. It also occupies undisturbed stands of virgin forest where old trees have diseased or decayed hearts. About three-fourths of the food of both three-toed species consists of wood-boring beetle larvae, with caterpillars being of secondary significance. The birds often strip bark from large areas of the trees as they search for woodborers. They also are relatively tame, so their presence is usually easy to detect. Almost nothing is known of their social behavior, but it is probably little different from that of the black-backed woodpecker. The clutch size is usually three to four eggs, and the incubation period lasts 13 days. Both sexes are known to assist in incubation and in rearing the young, with the male often taking on most of these responsibilities. Fledging occurs at 26 days. They are rather sedentary birds and probably maintain their foraging territories throughout the year.

Suggested viewing locations: In Wyoming, recently burned areas and old-growth spruce-fir forests are likely sites for finding this species (Dorn and Dorn, 1990). It is widespread in mountain ranges across the state, but its status in the Black Hills National Forest is still uncertain (Faulkner, 2010).

Suggested reading: BNA 588 (J. A. Tremblay, D. L. Leonard, Jr., and L. Imbeau, 2018); Gibbon, 1966.

Black-backed Woodpecker. *Picoides arcticus.* HL; Un, CC3, CC4; Res.

Status: The black-backed woodpecker is an uncommon resident in wooded areas of northwestern Wyoming south to Jackson Hole and local in the Black Hills. It is a rare breeder in both national parks (Grand Teton and Yellowstone). It is a conservation priority species in Wyoming and a US Forest Service regional Sensitive Species.

Habitats and ecology: This species is nearly identical to the American three-toed woodpecker as to its general adaptations to feeding and breeding in recently burned coniferous forest areas, especially those of lodgepole pine and subalpine spruce-fir. As such, it is highly local and eruptive. Apparently the birds feed exclusively on beetles in dead or dying conifers, and they sometimes nest in nearby aspens. In the interior Pacific Northwest, this species begins breeding in mixed coniferous forests during the young forest stage of succession (Sanderson, Bull, and Edgerton, 1980).

Breeding biology: This is a relatively eruptive species, coming into areas shortly after logging or forest fires, breeding for a few seasons, and then disappearing again. The birds seem to feed exclusively on conifers and usually nest in them but have at times nested in aspens. Dead tamarack and spruce swamp areas are favored foraging areas for this species in northern Minnesota, and the birds are usually found in pairs or probably family groups of three or four individuals. Little is known of their pair-forming behavior, but both sexes help in excavating the nesting hole and, as might be expected, both sexes incubate. The clutch size is usually four to five eggs. Incubation lasts 13 days and fledging occurs at 24 days. Probably the male does most of the feeding of the young; they eat mainly the larvae of wood-boring beetles. Thus, this bird is extremely beneficial in controlling this serious enemy of coniferous forests.

Suggested viewing locations: The same areas mentioned for the American three-toed woodpecker are likely to support this species, although the black-backed is generally rarer and more habitat-limited. Relatively dense regional breeding populations probably occur in central Idaho and northwestern Wyoming, at least where forest fires have occurred recently. This bird too should be searched for in localities where forest fires or other forest disruptions, such as logging or infestations by boring or bark beetles, have occurred within the past four to five years.

Suggested reading: BNA 509 (J. A. Tremblay et al., 2000).

Northern Flicker. *Colaptes auratus.* Ub; Co; Res.

Status: The northern flicker is a common resident or migrant throughout the state, more abundant in wooded areas but also present in open country where trees are scattered. It is common in both national parks (Grand Teton and Yellowstone).

Habitats and ecology: Flickers are unusual among woodpeckers in that much of their food consists of insects such as ants that are obtained by probing in the ground rather than by excavating trees. However, they do excavate holes in trees for nesting, usually those that are already dead or have decaying interiors, especially in relatively softwood species such as cottonwoods and aspens. Open woodlands, such as orchards, parks, and similar areas offering foraging opportunities on grassy areas nearby are preferred over dense forests.

In the interior Pacific Northwest, this species begins breeding in mixed coniferous forests during the young forest stage of succession (Sanderson, Bull, and Edgerton, 1980). In Colorado, nests have been found at elevations from 5,400 to 11,480 feet. There red-shafted forms were mostly found in coniferous and aspen habitats, while yellow-shafted birds were most common in cottonwood riverbottoms (Kingery, 1998). The highest population densities found by Hutto and Young (1999) in 566 transects of 11 unaltered forest, grassland, shrub, and wetland habitats of Idaho and Montana were in cottonwood-aspen communities.

Breeding biology: This species is relatively migratory over most of the region, and when returning to the nesting area both sexes seek out their old territories and nest sites. Males tend to arrive a few days before females and soon begin uttering location calls and drumming as a territorial advertisement. Recognition of previous mates is apparently site-induced, and sex recognition is based on the "moustache" markings of the male. Courtship displays include exposing the undersides of wing and tail, bobbing, and billing ceremonies. Males apparently select the nest site, often using a previous year's nest or starting to excavate a new one. Most of the excavation is done by the male, and copulation typically occurs just before the nest is finished. The clutch size is usually six to eight eggs. The incubation period lasts 11 to 13 days. The eggs are laid at daily intervals, and both sexes share incubation, with the male assuming most of the nocturnal responsibilities. Both sexes care for and brood the young, feeding them by regurgitation, with the males, again, taking most of the responsibility. The nestling period is about 26 days, but the parents continue to feed their offspring for some time after they leave the nest.

Suggested viewing locations: This woodpecker is ubiquitous in wooded and parkland areas throughout the region, with the yellow-shafted form most abundant in riverbottom woods along its eastern boundary and the red-shafted form mostly present elsewhere from the foothills to subalpine forests.

Suggested reading: BNA 166 (K. L. Wiebe and W. S. Moore, 2017).

Family Falconidae (Falcons)

American Kestrel. *Falco sparverius*. Ub; Co; Su; Res.

Status: The American kestrel is a common summer or permanent resident over nearly all the state and a common breeder in both national parks (Grand Teton and Yellowstone).

Habitats and ecology: This is an open-country falcon that occurs in agricultural areas, grasslands, sagebrush, and desert scrub and nests in tree cavities, rock or building crevices or cavities, and, rarely, in earthen holes. It avoids forests but sometimes forages as high as mountain meadows. In Wyoming, nesting occurs up to about 8,500 feet. The highest population densities found by Hutto and Young (1999) in 566 transects of 11 unaltered forest, grassland, shrub, and wetland habitats of Idaho and Montana were in sagebrush communities.

Breeding biology: American kestrels are perhaps the most sociable of the falcons and until pair formation may associate in small groups. During the courtship period males perform aerial dive displays, whining vocalizations, and courtship feeding. Courtship feeding serves to maintain the pair-bond and also provides food for the female and her young. Sometimes the female begs for food in flight by performing a distinctive "flutter-glide" display. Copulations reach a peak just before egg-laying, but courtship feeding peaks during the egg-laying period and continues through the brooding period. The clutch size is usually four to five eggs, and the incubation period lasts 29 or 30 days. The female does nearly all the incubating, and the young often hatch at intervals of about a day, somewhat less than the egg-laying interval. During the fledging period of approximately 30 days, the male continues to do most of the food gathering while the female broods and directly feeds the young. After about 20 days the adults bring in entire prey animals and place them in the nest for the young birds to tear apart and feed themselves. The family typically remains together for some time after fledging.

Suggested viewing locations: For most of the year this species can be found in any fairly open-country habitat, especially where some trees with woodpecker holes provide for nesting and overhead lines offer perching sites. In Wyoming, the Torrington and Lovell areas attract good numbers of kestrels (Dorn and Dorn, 1990).

Suggested reading: BNA 602 (J. A. Smallwood and D. M. Bird, 2002); Johnsgard 1990.

Merlin. *Falco columbarius*. Di; Un; Ch; Res.

Status: The merlin is an uncommon resident in at least the eastern parts of the state, mainly in montane wooded areas, and probably breeding locally, but there are few recent breeding records (Faulkner, 2010). It is generally rare in the national parks with no definite breeding records in Grand Teton or Yellowstone.

Habitats and ecology: This species is a forest and woodland-adapted falcon, usually breeding in clumps of open woodlands, often in bottomlands or valleys. In the Pacific Northwest, merlins begin breeding in mixed coniferous forests during the mature forest stage of succession (Sanderson, Bull, and Edgerton, 1980). In one Wyoming study, merlins nested preferentially in old magpie nests near major river systems and areas of mixed-grass and ponderosa pine savannas (Faulkner, 2010). During the nonbreeding season they also forage in grasslands, agricultural lands, desert scrub, and marshes or shorelines.

Breeding biology: Typically, males return to the breeding ground prior to females and begin calling while flying from one perch to another. Little actual nest building is done since the birds typically take over an already constructed nest of another species. The clutch size is usually five to six eggs. The eggs are laid at two-day intervals, and the female begins to incubate when the clutch nears completion. Probably the only time the male takes over incubation is during the short periods the female is off the nest eating food he has brought. The incubation period lasts 28 to 32 days. The eggs hatch at intervals, resulting in marked size differences among the young, which develop rapidly. Early in the brooding period all the food is brought in by the male, passed on to the female, and then torn up and divided among the young. Later the female assists in hunting and bringing in food, which mostly consists of small birds. The young fledge in 25 to 30 days but remain in the vicinity of the nest for some time. They initially begin to hunt by catching insects but soon learn to chase and capture young birds.

Suggested viewing locations: In Wyoming, this species breeds in woodlands but moves to open grasslands during winter, especially where horned larks are abundant (Dorn and Dorn, 1990). Merlins are notably common in Wyoming's Green River valley above the town of Green River (Scott, 1993).

Suggested reading: BNA 44 (I. G. Warkentin et al., 2005); Johnsgard 1990.

Peregrine Falcon. *Falco peregrinus.* Lo; Un, CC3, Ch; Su; Res.

Status: The peregrine is currently an uncommon summer resident. It once nested throughout the montane areas of Wyoming, including at least Yellowstone National Park, but a population crash after the introduction of DDT and related pesticides in the 1940s eliminated most breeding populations. Efforts have been underway since the 1980s to reestablish the species in some areas by releasing hand-reared birds, and they have proven to be very successful in the Rocky Mountain region. This falcon is still a conservation priority species in Wyoming; national populations have been increasing nearly 3 percent annually in recent decades (Sauer et al., 2017).

Habitats and ecology: This species is largely a cliff-nesting species, typically in woodland habitats. Nonbreeders occur over a wide habitat range, from mountain meadows to grasslands, marshes, and riparian habitats. McEneaney (1988) judged the Yellowstone National Park population to be about 12 birds during the late 1980s, but Rocky Mountain populations have greatly increased since then. In Wyoming, restoration efforts resulted in the first documented nestling in 1984 (Faulkner, 2010).

Breeding biology: After a pair has returned to its nesting area, the male or both birds perform courtship flights consisting of diving, swooping, and sometimes passing food in the air. Mating is frequent during this period and on into the period of egg-laying, with eggs laid at intervals of about two to three days. The clutch size is usually three to four eggs, and the incubation period lasts 28 to 29 days. The female does most of the incubation, with the male bringing prey to his mate and also occasionally relieving her. The female typically eats away from the nest, on a nearby "plucking post." For the first two weeks after hatching the female does nearly all the brooding and feeding of the young, but later both adults hunt extensively and simply drop their prey into the nest, letting the young birds compete for it and tear it up. Usually only two or three young fledge from each brood; fledging occurs about 35 to 42 days after hatching. With the advent of modern pesticides and before they were outlawed, virtually no young were fledged at most nests because the thin-shelled eggs laid by pesticide-poisoned females failed to hatch or the young did not survive to fledging.

Suggested viewing locations: Hayden Valley is an excellent location for finding these birds in Yellowstone National Park (McEneaney, 1988), as is Clarks Fork Valley in Park County (Dorn and Dorn, 1990).

Suggested reading: BNA 660 (C. M. White et al., 2002); Johnsgard 1990.

Prairie Falcon. *Falco mexicanus.* Ub; Un, Ch; Res.

Status: The prairie falcon is a widespread uncommon summer resident or year-round resident throughout the state, mainly in mountain or rimrock areas that offer open country for hunting. It is an uncommon breeder in Yellowstone National Park and occasional in Grand Teton.

Habitats and ecology: Breeding birds are largely associated with plains, sagebrush, or desert scrub habitats with steep cliffs nearby for nesting; sometimes tundra areas also support breeders, and foraging may be done on mountain meadows or similar alpine habitats.

Breeding biology: Prairie falcons probably first nest when two years old; yearlings normally wander during the breeding season. The birds arrive on the Wyoming and Colorado nesting grounds in late February or early March, and the male engages in aerial courtship for about a month while the pair examines potential nest sites. Frequently, nest sites of the previous year are used, even if the female is mated to a new male. The male begins to do most of the hunting for the pair during the courtship period, and the female later does nearly all the incubation. The clutch size is usually four to five eggs, and the incubation period lasts 29 to 31 days. Only when the female is eating food brought by her mate does he incubate, but the male performs the major role in nest defense. The young begin to acquire their flight feathers at about 30 days and fledge at about 40 days of age. Evidently a large portion of the young survive their first autumn, but there is an overall mortality rate of about 80 percent by the end of the first year of life.

Suggested viewing locations: In Wyoming, good locations during the breeding season include Flaming Gorge Reservoir and buttes around Rock Springs and Green River (Sweetwater County), Beaver Rim (Fremont County), Wind River Canyon (Washakie County), and the western slope of the Bighorn Mountains (Dorn and Dorn, 1990).

Suggested reading: BNA 346 (K. Steenhof, 2013); Hickey, 1969; Snow, 1974b; Johnsgard 1990.

Family Tyrannidae (New World Flycatchers)

Olive-sided Flycatcher. *Contopus cooperi*. Di; Un, CC1, CC2, CC3; Su; Res.

Status: The olive-sided flycatcher is an uncommon but widespread summer resident in Wyoming east to the Bighorn, Laramie, and Medicine Bow ranges, primarily in coniferous forests but also in riparian woodlands. It breeds in both of the national parks—common in Grand Teton and uncommon in Yellowstone. This flycatcher is a nationally declining species and a US Forest Service Sensitive Species in the Rocky Mountain region.

Habitats and ecology: This species is associated with coniferous montane forests (especially Douglas-fir and lodgepole pine), burned-over or logged forests with standing snags, and muskeg areas in the region. Typically tall conifers and open, often boggy or meadow-like areas are present in territories. In Colorado, it has been found breeding from 7,000 to 11,000 feet elevation, most often in montane coniferous forests with closed canopies (Kingery, 1998). The highest population densities found by Hutto and Young (1999) in 566 transects of 11 unaltered forest, grassland, shrub, and wetland habitats of Idaho and Montana were in sagebrush communities, a favorite foraging habitat.

Breeding biology: Males of this species establish territories along the edges of tall coniferous forests, where they can sally out to obtain insect prey. There they can also sit on some high, exposed perch, such as a dead tree or branch, uttering their loud and distinctive three-syllable song, *whip-whee'-peooo*, variously interpreted as "look, three deer," or "quick, three beers." Courtship and pair formation have not been extensively studied in this species, but they probably closely resemble those of the other tyrant flycatchers. The clutch size is usually three eggs, and the incubation period lasts 16 to 17 days. The nestling period is usually 15 to 19 days, but in one instance the young remained attached to the nest for 23 days, flying to and from it with their parents during the last few days.

Suggested viewing locations: In Wyoming, good viewing areas include the forested uplands of Fossil Butte National Monument (Lincoln County), Pine Mountain (Sweetwater County), and the Brooks Lake area (Fremont County) (Dorn and Dorn, 1990). Breeding occurs from the Greater Yellowstone ecosystem east at elevations of 7,500 to 10,000 feet to the Bighorns and southeast to the Laramie and Medicine Bow ranges (Faulkner, 2010).

Suggested reading: BNA 502 (B. Altman and R. Sallabanks, 2012).

Western Wood-Pewee. *Contopus sordidulus*. Ub; Co; Su; Res.

Status: The western wood-pewee is a common summer resident throughout the entire state in summer. Breeding occurs in both national parks—common in Grand Teton and uncommon in Yellowstone.

Habitats and ecology: This wood-pewee breeds in most coniferous forest types and also to varying extent in aspens, riparian forests, and various open deciduous or mixed woodland habitats. Open forests are favored, especially those dominated by conifers. In the interior Pacific Northwest, this species begins breeding in mixed coniferous forests during the young forest stage of succession (Sanderson, Bull, and Edgerton, 1980). The highest population densities found by Hutto and Young (1999) in 566 transects of 11 unaltered forest, grassland, shrub, and wetland habitats of Idaho and Montana were in cottonwood-aspen communities. In Colorado, these birds nest from 3,800 to 10,000 feet elevation, in trees with dead tops or exposed branches (Kingery, 1998).

Breeding biology: A rather late arrival among the flycatchers, pewees usually reach the northern states in late May, near the end of the spring migration period. As is typical in the Tyrannidae, only the female constructs the nest, which is usually on the same branch year after year. In one case, a fork of an elm tree was used as a nest location by this species every year for 35 years. The clutch size is usually of three eggs. The incubation period lasts 12 to 13 days. Incubation is done by the female, but the male occasionally feeds her and remains near the nest to help feed the young when they hatch. Like the nest itself, the juveniles closely resemble the surrounding bark and lichens. By 15 to 18 days after hatching they are ready to leave the nest. They are probably dependent on the parents for food for some time after fledging, until they have become skilled in catching flying insects.

Suggested viewing locations: In Wyoming, this abundant species is ubiquitous statewide in wooded areas up to about 9,000 feet, especially in ponderosa pine or mixed conifers (Faulkner, 2010).

Suggested reading: BNA 451 (C. Bemis and J. D. Rising, 1999); Eckhardt, 1976.

Willow Flycatcher. *Empidonax traillii*. Di; Un, CC1; Su; Res.

Status: The willow flycatcher is an uncommon summer resident in woodlands from western Wyoming east to the Bighorns, Laramie, and Medicine Bow ranges, and local in the Black Hills. Breeds in both national parks—common in Grand Teton and uncommon in Yellowstone.

Habitats and ecology: This flycatcher is especially associated with riparian or wetland habitats in the region, including willow thickets, low gallery forests along streams, prairie coulees, and farther north in woodland edge habitats such as muskegs and boggy openings. The highest population densities found by Hutto and Young (1999) in 566 transects of 11 unaltered forest, grassland, shrub, and wetland habitats of Idaho and Montana were in wetland communities. In Wyoming, this species breeds generally west of a line from Sheridan to western Laramie County and increases in abundance in the western mountains. A small local population also exists in the western Black Hills (Faulkner, 2010).

Breeding biology: In southern Michigan, males arrive at the nesting areas somewhat before females and begin to establish territories that average about two acres, always including shrubs and small trees as well as clearings. Water is present either in the territory or very close to it. Birds usually sing from the highest point in the territory, up to 30 songs per minute. Nests are built by the females, usually in upright crotches of shrubs that the returning bird can fly to directly. The clutch size is usually three to four eggs. The incubation period lasts 12 or 13 days. Only the female is known to incubate. The young are fledged in 12 to 16 days, and they remain in their parents' territory until fall. They continue to beg for food until they are about 24 to 25 days old, when they have become fairly adept at catching insects.

Suggested viewing locations: Widespread and abundant throughout the region in thickety areas and woodland edges.

Suggested reading: BNA 533 (J. A. Sedgwick, 2000).

Least Flycatcher. *Empidonax minimus*. SD; Co; Su; Res.

Status: The least flycatcher is a common summer resident in northeastern Wyoming, west to the Bighorn Mountains, and south to perhaps the Laramie Mountains, but the southern and western breeding limits are very uncertain.

Habitats and ecology: Associated with open and edge-dominated habitats such as mature deciduous floodplain forests with shrubby understories in prairie areas, scattered prairie grovelands, shelterbelts, woody lake margins, and urban parks or gardens.

Breeding biology: Shortly after returning to their nesting areas, male least flycatchers establish breeding territories that are surprisingly small, averaging only 0.18 acre in one study. The territory usually also includes exclusive foraging areas as well as a nest site; sometimes communal foraging areas are shared. Territories are advertised by the males' simple *cha-beck* song and are primarily defended by males. Females defend only a small area around the nest. The nest is built by the female in six to eight days, and the clutch size is usually four but may range from three to six. The incubation period lasts 14 to 16 days. The female also does all the incubating and brooding, although the male remains near the nest and occasionally feeds her. He also feeds the chicks when they hatch and, at least initially, provides most of their food, The juveniles leave the nest at 23 to 25 days of age and may leave the territory a few days after fledging, but they remain dependent on their parents until about three weeks of age.

Suggested viewing locations: This species is common in the riparian forests of northeastern Wyoming, such as the upper canyon of the North Fork of Crazy Woman Creek (Johnston County) or in mature woods above LAK Reservoir (Weston County) (Dorn and Dorn, 1990).

Suggested reading: BNA 99 (S. Tarof and J. V. Briskie, 2008); Davis, 1959.

Dusky Flycatcher. *Empidonax oberholseri*. Di; Co; Su; Res.

Status: The dusky flycatcher is a common summer resident in woodlands from western Wyoming east to the Bighorns, Laramie, and Medicine Bow ranges and is local in the Black Hills. This species is common in both national parks (Grand Teton and Yellowstone).

Habitats and ecology: The dusky flycatcher is associated with open woodland with dense shrubby understories, ranging from riparian edges through oak–mountain mahogany woodlands to aspens and open ponderosa pine woods. In Montana, brushy logged-over slopes seem to be favored habitats, while in Colorado areas dominated by shrubs or aspens are preferred nesting habitats (Kingery, 1998). In the interior Pacific Northwest, this species begins breeding in mixed coniferous forests during the shrub-seedling stage of succession (Sanderson, Bull, and Edgerton, 1980). The highest population densities found by Hutto and Young (1999) in 566 transects of 11 unaltered forest, grassland, shrub and wetland habitats of Idaho and Montana were in cottonwood-aspen communities.

Breeding biology: Dusky flycatchers closely resemble other small flycatchers such as willow flycatchers and western flycatchers, and in the Black Hills they occupy drier habitats than the cordilleran flycatcher. The dusky flycatcher has a three-syllable territorial advertisement that sounds like *prillit, prrddrt, pseet*, with the second syllable low and burred. Territories are marked by this song, pursuit flights, and trill calls, the last given as the male perches above the female. Trill-calling plays an important role in pair formation. In Montana, the nests are typically built in crotches of small bushes, and the eggs are laid at the rate of one a day. The clutch size is usually three or four eggs. The incubation period lasts 15 to 16 days. Only the female incubates, starting incubation after the second egg is laid. The eggs hatch over a period of two to three days, and both sexes feed the young, although only the female broods. The nestling period is 15 to 17 days.

Suggested viewing locations: Brushy foothill forests are good places in Wyoming to search for this species (Dorn and Dorn, 1990), which is widespread throughout all the major mountain ranges and the western Black Hills (Faulkner, 2010). Dense regional breeding populations probably occur in the Greater Yellowstone ecosystem and northeastern Wyoming.

Suggested reading: BNA 78 (M. E. Pereyra and J. A. Sedgwick, 2015).

Cordilleran Flycatcher. *Empidonax occidentalis*. Wi; Co; Su; Res.

Status: The Cordilleran flycatcher is a common summer resident over much of the state, mainly in woods near streams and in montane forests. It is present in both national parks: a common breeder in Grand Teton but rare in Yellowstone.

Habitats and ecology: A widespread and adaptable small

flycatcher, the Cordilleran ranges from riparian woodlands through aspens into the coniferous forest zones, all the way to the upper spruce-fir zones. Shrubby riparian areas with cottonwoods are favored. In Colorado, nesting occurs from about 6,000 to 10,000 feet in montane and subalpine forests. In the Pacific Northwest, this species begins breeding in mixed coniferous forests during the young forest stage of succession (Sanderson, Bull, and Edgerton, 1980). The highest population densities found by Hutto and Young (1999) in 566 transects of 11 unaltered forest, grassland, shrub, and wetland habitats of Idaho and Montana were in cottonwood communities. Nests are often placed on ledges, so they favor ravines with rock outcrops, steep banks, and similar sites suitable (including bridges and vacant buildings)for such nest placement.

Breeding biology: These flycatchers are highly aggressive, and their territorial aggression is directed not only toward their own species but also toward other species of similar size. Unmated males sing their advertising *ps-seet"-ptsick see!* notes throughout most of the day, whereas mated males sing only at dawn. Nests are constructed by one of the pair, presumably the female, over a period of four to five days, and the first egg is laid within a day or two of the completion of the nest. The clutch size is usually four eggs, and the incubation period lasts 14 to 15 days. Incubation is performed by a single bird, presumably the female, while the mate occasionally feeds her on the nest. During the first few post-hatching days only one of the pair does all the brooding and most of the feeding, again most probably the female, but soon both parents are kept busy feeding the growing brood. The nestling period lasts from about 14 to 18 days, and for a period after the young depart from the nest the adults continue to feed them at a rate even greater than when they were in the nest. After about four days the female may stop feeding the brood and begin her second nest. As the young birds grow, they become more independent and gradually drift away from the territory.

Suggested viewing locations: In Wyoming, this is a common species in woodlands above the juniper zone, and it is usually found along watercourses. It is widespread throughout all the major mountain ranges and the western Black Hills (Faulkner, 2010) and is especially common in the Laramie range (Scott, 1993).

Suggested reading: BNA 556 (P. E. Lowther, P. Pyle, and M. A. Patten, 2016).

Say's Phoebe. *Sayornis saya*. Ub; Co; Su; Res.

Status: Say's phoebe is a common summer resident over most of the state but infrequent in the mountains and rare in the national parks (Grand Teton and Yellowstone).

Habitats and ecology: This phoebe is generally associated with grasslands, sagebrush, and agricultural areas in the region, especially prairie coulees and steep, eroded riverbanks.

They sometimes reach foothill areas but do not breed in the wooded mountain zones. In Wyoming, they are widespread throughout except at high elevations. In Colorado, the birds usually nest below 9,000 feet, with the highest densities under 8,000 feet on the Western Slope (Kingery, 1998).

Breeding biology: Male phoebes arrive on the nesting grounds before females, and when the females return, pairs form or reform rather rapidly. Nest building or the repair of a previous nest may begin only a week or two after the birds arrive and is presumably done by the females. Often the same nest is used in subsequent years or for successive clutches. At favored nest sites the loss of one or both members of a pair brings a rapid replacement. The clutch size is usually four to five eggs. The incubation period lasts 12 to 14 days. Males remain nearby on a convenient lookout post. However, the males guard their nests and feed their mates, and then feed the brood virtually alone during the first week or so of their lives. When a brood is ready to leave the nest at about two weeks of age, the male takes over and teaches the young birds to capture insects while the female prepares to produce a second clutch of eggs. Apparently she assumes the entire job of feeding herself and her second clutch, although reportedly the male may again appear to take care of the brood when it fledges, freeing the female for a possible third brood.

Suggested viewing locations: In Wyoming (and elsewhere), this species is abundant in open country. Good Wyoming viewing sites include the badlands area north-northwest of Baggs, Fossil Butte National Monument (Lincoln County), the Newcastle area (Weston County), and around Boysen Reservoir (Fremont County) (Dorn and Dorn, 1990).

Suggested reading: BNA 374 (J. M. Schukman and B. O. Wolf, 1998).

Western Kingbird. *Tyrannus verticalis*. Wi; Co; Su; Res.

Status: The western kingbird is a common summer resident in most of the state but infrequent in montane areas and rare in the national parks (Grand Teton and Yellowstone).

Habitats and ecology: This species is always associated with edge habitats near open country, such as shelterbelts, hedgerows, margins of forests, tree-lined residential districts, and riparian forests. It occupies more open country and occurs at somewhat lower elevations than the Cassin's kingbird; in Colorado it is usually found below 7,000 feet and most often in rural or residential areas, or in riparian woodlands (Kingery, 1998).

Breeding biology: Like the eastern kingbird, this species is highly territorial and generally is extremely intolerant of larger birds, such as crows and hawks, in the vicinity of its nest. The birds are also at least as noisy as the eastern kingbird, and during the early stages of territorial establishment and pair formation they are particularly conspicuous for their calling and singing. Yet, in spite of their overt aggressiveness, there are reported cases of several pairs occupying the same tree or sharing a small grove for nesting. A traditional return to a previous year's nesting territory is common, as in eastern kingbirds. The clutch size is usually four eggs. The incubation period lasts 12 to 14 days. The female does most of the incubating, but males have been seen on the nest as well. Likewise, both parents actively feed the young, which remain in the nest for about two weeks. In one observed case in Oklahoma, a pair began a new nest only four days after its first brood fledged, but presumably the young birds remain at least partially dependent on their parents for some weeks after fledging.

Suggested viewing locations: This kingbird is a nearly ubiquitous breeder in open country throughout the entire region, especially along riparian woodlands in otherwise tree-free habitats. In Wyoming, the birds occur statewide at lower elevations, especially where flying insects are abundant (Faulkner, 2010).

Suggested reading: BNA 227 (L. R. Gamble and T. M. Bergen, 2012).

Eastern Kingbird. *Tyrannus tyrannus*. Ub; Co; Su; Res.

Status: The eastern kingbird is a common summer resident throughout the state but is infrequent in montane areas and rare (Yellowstone) or occasional (Grand Teton) in the national parks.

Habitats and ecology: Eastern kingbirds are associated with open areas with scattered trees or tall shrubs, such as forest edges, fencerows, riparian areas, agricultural lands, and farmsteads. In Wyoming, the birds occur statewide at lower elevations, especially where scattered mature trees or utility poles occur in otherwise open landscapes (Faulkner, 2010). In Colorado, they occupy many of the same habitats as western kingbirds but are more likely to occur in woodlands adjacent to open water. Cottonwoods and elms are favored nesting trees there (Kingery, 1998).

Breeding biology: Eastern kingbirds arrive on their breeding grounds when insect populations begin to become noticeable. They soon become extremely conspicuous as the males begin territorial and associated courtship behavior. Aerial displays are common, with the bird flying erratically in a series of swoops and dives not far above the ground and uttering harsh screams. Chases and fights between birds on adjacent territories are also prevalent at this time. Kingbirds have a strong tendency to return to the same nesting territory in subsequent years, although the specific nest site varies. Males help build the nest, but the female typically does the incubating. The clutch size is usually three to four eggs. The incubation period lasts 12 to 14 days. After the eggs hatch both sexes are kept constantly busy bringing food to the young, the female generally being more active in feeding and brooding than her mate. The young are in the nest for approximately two weeks. Thereafter they remain as a group, often on a wire or an exposed tree branch, waiting for their parents to come and feed them. The young begin to catch flies after they are about eight days out of the nest, and they continue to improve in flying ability for the next month. They are not fed by the adults beyond about 35 days after fledging.

Suggested viewing locations: This species is widespread and common in Wyoming and elsewhere throughout the region, with a somewhat more easterly (more woodland adapted) distribution than the western kingbird. Like the western kingbird, riparian woodlands in eastern Wyoming attract large numbers of eastern kingbirds.

Suggested reading: BNA 253 (M. T. Murphy and P. Pyle, 2018).

Family Laniidae (Shrikes)

Loggerhead Shrike. *Lanius ludovicianus*. Ub; Un, CC3, CC4; Su; Res.

Status: The loggerhead shrike is an uncommon summer resident nearly throughout the state but is rare in the two national parks and a reported breeder only in Grand Teton National Park. This species is declining nationally and is a US Forest Service Sensitive Species in the Rocky Mountain region.

Habitats and ecology: Like the northern shrike, the loggerhead is associated with open habitats that have scattered perching sites, and it ranges altitudinally from agricultural lands on the prairies to montane meadows. Sagebrush areas, desert scrub, and pinyon-juniper woodlands offer ideal nesting and foraging areas, but some nesting also occurs in woodland edge situations, farmlands, and similar habitats. In Wyoming, these birds occur statewide at elevations up to the foothills zone (Faulkner, 2010). In Colorado, rural areas account for most breeding-season sightings, with shortgrass prairies a distant second. Nests there are usually placed in small, scattered trees, at times at elevations as high as 8,000 feet and possibly to 8,900 feet (Kingery, 1998).

Breeding biology: On a study area in north-central Colorado, shrikes arrived in early April and had established territories by the first of May. Both sexes help build the nest. Nests in Colorado were separated by at least 400 meters (1,300 feet), and nest sites of previous years were often used. The clutch size is usually four to five eggs. The incubation period lasts 14 to 16 days. The female incubates, but her mate feeds her, and both sexes bring food to the young. The fledgling period is normally 17 days but may require up to 20 days. The young birds gradually learn how to wedge or impale food items in forks or on thorns or other sharp objects. Impaling is generally recognized as a means of short-term food storage and is thus most likely to be done by shrikes that are not hungry.

Suggested viewing locations: This species can be found in the same open-country habitats as the northern shrike and is usually seen perching on electrical or fence wires along country roads. The densest regional breeding populations probably occur in eastern Wyoming (such as at Thunder Basin National Grassland).

Suggested reading: BNA 231 (R. Yosef, 1996).

Family Vireonidae (Vireos)

Plumbeous Vireo. *Vireo plumbeus*. SD; Un; Su; Res.

Status: The plumbeous vireo is an uncommon summer resident in forested areas over much of the state, from the Black Hills and Bighorns south and west through the Laramie and Medicine Bow ranges to the juniper woodlands of southwestern Wyoming.

Habitats and ecology: Open, coniferous, or mixed forests with considerable undergrowth seem to be the plumbeous vireo's favored habitat, especially those that offer open branches for foraging at low to medium tree levels. Fairly dry and warm forests are favored by the plumbeous vireo over those that are moist and cool, and breeding extends from the open oak or aspen and ponderosa pine zones upward through the lower coniferous communities, to about 8,000 feet in Colorado and reportedly to 10,000 feet elsewhere.

Breeding biology: Like all vireos, males of this species are persistent singers, although in this species the songs tend to be a series of nasal phrases and pauses, *churech, ch-reet; churech, ch-reet*, with no unique rhythm. Singing by the male also is stimulated when the nest is approached and, like the red-eyed vireo, may even occur while the male is sitting on the nest. Like other vireos, the nests of this species are built across a horizontal fork of two branches. The nest is suspended between them and supported only at its upper edges. It is composed of a mixture of many plant materials, bound together with spider webbing, and lined with grasses, mosses, or fur. It is built mostly by the female, over a period of about seven days. The clutch size is usually four eggs but varies from three to five, and incubation by both sexes requires 14 to 15 days.

Suggested viewing locations: Plumbeous vireos breed in all the major mountain ranges of Wyoming except in the Greater Yellowstone ecosystem but are especially common in the western Black Hills, Laramie Mountains, and eastern Bighorns (Dorn and Dorn, 1990; Faulkner 2010).

Suggested reading: BNA 366 (C. Goguen and D. R. Curson, 2012).

Warbling Vireo. *Vireo gilvus*. Ub; Co; Su; Res.

Status: The warbling vireo is a very common summer resident in forests throughout the entire state and a common to abundant breeder in both national parks (Grand Teton and Yellowstone). It is the most common and widespread vireo of Wyoming.

Habitats and ecology: Fairly open woodlands, especially of deciduous trees, are favored by this species. It is probably most common along riparian forests that support tall trees, but it also occurs in aspen groves and well-wooded residential or park areas, especially where tall cottonwoods are present. In coniferous forest areas, the birds favor areas where single or clumped broad-leaved trees such as aspens or birches occur. Foraging is done near the crowns of fairly densely leaved trees, and nests are sometimes located as high as 90 feet above ground in very tall forests. In Colorado, most breeders occupy aspen woodlands (Kingery, 1998). Relatively few nest in eastern riparian woodlands of the prairie rivers, whereas in Nebraska these constitute a primary nesting habitat.

Breeding biology: In spite of its widespread occurrence, this species has not been extensively studied, perhaps because many of its breeding activities take place high in trees. An early observation by Audubon indicated that both sexes helped build the nest, which is unusual in vireos, and that eight days were required to complete it. The clutch size is usually four eggs. The incubation period lasts 12 to 13 days. Both sexes incubate, and males often sing while on the nest. The fledging period is believed to be 16 days. Adult birds continue to sing well into the summer, and young males learn to sing fairly well before they leave in fall. In this respect, this species resembles the red-eyed and yellow-throated vireos.

Suggested viewing locations: In Wyoming, this species is likely to be found in almost every aspen stand in the mountains (Dorn and Dorn, 1990); at lower elevations riparian willow and cottonwood stands are equally attractive. The densest regional breeding populations probably occur in the Greater Yellowstone ecosystem and the Black Hills region. The male's loud and distinctively syncopated songs, uttered even while he is sitting on the nest, makes locating the birds very easy.

Suggested reading: BNA 551 (T. Gardali and G. Ballard, 2000); Dunham, 1964.

Family Corvidae (Crows, Jays, and Magpies)

Gray (Canada) Jay. *Perisoreus canadensis*. Dis; Un; Ch; Res.

Status: The gray jay is an uncommon resident in montane forests throughout the state; it is present in both Grand Teton and Yellowstone National Parks and common in both.

Habitats and ecology: This species is associated with a wide variety of boreal and montane coniferous forest types and occasionally extends into aspens and riparian woodlands outside the breeding season. Nesting almost always is in coniferous vegetation. In the interior Pacific Northwest, this species breeds in mixed coniferous forests from the pole-sapling to the mature forest stage of succession (Sanderson, Bull, and Edgerton, 1980). In Colorado, nesting occurs from about 8,500 feet to timberline, usually in spruce-fir forests but sometimes in lodgepole pines. Territories are large and permanent, and almost everything edible is consumed (Kingery, 1998).

Breeding biology: Like many corvids, gray jays regularly cache food, but in contrast to other species, these birds produce a special saliva that helps bind food particles together so the mass can be firmly held in position among conifer foliage. Breeding in this boreal species begins very early, with nest building sometimes beginning in February. Both sexes help build the nest, the female doing most of the actual construction in one observed case. The clutch size is usually three to four eggs. The incubation period lasts 16 to 18 days. As with other corvids, only the female incubates, and she typically sits on the nest from the time the first egg is laid, although incubation does not begin immediately. Both sexes feed the young, but the male brings most of the food during the first few days after hatching. Fledging occurs approximately 15 days after hatching, and it is likely that the young birds remain with their parents through the first winter of life.

Suggested viewing locations: In Wyoming, gray jays occur in all the mountain ranges and are most often and easily seen at higher elevations at tourist-frequented sites of the Teton, Wind River, and Medicine Bow ranges (Dorn and Dorn, 1990). McEneaney (1988) judged the best observation areas of the Yellowstone population to be around Yellowstone River and Bridge Bay. Other montane parks and national forests in the region are also favorite places for finding this common species.

Suggested reading: BNA 40 (D. Strickland and H. R. Ouellet, 2018); Dow, 1965.

Pinyon Jay. *Gymnorhinus cyanocephalus*. SD; Un, CC1, CC2; Res.

Status: The pinyon jay is an uncommon resident over most of Wyoming except the northwestern and southeastern corners, especially at lower elevations. It is a vagrant in Grand Teton National Park.

Habitats and ecology: This jay species is generally associated with pine forests growing on dry substrates, especially the pinyon-juniper association, but extending during the nonbreeding period into oak–mountain mahogany, sagebrush, and desert scrub habitats. In Colorado, the birds strongly prefer pinyon-juniper habitats, but where pinyon pines are lacking they use juniper woodlands and consume juniper berries rather than pinyon nuts. On the eastern slope of the Continental Divide only about 2 percent of the breeding-season observations were in ponderosa pines (Kingery, 1998).

Breeding biology: Pinyon jays are highly gregarious, usually gathering in flocks of up to 50 birds for much of the year. Pair-bonds are probably rather permanent in these flocks, and as early as mid-November males begin feeding their mates by transferring pine seeds or other morsels. This behavior is also performed by first-year birds, although initial breeding may not occur until they are at least another year older. Later, females actively solicit feeding by courtship begging, a display that continues through nesting and stimulates the male to feed his incubating mate. In late stages of courtship, a male may pick up a bit of vegetation, present it to his mate, and then fly up into a nearby tree, as if to lure her away from the flock. In this way, specific courtship crotches or branches are established, although the actual nest is often built in another

location. Nests are usually placed on the south side of trees, probably for warmth. They are built by both members of a pair, usually over about a week, and the first egg is laid about two days later. The clutch size is usually three to four eggs. The incubation period lasts 17 days. Most birds in a colony begin and complete their nests at nearly the same time; the colony's location is dependent on the caches of pine seeds from the previous fall. Fledging occurs about three weeks after hatching, and the parents remain with their young for a prolonged period, continuing to feed them well after they are fledged.

Suggested viewing locations: In Wyoming, the best locations for finding this southwestern species are at Flaming Gorge in Sweetwater County, the drier slope of the Bighorn Mountains, around Alcova Reservoir (Natrona County), and in the Guernsey area (Dorn and Dorn, 1990).

Suggested reading: BNA 605 (R. P. Balda, 2002).

Steller's Jay. *Cyanocitta stelleri*. SD; Co; Res.

Status: The Stellar's jay is a common resident in coniferous woodlands throughout the state except for the Bighorn Mountains and Black Hills. It is common in Grand Teton National Park but uncommon in Yellowstone.

Habitats and ecology: This species is centered in the ponderosa pine zone but also extends down into the pinyon-juniper zone and as high as the Douglas-fir zone. In the interior Pacific Northwest, this species breeds in mixed coniferous forests from the pole-sapling to the old-growth forest stage of succession (Sanderson, Bull, and Edgerton, 1980). In Colorado, it occurs in foothills and lower mountains from 6,000 to 9,000 feet, most often in ponderosa pine forests but also in other conifers and aspen forests. It occupies an ecological zone above that of the blue jay and western scrub-jay and below the gray jay's (Kingery, 1998).

Breeding biology: Steller's jays are usually found in pairs, but they sometimes form small parties when aggregating at a food source or mobbing a predator. Paired birds tend to remain in or near their breeding territories throughout the year, but immature birds may wander about in the winter. Little has been written on nesting biology, but both sexes help build the nest and probably only the female incubates, although it has been suggested that both sexes incubate in Alaska. The clutch size is usually four eggs, and the incubation period lasts about 16 days. The fledging period is about 18 days, and the young birds are fed for a month or more after they fledge and presumably remain with their parents for most of their first year.

Suggested viewing locations: This species is common throughout most of the coniferous forests of Wyoming but are curiously absent from the Black Hills and are only rare vagrants in the Bighorns. More widely, they are generally common throughout the Rocky Mountains proper.

Suggested reading: BNA 343 (L. E. Walker et al., 2016); Brown, 1964.

Clark's Nutcracker. *Nucifraga columbiana*. Wi; Un, Ch; Res.

Status: Clark's nutcracker is an uncommon resident in wooded areas over much of the state, common to abundant in both national parks (Grand Teton and Yellowstone).

Habitats and ecology: This bird is widespread in coniferous habitats, from the ponderosa pine zone to timberline. In the interior Pacific Northwest, it breeds in mixed coniferous forests in the mature and old-growth forest stages of succession (Sanderson, Bull, and Edgerton, 1980). It's more common in the higher coniferous zones in summer but descends during winter to the pinyon pine zone and sometimes out onto the plains. In Colorado, the birds range from 8,000 feet to timberline during the breeding season and are especially common in spruce-fir and mixed coniferous forests. The highest population densities found by Hutto and Young (1999) in 566 transects of 11 unaltered forest, grassland, shrub, and wetland habitats of Idaho and Montana were in ponderosa pine communities. During their breeding period they continue to eat pinyon seeds, obtained from stored underground caches (Kingery, 1998).

Breeding biology: This species, ecologically speaking, is one of the most important birds of the western mountains. Nutcrackers find and cache thousands of pine nuts and other seeds every year, often resorting to stealing pine cones from squirrels or "panhandling" from humans. They are remarkably able to remember where hundreds of buried foods can be retrieved months later. A nutcracker can ingest and carry 70 to 95 pine nuts in a pouch below its tongue (sublingual pouch), and certainly many of its caches are never used, with the seeds left to sprout and start a new generation of trees. However, the birds can locate cached foods hidden under as much as eight inches of snow with remarkable accuracy. Nutcrackers are monogamous, but like other corvids they often forage in small groups, probably made up of closely related individuals. Their stick and bark nests are often built on a horizontal branch of a pine or juniper, with both sexes participating in its construction. The usual clutch is of two or three eggs. The incubation period lasts 16 to 18 days, with incubation by both sexes and later brooding by both. The fledging period is 22 days.

Suggested viewing locations: In Wyoming, this conspicuous species occurs in all the higher mountain ranges. High-elevation picnic areas and scenic viewpoints in national or state parks are good places to find nutcrackers, which often are very tame at such locations in hopes of getting fed. McEneaney (1988) judged the best Yellowstone observation areas to be around Dunraven Pass and the Upper Terrace Drive.

Suggested reading: BNA 331 (D. F. Tomback, 1998); Vander Wall and Balda, 1977; Tomback, 1983.

Black-billed Magpie. *Pica hudsonia*. Ub; Co; Ch; Res.

Status: The black-billed magpie is a common resident throughout the state in most habitats and a common breeder in both national parks (Grand Teton and Yellowstone).

Habitats and ecology: This magpie species is of widespread occurrence, especially common in riparian areas with thickety vegetation, agricultural areas with scattered trees, sagebrush, aspen groves, and the lower levels of the coniferous forest zones. Small, thorny trees are especially favored nest sites, but junipers and similar trees are also used. In Colorado, the birds range up to 10,000 feet of elevation during the breeding season and tend to favor riparian, rural, and pinyon-juniper areas for nesting. Conifers appear to be avoided, except for open ponderosa woodlands (Kingery, 1998).

Breeding biology: At least in some areas, pairs often remain in the general vicinity of their breeding areas through the winter period, and many use old nests for nighttime brooding during cold weather. Rarely, however, are old nests used again for nesting; new ones are typically built each year and for each breeding attempt. In eastern Wyoming, birds may begin carrying mud to anchor nest bases in late February, but intensive nest-building does not occur until mid-March or April. Both sexes gather materials, the male bringing more sticks than the female, and rarely each partner will begin a nest at a different location. A surprisingly long average period of 43 days is required to complete a nest, and during the latter part of this time intensive displaying also occurs, especially courtship feeding of the female. The clutch size ranges from five to nine eggs. The female does all the incubating, but her mate feeds her throughout the 17- to 22-day incubation period. Both sexes feed the nestlings equally, and they remain in the nest for an average of nearly four weeks. After the young are able to fly well, the family gradually wanders out of its nesting area. Although it is known that birds sometimes acquire territories and breed when a year old, it is likely that most initial breeding occurs in the second year.

Suggested viewing locations: In Wyoming, this species might be seen almost anywhere in open country at lower elevations, especially in those areas with scattered small trees and that are fairly close to water. The same rule applies elsewhere in the Rocky Mountain region.

Suggested reading: BNA 389 (C. H. Trost, 1999).

American Crow. *Corvus brachyrhynchos*. Wi; Co; Res.

Status: The American crow is a common summer or year-round resident in wooded habitats throughout the region, but it is more common at lower elevations and varies in abundance from common in Grand Teton National Park to rare in Yellowstone.

Habitats and ecology: Forested areas, wooded riverbottoms, orchards, woodlots, large parks, and suburban areas are all used by this species, but it is often replaced by the common raven in rocky canyons and higher montane areas, as seems to be the case in Yellowstone National Park. In Colorado, crows breed at elevations up to 10,000 feet, with rural areas and croplands favored habitats. Trees, shrubs, and utility poles may serve as nest sites (Kingery, 1998).

Clark's nutcracker, adult

Breeding biology: Crows generally begin to flock after the breeding season, and at least in northern areas tend to migrate some distance southward, where they develop massive roosting flocks. It is likely, however, that pairs are maintained within these large flocks, and shortly after returning to the breeding areas the birds typically become well spaced and territorial. Crows utter a surprisingly broad range of notes, including more than a dozen distinct calls, and in addition they commonly mimic other species. Both sexes help build the nest, and although it has been reported that both sexes incubate, this seems unlikely in light of what is known of related species. The clutch size is usually four to five eggs. The incubation period lasts 18 days. The young birds fledge in about 36 days but remain with their parents for a protracted period. Like jays, crows often move about in family groups, with members quickly alerting other members of finding food or observing possible danger.

Suggested viewing locations: Crows are widespread in Wyoming wherever they are not displaced by ravens, and unlike most ravens they will commonly frequent towns and villages. They are very common in the Black Hills as well as around Lovell, Torrington, and Riverton (Dorn and Dorn, 1990).

Suggested reading: BNA 647 (N. A. Verbeek and C. Caffrey, 2002).

Common Raven. *Corvus corax*. Di; Co; Res.

Status: The common raven is an uncommon resident throughout much of the state, mainly in western mountainous regions and east through the Bighorn, Laramie, and Medicine Bow ranges. It is present in both national parks, varying in abundance from common in Grand Teton to abundant in Yellowstone.

Habitats and ecology: Ravens are generally associated with wilderness areas of mountains and forests, especially where bluffs or cliffs are present for nesting. Where these are unavailable, tall coniferous trees are used for nesting, as in Jackson Hole, where more than 90 percent of the nests are in trees. Ravens are often found up to timberline in late summer, or even in alpine tundra. They also extend out into sage and grassland areas, scavenging for road-killed mammals and birds. In Colorado, ravens breed widely in mountains and forests, nesting most commonly in pine-juniper woodlands and upland coniferous forests, with foraging territories often covering several square miles and often encompassing an entire valley, ridge to ridge. There they often nest on cliffs that have a southern aspect (Kingery, 1998).

Breeding biology: The largest of all passerine birds, ravens also have the broadest worldwide distribution of any of the Corvidae. But apart from their size, the birds are typical crows. Pairs form rather large territories and largely remain within them, whereas immature birds and adults that lack territories tend to roam about in flocks. In agonistic and sexual situations, ravens perform a "self-assertive" display in which they raise the feathers above the eyes and all the head and throat feathers, producing a very shaggy-headed appearance, often followed by bowing and crowing while spreading the tail. The nest is built in an inaccessible location by both members of the pair, and the same site is sometimes used in successive years. From one to four weeks are spent in nest building, and an additional day is required for each egg that is laid. Although the female spends most of her time on the nest as soon as the first egg is laid, incubation does not begin until the clutch is complete or nearly complete. The clutch size is usually four to six eggs. The incubation period lasts 20 to 21 days. Only the female incubates, but she is fed by the male, and both sexes feed the young. The female broods them for about 18 days, but they do not fledge until they are about six weeks old. Thereafter they remain in the care of their parents for nearly half a year.

Suggested viewing locations: In Wyoming, ravens are almost entirely found west of a line extending from Sheridan to Casper and Laramie (Faulkner, 2010) and are common in all the major mountain ranges. They are especially abundant in Jackson Hole, where they are attracted to carrion and wastes at places like the National Elk Refuge or at wolf kills in Yellowstone Park. McEneaney (1988) judged the best Yellowstone observation areas to be around Artist Point and Old Faithful, both heavily visited by humans. The densest regional breeding populations probably occur in the Greater Yellowstone ecosystem.

Suggested reading: BNA 476 (W. I. Boarman and B. Heinrich, 1999); Heinrich, 1979.

Family Alaudidae (Larks)

Horned Lark. *Eremophila alpestris*. Ub; Co; Res.

Status: The horned lark is a common resident throughout the state, although some migratory movements into the state from more northerly populations do occur. It is present in both national parks, varying in abundance from common in Grand Teton National Park to uncommon in Yellowstone.

Habitats and ecology: Open-country habitats, ranging from short-grass plains through agricultural lands such as pastures, desert scrub, mountain meadows, and alpine tundra, are the basic requirements for this species, which has an enormous ecological and geographic range in North America. The highest population densities found by Hutto and Young (1999) in 566 transects of 11 unaltered forest, grassland, shrub, and wetland habitats of Idaho and Montana were in grassland communities, followed by sagebrush.

Breeding biology: In the midwestern states, horned larks begin to establish and defend territories in January and February, while pairs are being formed. Territories are large (averaging about four acres in two studies) and are defended by males but only against other males. Two advertisement songs are uttered, either on the ground or in the air. These songs seem related to courtship rather than to territorial defense and are most common after losing a mate or fledging a brood. Courtship feeding and other displays are also performed at this time. The female selects a nest site almost anywhere in the territory and constructs the nest alone. She digs a cavity with her bill and feet, often "paving" it on one side with various objects, for still uncertain reasons. The paving may cover and hide the fresh dirt that has been dug out, or may help keep the nest lining from blowing away during early stages. The female sometimes begins incubation slightly before the clutch is completed but more often begins when the last egg is laid. The clutch size is usually four eggs. The incubation period lasts 12 to 14 days. There is a relatively short nestling period of 9 to 12 days, averaging about 10, and the young birds leave the nest when their flight feathers are only about one-third to one-half grown. Until they are at least 15 days old, the young are able to fly only a few yards. The young birds begin to flock soon after leaving the nest, and in many areas the female shortly thereafter begins a second nesting.

Suggested viewing locations: In Wyoming, horned larks occur statewide ands may easily be found during almost any season in open, grassy country up to the alpine zone. The same applies elsewhere in the Rocky Mountain region. The densest regional breeding populations probably occur in the short-grass plains of southern Wyoming (Seedskadee and Hutton Lake National Wildlife Refuges).

Suggested reading: BNA 195 (R. C. Beason, 1995).

Family Hirundinidae (Swallows)

Tree Swallow. *Tachycineta bicolor*. Ub; Co; Su; Res.

Status: The tree swallow is a common summer resident throughout the region but absent from treeless areas and the high montane communities. It is present in both national parks (Grand Teton and Yellowstone) and abundant in both.

Habitats and ecology: Breeding in the region extends from riparian woodlands through the aspen zone and into the lower levels of the coniferous forest zone. In the Pacific Northwest, this species breeds in mixed coniferous forests from the young to the old-growth forest stages of succession (Sanderson, Bull, and Edgerton, 1980). Outside the breeding season, tree swallows are often seen over lakes or rivers as well as over other open habitats. Nesting is especially prevalent in aspen stands where old woodpecker holes are available, but pairs will also at times use birdhouses erected for bluebirds. In some areas, such as along plains rivers where aspens are lacking, cottonwoods may serve for nesting; with either aspen or cottonwood nest sites, easy access to open foraging areas is important (Kingery, 1998).

Breeding biology: One of the earliest spring migrants of the eastern species of swallows, these birds reach Minnesota and the Dakotas in late April, at least a month before nesting gets underway. Much of the courtship apparently occurs in the air, and it includes synchronized flying by a pair. In one reported

case, a male grasped the female's breast in midair and the two birds tumbled downward until they almost reached the ground. The female then flew to the vicinity of the nest and perched, whereupon the male glided above her and landed on her back. The female constructs the nest with little or no assistance from the male, at times carrying in more than 100 feathers for nest lining. Evidently the male brings food to his incubating mate only rarely; instead, she leaves the nest a few times during daylight to forage for herself. The males often spend the evening perched near the nest, leaving it for their own roosting sites only after dark. The clutch size is usually four to six eggs. The incubation period lasts 13 to 16 days. The nestling period varies considerably, depending on brood size and thus on the rate of feeding, so that the young may spend as few as 16 days or as many as 24 days in the nesting cavity. Once the young leave the nest, however, they rarely return to it.

Suggested viewing locations: Tree swallows in Wyoming and elsewhere are often associated with aspen stands that have good woodpecker populations. They are also very common in the riparian woodlands of eastern Wyoming, such as along the North Platte River. They occur statewide and are especially common in aspen groves near water (Faulkner, 2010). The densest regional breeding populations probably occur in the Greater Yellowstone ecosystem.

Suggested reading: BNA 11 (D. W. Winkler et al., 2011).

Violet-green Swallow. *Tachycineta thalassina*. Ub; Co; Su; Res.

Status: The violet-green swallow is a common summer resident in mountainous areas, occurring locally throughout the entire state. It is common in both national parks (Grand Teton and Yellowstone).

Habitats and ecology: Generally associated with open coniferous forests, such as ponderosa pines, this species also breeds in aspen groves, riparian woods, and sometimes also in urbanized areas. Nesting sites are rather variable and include old woodpecker holes, natural tree or cliff cavities, and occasionally birdhouses. Cliff nest sites in Colorado are second only to upland deciduous forest habitats in nest-choice frequency (Kingery, 1998).

Breeding biology: The violet-green swallow is an unusually early migrant, typically arriving before other swallow species. It thus may not begin nesting for nearly a month after arrival. It spends some time seeking out a suitable cavity, and apparently the female makes the choice with the male playing a minor role. But once the site is chosen, both sexes begin to bring in nesting materials, the female doing most of the carrying. About six days are spent in building the nest, and the female roosts on it at night. The clutch size is usually four to five eggs. The incubation period lasts 13 to 15 days. Eggs are laid at daily intervals, and the female may begin incubating before the clutch is completed, though this does not happen

Overleaf: Common raven, adults

often. Thus, the period of hatching is sometimes rather staggered and has been noted to require as long as five days. The female does most of the feeding and also broods during the first ten days or so after hatching. The fledging period is somewhat variable but averages about 23 to 25 days. After leaving the nest, neither the adults nor the young return to it.

Suggested viewing locations: In Wyoming and elsewhere, violet-green swallows occupy deep canyons during the breeding season. They occur in all the major mountain ranges and the Black Hills, and may be easily seen in the Wind River Canyon (Washakie County), Yellowstone Canyon, and canyons in the Bighorn Mountains as well as at Beaver Rim (Fremont County) and LAK Reservoir in Weston County (Dorn and Dorn, 1990).

Suggested reading: BNA 14 (C. R. Brown, A. M. Knott, and E. J. Damrose, 2011); Edson, 1943.

Northern Rough-winged Swallow. *Stelgidopteryx serripennis*. Ub; Co; Su; Res.

Status: The northern rough-winged swallows is a common summer resident throughout the state, mainly at lower elevations and in open habitats. It is present in both national parks, varying in abundance from occasional in Grand Teton to uncommon in Yellowstone.

Habitats and ecology: This swallow is associated with open areas, including agricultural lands, rivers and lakes, and grasslands near water; it breeds almost exclusively in cavities dug in earthen banks of clay, sand, or gravel. Ready-made cavities in rock are also used. In Colorado, the birds nest at elevations up to 7,500 feet, mostly in streamside banks in prairies, foothills, and deserts and typically as solitary pairs (Kingery, 1998).

Breeding biology: Almost as soon as they arrive on their nesting grounds, these swallows begin to show interest in suitable nesting sites, and they may seek out old kingfisher or bank swallow excavations that are still usable. Males establish a limited territory around a potential nest site, perching near it and pursuing females from it. Females carrying nesting materials are especially pursued, although this behavior may be associated more with copulation than with courtship. Copulation has not been described and presumably occurs in the nesting cavity. Evidently only the female gathers and carries nest-lining material; apparently neither bird does any excavating. About six days are needed to construct the rather bulky nest, but it may rarely take as long as 20 days. The clutch size is usually six to seven eggs. The incubation period lasts 15 to 16 days. The female usually starts incubating with the laying of the next-to-the-last egg, and hatching may extend a few hours or as long as several days. Brooding is done primarily if not exclusively by the female, but both sexes feed the young. The young birds are able to fly a few days before they leave the nest, which usually occurs at 18 to 21 days of age. Young birds rarely return to the nest after they leave it, and there is no evidence on how long the young remain dependent on their parents for food after fledging.

Suggested viewing locations: This swallow is found throughout Wyoming except at high elevations and is especially common at Flaming Gorge Reservoir (Sweetwater County).

Suggested reading: BNA 234 (M. J. De Jong, 1996).

Bank Swallow. *Riparia riparia*. Wi; Co; Su; Res.

Status: The bank swallow is an uncommon summer resident in suitable habitats throughout the state, mainly at lower elevations. It is present in both national parks, varying in abundance from common in Grand Teton National Park to uncommon in Yellowstone.

Habitats and ecology: Breeding almost always occurs near water, such as in steep banks along rivers, road cuts near lakes, gravel pits, and similar areas with steep slopes of clay, sand, or gravel. Outside the breeding season the birds are of broader distribution, sometimes foraging over agricultural lands.

Breeding biology: Shortly after bank swallows arrive in a nesting area, they begin to gather near the breeding site. Unpaired birds apparently select a burrow site, which may be the same burrow they used the previous year. Thereafter they defend the area from intrusion, although potential mates continue to return to a defended spot until one is eventually tolerated and accepted. Sexual chases of the female by the male are a common feature of pair formation, accompanied by male song. Another vocalization, the mating song, is uttered by both members of a pair as they sit side by side or facing each other in the burrow opening. This behavior may be a preliminary to copulation, which probably occurs in the nest chamber. When a burrow needs to be dug or deepened, both sexes share equally in the task, then gather materials such as feathers and grass for nest lining. The clutch size is usually four to five eggs. The incubation period lasts 12 to 15 days. Incubation is by the female and may begin before the clutch is completed. Thus some eggs may hatch as early as 13 days after the clutch has been completed. Both parents alternate at brooding the young, and both feed the young and keep the nest clean. Birds as young as 20 days of age may be able to fly, but they often do not leave the nest for some time thereafter.

Suggested viewing locations: In Wyoming, this is a widespread species at lower elevations in summer, where it nests along cuts in clay, sand, or silt banks near water. Where these banks are unstable, the colonies may last for only a year or two and then disappear. The densest regional breeding populations probably occur in the Greater Yellowstone ecosystem.

Suggested reading: BNA 414 (B. A. Garrison, 1999).

Cliff Swallow. *Petrochelidon pyrrhonota*. Ub; Co; Su; Res.

Status: The cliff swallow is a common summer resident throughout the state, present in both national parks (Grand Teton and Yellowstone) and abundant in both.

Habitats and ecology: A wide variety of nesting areas are used by this species, but in Wyoming vertical cliffsides and the sides or undersides of bridges are perhaps most commonly used.

The nests are gourd-like structures of dried mud, made of small mud globules that are gathered by the birds and carried back in their beaks. In Colorado, the birds nest as high as 12,000 feet, but are mostly found at much lower elevations (Kingery, 1998).

Breeding biology: At least in the northern states, cliff swallows begin to pair immediately upon arrival of their nesting grounds. This activity takes place at or near the nest, and the pair-bond apparently consists primarily of mutual tolerance at the nesting site. Male "primary squatters" persistently return to specific perching places, and their singing attracts secondary visitors to that location, some of which are unpaired females. Both sexes defend the nest site, and both bring mud to construct the nest, which requires nearly two weeks of effort. When the nest is nearly completed, copulation occurs in the nest cup, and copulatory behavior continues until the middle of the laying period. Many cliff swallows occupy old nests if they are still usable; otherwise they construct entirely new ones. The clutch size is usually four to five eggs. The incubation period lasts 12 to 14 days. Incubation may begin before the clutch is complete, and males regularly participate. There is a relatively long nestling period in this species, averaging about 24 days, and a relatively low proportion of females (27 percent in one study) attempt a second clutch. In at least some cases, females change mates for their second nesting, and a considerable amount of courtship activity is evident between broods.

Suggested viewing locations: This species occurs throughout the state except at high elevations, especially where there are sheer rock cliffs, concrete bridges, or even large metal culverts. The densest regional breeding populations probably occur over much of the Greater Yellowstone ecosystem.

Suggested reading: BNA 149 (C. R. Brown et al., 2017).

Barn Swallow. *Hirundo rustica.* Pan; Co; Su; Res.

Status: The barn swallow is a common summer resident throughout the state. It is present in both national parks, varying in abundance from common in Grand Teton to uncommon in Yellowstone.

Habitats and ecology: Except for the purple martin, this species is the swallow that is most closely associated with humans in the Rocky Mountain region. Although it may still occasionally nest on cliff or cave walls, its normal current nesting sites are the horizontal beams or upright walls of buildings and similar structures. In Colorado, the birds nest at elevations up to 10,000 feet, most often at rural sites such as farmhouses, barns, and sheds (Kingery, 1998).

Breeding biology: Within about two weeks after their arrival in nesting areas, most barn swallows have formed pair-bonds. Pair formation takes place on fences and utility lines near nesting areas, with unpaired birds perching alone and singing, and perching or flying between paired birds. Both sexes gather mud for nests; when available, horsehair is added to the mud cup, and feathers are added later for lining. Many times an old nest is used, with new materials added as necessary. An average of about six days is needed to build a nest, and eggs are not laid until the nest is completed. Five eggs are a typical clutch size. The incubation period lasts 14 to 16 days. Only the female incubates in most nests, but in some cases males also participate. An average of 21 days, with a range of 18 to 27 days, has been observed as the nestling period in this species. Courtship behavior soon begins again, such as "song flights" given by flocks of swallows flying high and chasing each other. Partners are not changed between broods, and the same nest is usually used again, often with more mud and feathers added to it. A gap of about one month lapses between nesting cycles, and only about a third of the swallows in one New York study raised second broods. Second clutches most often have four rather than five eggs, but egg and nesting mortality rates are similar in the two nesting cycles.

Suggested viewing locations: In Wyoming, this species is abundant at elevations up to about 8,000 feet, especially around farmyards or abandoned buildings as well as bridges and culverts.

Suggested reading: BNA 452 (C. R. Brown and M. B. Brown, 1999).

Family Paridae (Chickadees and Titmice)

Black-capped Chickadee. *Poecile atricapillus.* Ub; Co; Su; Res.

Status: The black-capped chickadee is a common resident in deciduous and coniferous forests throughout the state. It is present in both national parks, varying in abundance from common in Grand Teton to uncommon in Yellowstone.

Habitats and ecology: This chickadee is associated with a wide variety of wooded habitats of coniferous and hardwood types, and it breeds wherever suitable nesting cavities exist. These typically consist of old woodpecker holes, but sometimes the birds excavate their own nest cavities in the rotted wood of dead stumps. Bird houses are also occasionally used. Aspen groves and riparian woodlands are favored nesting areas in the Rocky Mountain region. In Colorado, most chickadees nest at from 5,000 to 9,000 feet elevation, with large aspens stands their favorite breeding habitat, and lowland riparian woods with cottonwoods their second most frequent choice (Kingery, 1998).

Breeding biology: Chickadees are largely nonmigratory, but winter flocking does occur. Pair-bonds are weak or absent during this time, although there is enough contact to allow frequent re-pairing with past mates. Courtship is apparently simple, consisting mainly of the loud *phoe-be* song by males. Territories are not established until later, when the pair begins to excavate a nest site. Both sexes excavate, the female taking the lead, and both birds work intermittently during daylight. Eggs are laid daily and are covered with nesting material by the female, who also sleeps in the nest but does not begin incubating until the clutch of six to eight eggs is complete. The incubation period lasts 12 to 13 days. Only the female incubates, but the male feeds her at intervals. During the first week after hatching the behavior of adults is similar to their behavior during incubation, but the male stops feeding the female and both parents feed the young. After brooding is terminated, both sexes feed the young at about an equal rate, and the young birds leave the nest at 16 or 17 days of age. Fledglings are able to forage for themselves about ten days after leaving the nest, but they remain with their parents for three to four weeks. A small proportion of adults attempt a second brood.

Suggested viewing locations: In Wyoming, this eastern-oriented chickadee is most common in eastern areas, such as the western Black Hills, but also the Tongue River Canyon near Dayton, along the North Platte River, and also west to the Wind River near Dubois (Dorn and Dorn, 1990). It is usually found below 8,500 feet but may reach 10,000 feet along riparian corridors (Faulkner, 2010). The densest regional breeding populations probably occur in the western Black Hills.

Suggested reading: BNA 39 (J. R. Foote et al., 2010).

Mountain Chickadee. *Poecile gambeli*. Wi; Co; Su; Res.

Status: The mountain chickadee is a common resident in mountain forests of the region, occurring in both national parks (Grand Teton and Yellowstone) and common in both. It is more montane in distribution than the black-capped chickadee.

Habitats and ecology: Largely limited to montane coniferous forests, this chickadee is usually absent from deciduous stands, although aspens are frequently used for nesting. This species prefers open coniferous forests, especially pines, including both ponderosa pines and also pinyons. In the interior Pacific Northwest, it breeds in mixed coniferous forests from the pole-sapling to the old-growth forest stages of succession (Sanderson, Bull, and Edgerton, 1980). Woodpecker holes or self-excavated cavities in rotted wood are used for nesting. In Colorado, this species ranges up to about 9,000 feet elevation, reaching its greatest abundance in old-growth spruce-fir, lodgepole pine, and ponderosa pine forests, and is less common in pinyon-juniper woodlands (Kingery, 1998). The highest population densities found by Hutto and Young (1999) in 566 transects of 11 unaltered forest, grassland, shrub, and wetland habitats of Idaho and Montana were in ponderosa pine communities.

Breeding biology: This species differs slightly, if at all, in its social behavior from that of the better-known black-capped chickadee. Its primary call is a more throaty or hoarser version of that species' familiar *chick-a-dee-dee* vocalization, and that species' *phoe-be* call is rendered as *phoe-be-be*. It nests in pre-excavated tree cavities most often, usually at a height of less than 15 feet above ground but rarely as high as 80 feet. It also has been found nesting in bird boxes, pine stumps, and even under rocks. There is one record of a mountain chickadee sharing a nest box with a juniper titmouse, with the chickadee doing the incubating. Clutch sizes usually range

from 6 to 9 but have been reported to be as large as 12, a remarkable number of eggs for such a tiny bird. Incubation is performed by the female alone and lasts 14 days. When disturbed at the nest, females utter a hissing sound, which perhaps mimics something like a snake's hiss and possibly helps to deter predators. Both parents feed the young, which reportedly remain in the nest for a surprisingly long period of 21 days. This compares with a fledging period of 14 to 18 days reported for the black-capped chickadee.

Suggested viewing locations: In Wyoming, this chickadee is common in all the mountain ranges except for the Black Hills (Dorn and Dorn, 1990) and is largely limited to mature coniferous forests from 7,500 feet to timberline (Faulkner, 2010). McEneaney (1988) judged the Yellowstone population to be in the thousands during the late 1980s, with the best observation areas being around Fishing Bridge and on the Blacktail Plateau.

Suggested reading: BNA 453 (D. A. MacCallum, R. Grundel, and D. L. Dahlsten, 1999).

Juniper Titmouse. *Baeolophus ridgwayi*. HL; Un; Res.

Status: The juniper titmouse is an uncommon resident in the woodlands of southwestern Wyoming.

Habitats and ecology: In this region, the juniper titmouse is associated almost exclusively with the pinyon-juniper association; in some other areas they also extend into oak woodlands. Nesting is usually done in cavities of partially decayed and split-open trunks of junipers, the cavity typically being at least partially excavated or modified by the birds. They sometimes also use woodpecker holes and occasionally will nest in birdhouses. In Colorado, the pinyon-juniper habitat accounted for 96 percent of breeding-season observations (Kingery, 1998). In Wyoming, nesting is confined to Utah juniper woodlands in Sweetwater, Uinta, and Carbon Counties (Faulkner, 2010).

Breeding biology: At least in the interior populations of this species, a certain amount of winter flocking seems typical. Apparently pairs are formed during the flocking period and before territories are established. Pair formation is marked by singing and males' making "approach threats" toward females and chasing them in a sexual flight that represents attempted copulation. A submissive display by females, involving wing-quivering and a soft call, stimulates feeding by the male and helps to establish a pair-bond. Apparently only the female searches for a nest site, and she also is the only one that gathers materials to fill the chosen cavity. The clutch is usually six to eight eggs, and the male continues to feed the female during incubation, which lasts 12 to 14 days. Both sexes help feed the young about equally, at least after the female terminates her brooding. The nestling period is approximately 20 days, and after leaving the nest the young birds continue to forage within a narrow radius of their nest for some time. By the time they are five weeks old, they are foraging for themselves, and they gradually begin to disperse from their parental territory. In some cases, they establish temporary territories of their own while still juveniles, but they may be nearly a year old before they successfully obtain suitable breeding territories.

Suggested viewing locations: A good location for finding this highly localized species is the Minnie's Gap area in Sweetwater County and among scrub junipers five to ten miles west of Baggs.

Suggested reading: BNA 485 (C. Cicero, P. Pyle, and M. A. Patten, 2000).

Family Aegithalidae (Long-tailed Tits and Bushtits)

Bushtit. *Psaltriparus minimus.* HL; Un, Ch; Res.

Status: The bushtit is an uncommon resident along the southwestern edge of Wyoming, where it is most likely to be found in pinyon pine–juniper woodlands and similar scrub oak–mountain mahogany habitats.

Habitats and ecology: Commonest in open woodlands such as pine and juniper habitats, these birds also at times occur in sagebrush or even aspen-covered hillsides. Nests are usually in small pines or junipers and are beautiful soft, woven hanging structures made of mosses, spider webs, and hair or feathers, with lateral entrances most often less than ten feet above ground. In Colorado, 85 percent of breeding bushtits were recorded in pinyon-juniper woodlands, with upland and riparian shrubland also used to a small degree (Kingery, 1998). In Wyoming, known nesting is confined to Utah juniper woodlands in southern Sweetwater and Uinta Counties, but a local population also exists and presumably nests around Coal Mountain in Natrona County (Faulkner, 2010).

Breeding biology: Bushtits are found in small flocks during the nonbreeding period, moving about in close-knit groups from about mid-September until the first of April. Courtship begins in the flocks and consists of sexual posturing, trills, and excited location notes. Territories are poorly defined and rather variably defended, probably depending on the abundance of nesting materials and food they contain. Nests are built by both members of a pair, and in one case a third bird of unknown sex was seen helping with nest-building and incubation. Nest-building is a long and intricate process, requiring from 13 to 51 days in eight observed instances. Rarely, nests from the past year are usable, and even in renesting efforts a new nest might be built, often using materials from the first nest. In renesting efforts, the pairs often dissolve and the members may take new mates. The clutch size is usually five to seven eggs. The incubation period lasts 12 to 13 days. Incubation apparently is equally shared by the two sexes, and both birds sleep in the nest at night. Both sexes feed the young and remain in the nest at night during the entire nestling period, which lasts about 14 days. After leaving the nest the family forms a small flock, with the adults initially doing all the foraging

for the young, but about 14 days after leaving the nest the young birds are independent.

Suggested viewing locations: In Wyoming, Seedskadee National Wildlife Refuge and Flaming Gorge National Recreation Area of Sweetwater County are good locations for finding this tiny and elusive species.

Suggested reading: BNA 598 (S. A. Sloane, 2001).

Family Sittidae (Nuthatches)

Red-breasted Nuthatch. *Sitta canadensis.* Ub; Un; Res.

Status: The red-breasted nuthatch is an uncommon resident in montane coniferous forests throughout the entire state. It is a common resident in both national parks (Grand Teton and Yellowstone).

Habitats and ecology: This nuthatch is limited largely, but not entirely, to coniferous forests, primarily those of relatively tall firs, where much of the foraging occurs at rather high portions of the trees. To a much more limited degree, aspens and riparian woodlands are sometimes also used. In Colorado, this species is associated with coniferous forests and aspens during the breeding season, at elevations from 6,500 to 11,500 feet (Kingery, 1998). In the Pacific Northwest, this species breeds in mixed coniferous forests from the young to the old-growth forest stages of succession (Sanderson, Bull, and Edgerton, 1980). The highest population densities found by Hutto and Young (1999) in 566 transects of 11 unaltered forest, grassland, shrub, and wetland habitats of Idaho and Montana were in ponderosa pine communities.

Breeding biology: During the winter, red-breasted nuthatches may remain paired if the food supply is good or if the birds are close to a feeding station. At this season the birds maintain contact by uttering location calls, but by late winter unpaired males begin singing a series of plaintive *waa-aan*s from tall trees, which probably serve both territorial and courtship functions. A major behavior during pair formation is courtship feeding of the female by her mate, which continues through the incubation period. Breeding occurs in the trunks of dead trees or the rotting portions of live trees, with the birds typically excavating their own nesting holes. Pairs seek out nesting sites together, with the female making the final choice and also doing the initial excavating. Courtship chases of the female are frequent during nest building and may end in copulation. When the nest excavating is nearly finished, both members of the pair bring resin in their bills, land in the nest entrance, and spread the resin above and below the hole. This sticky material probably deters other animals from entering the hole. The clutch size is usually five to six eggs. The incubation period lasts 12 days. The female does all the incubation, but both sexes feed the young, which fledge within periods that have been estimated to range from 14 to 21 days.

Suggested viewing locations: In Wyoming, and elsewhere throughout the entire region, this nuthatch is common in coniferous forests. It occurs statewide in all of Wyoming's mountain ranges, at middle to higher elevations (Faulkner, 2010). The densest regional breeding populations probably occur in the western Black Hills and the Greater Yellowstone ecosystem.

Suggested reading: BNA 459 (C. K. Ghalambor and T. E. Martin, 1999); Kilham,1973.

White-breasted Nuthatch. *Sitta carolinensis*. Wi; Un; Res.

Status: The white-breasted nuthatch is an uncommon resident of deciduous forests and woodlands nearly throughout the state. It is present in both national parks, varying in abundance from common (Grand Teton) to uncommon (Yellowstone).

Habitats and ecology: This nuthatch is largely confined in the eastern states to deciduous forests, but in the Rocky Mountains it is associated with lower-elevation coniferous forests, especially the ponderosa pine zone and also the pinyon-juniper zone. Nesting occurs in old woodpecker holes or in self-excavated holes in rotted wood of dead or partially dead trees, often aspens. In Colorado, breeding birds extend up to 10,000 feet elevation (Kingery, 1998). In the interior Pacific Northwest, this species breeds in mixed coniferous forests from the mature to the old-growth forest stages of succession (Sanderson, Bull, and Edgerton, 1980). The highest population densities found by Hutto and Young (1999) in 566 transects of 11 unaltered forest, grassland, shrub, and wetland habitats of Idaho and Montana were in mixed conifer communities.

Breeding biology: Apparently white-breasted nuthatches maintain their pair-bonds throughout most of the year and perhaps permanently, although during winter the paired birds roost in different areas and maintain little contact with each other. The male begins to sing in late winter, uttering early-morning "rendezvous songs" from tall trees to attract the female. Males also sing and display directly to their mates when they arrive and may keep in touch with them during foraging by uttering a series of *wurp* notes. The female takes the initiative in choosing a nest site and does all the nest building, but both sexes participate in "bill-sweeping" in and around the nest. This behavior is of uncertain significance but consists of arclike movements of the bill near the tree or cavity surface, sometimes while holding an insect or other object. It has been suggested that the odors thus spread may repel squirrels, which often eat bird eggs. The clutch size is usually five to nine eggs. The incubation period lasts 12 to 14 days. The female does the incubating, but the male feeds her during egg-laying and incubation, and males later help feed the young. The fledging period is approximately two weeks.

Suggested viewing locations: Almost any deciduous or coniferous woodland in the entire region is likely to support this nuthatch, especially open ponderosa pine forests at lower elevations. In Wyoming, the species has a patchy distribution

statewide (Faulkner, 2010). The densest regional breeding populations probably occur in the western Black Hills.

Suggested reading: BNA 54 (T. C. Grubbs, Jr. and V. V. Pravosudov, 2008); Kilham, 1968.

Pygmy Nuthatch. *Sitta pygmaea*. Lo; Un; Res.

Status: The pygmy nuthatch is an uncommon resident in some portions of the region, mainly in the drier areas with ponderosa pine, especially the southern Bighorns, the Laramie and Medicine Bow ranges, and the hill country of Natrona and Platte Counties.

Habitats and ecology: Primarily associated with the ponderosa pine zone, but also occurring locally in the pinyon-juniper zone, the pygmy nuthatch generally forages fairly high in the tall pines but nests closer to the ground in snags or stubs that have rotted trunks providing excavation opportunities. In Colorado, two-thirds of breeding-season observations were in ponderosa pine woodlands, and most of the remainder were in other coniferous habitats. Mature trees with snags or rotting portions are preferred (Kingery, 1998). In the interior Pacific Northwest, this species breeds in mixed coniferous forests from the mature to the old-growth forest stages of succession (Sanderson, Bull, and Edgerton, 1980).

Breeding biology: Pygmy nuthatches are more or less permanently territorial; males hold small territories and limit most defense to the nest site. This site may be an existing cavity, or the pair may excavate a new one. Sometimes three or more birds have been seen excavating a single site, and at least in some cases the extra birds are males. Up to a month or more may be needed for excavation, and it takes one day to lay each egg. The clutch size is usually five to nine eggs. The incubation period lasts 15 to 16 days. Only females incubate, but both sexes sleep in the cavity at night, and in the observed threesomes all the birds roosted there. Females are fed on or off the nest by the male or males. The eggs typically hatch within a 24-hour span, and the young are fed by both adults and, when present, additional helpers. The young fledge in 20 to 22 days but do not gain independence from the adults until they are approximately 45 to 50 days old.

Suggested viewing locations: In eastern and central Wyoming this nuthatch can be found in mature ponderosa pine forests, such as those in Curt Gowdy State Park (Laramie County) and along Middle Crow Creek (Albany County) in the Pole Mountain Division of Medicine Bow National Forest (Dorn and Dorn, 1990). The birds are rare and of uncertain breeding status in the Greater Yellowstone ecosystem (Faulkner, 2010).

Suggested reading: BNA 567 (H. E. Kingery and C. K. Ghalambor, 2001).

Family Certhiidae (Treecreepers)

Brown Creeper. *Certhia americana*. SD; Un; Res.

Status: The brown creeper is an uncommon resident in mature forested areas almost throughout the state. It is present in both national parks, varying in abundance from occasional (Grand Teton) to rare (Yellowstone).

Habitats and ecology: Brown creepers are associated with forests throughout the year, including both deciduous and coniferous forests. Virtually all foraging is done on the trunks of fairly large trees, where the birds forage for insects in bark crevices and grooves. In Colorado, breeding birds most frequent occupy mature spruce-fir and lodgepole pine forests at 9,000 to 11,500 feet of elevation. Dense stands of mature trees are strongly favored, making them very sensitive for logging effects (Kingery, 1998). In the Pacific Northwest, this species breeds in mixed coniferous forests from the mature to the old-growth forest stages of succession (Sanderson, Bull, and Edgerton, 1980). The highest population densities found by Hutto and Young (1999) in 566 transects of 11 unaltered forest, grassland, shrub, and wetland habitats of Idaho and Montana, and that also occur in Wyoming, were in spruce-fir and mixed conifer communities.

Breeding biology: Although a few creepers may remain in the northern states through the winter, they are generally migratory, and in early spring small groups may be encountered foraging in loose flocks, maintaining contact by delicate *cree-cree-cree-ep* notes. At least in the European race of creeper it is known that during cold winter nights the birds frequently roost in cracks in tree trunks. Clinging woodpecker-like to the bark and supported by the tail, they can withstand subfreezing temperatures even when partly covered by snow. When spring comes the pair begins to work intermittently on the nest, which may require a month to finish. Both sexes bring materials, but only the female does the construction. The clutch size is usually six eggs. The incubation period lasts 14 to 15 days. Observations in North America suggest that only the female incubates, whereas in England it has been reported that the male participates in this to some extent. Both sexes also feed the young, which are ready to leave the nest in 13 to 14 days. Even though their short tail feathers do not provide them any support, young fledglings are able to cling to vertical branches and move about like adults.

Suggested viewing locations: In Wyoming, this inconspicuous bird occurs almost statewide and can often be found during winter in most towns and cemeteries. During summer it might be searched for in high-altitude coniferous forests statewide, such as the Togwotee Pass area (Teton-Fremont county line) or along the west side of Blackhall Mountain in Carbon County (Dorn and Dorn, 1990).

Suggested reading: BNA 669 (J-F. Poulin et al., 2013).

Family Troglodytidae (Wrens)

Rock Wren. *Salpinctes obsoletus*. Ub; Un, Ch; Su; Res.

Status: The rock wren is an uncommon summer resident throughout the state in rocky areas, especially in dry sagebrush-dominated localities. It is present in both national parks, varying in abundance from occasional in Grand Teton to common in Yellowstone.

Habitats and ecology: Closely associated with eroded slopes, badlands, rocky outcrops, cliff walls, talus slopes, and similar rock-dominated habitats at generally rather low elevations, this wren does sometimes occur in alpine areas, even as high as 13,000 feet on Colorado's Pike's Peak. Crannies in cliffs are favorite nesting sites, but the birds can nest among small as well big rocks, which allows them a broader breeding distribution than is true of canyon wrens. The highest population densities found by Hutto and Young (1999) in 566 transects of 11 unaltered forest, grassland, shrub, and wetland habitats of Idaho and Montana were in sagebrush communities.

Breeding biology: In Arizona, where rock wrens occur with canyon wrens, both species feed in a generalized fashion on similar foods, but rock wrens forage almost exclusively in open or relatively unvegetated situations, while canyon wrens forage mostly in secluded or covered habitats. The species also differ in favored nest sites, with this species using slopes of loose rocks and boulders rather than cliff or canyon walls. Eggs are laid at the rate of one a day, typically totaling five or six, with incubation starting when the clutch is complete. The incubation period lasts 12 to 14 days. Only the female incubates, but she is usually fed by the male, and both sexes feed the nestlings. When the young leave their nest (after about 14 days) the adults soon begin gathering nest material for their second brood, or may begin a second clutch in the same nest.

Suggested viewing locations: This species occurs statewide in Wyoming, from low elevations to above timberline. Good Wyoming birding locations include the badlands north-northwest of Baggs, the lower western slope of the Bighorn Mountains, rocky areas around Boysen Reservoir (Fremont County), the foothills of the Laramie Range west of Wheatland, and rocky hills south of Rock Springs (Dorn and Dorn, 1990).

Suggested reading: BNA 486 (P. E. Lowther, D. L. Kroodsma, and G. H. Farley, 2000).

Canyon Wren. *Catherpes mexicanus*. SD; Un, Ch; Res.

Status: The canyon wren is an uncommon resident in western Wyoming, with local breeding farther east in the Bighorns, Black Hills, and Laramie Mountains. It is present in both national parks (Grand Teton and Yellowstone).

Habitats and ecology: Rocky canyons, river bluffs, cliffs, rockslides, and similar topographic sites are favored by the canyon wren, especially those offering shady crevices. Wyoming populations include disjunctive groups in the Black Hills and Devils Tower regions, a large region centered in the Laramie Mountains, and much of western Wyoming except for the Yellowstone-Teton region (Faulkner, 2010). In Colorado, these birds breed at elevations of 3,900 to 8,500 feet in varied topography such as sandstone outcrops, rimrocks, and vertical canyon cliff sides. They are often found in stream-fed canyons but sometimes nest well away from water. Sometimes old buildings, fences, or other structures are used for nest sites (Kingery, 1998).

Breeding biology: In Arizona, where rock wrens occur with canyon wrens, both species feed in a generalized fashion on similar foods, but rock wrens forage almost exclusively in open or relatively unvegetated situations, while canyon wrens forage mostly in secluded or covered habitats. The species also differ in favored nest sites, with this species using slopes of loose rocks and boulders rather than cliff or canyon walls. Eggs are laid at the rate of one a day, with incubation starting when the clutch (usually of five to six eggs) is complete. The incubation period lasts 12 to 18 days. Only the female incubates, but she is fed by the male, and both sexes feed the nestlings. When the young leave their nest (after about 14 days) the adults soon begin gathering nest material for their second brood or may begin a second clutch in the same nest.

Suggested viewing locations: In Wyoming, good localities for finding this wren include canyons in the Bighorn Mountains, Wind River Canyon (Washakie County), and the cliffs on the west side of Flaming Gorge Reservoir (Sweetwater County) (Dorn and Dorn, 1990).

Suggested reading: BNA 197 (S. L. Jones and J. S. Dieni, 1995).

House Wren. *Troglodytes aedon.* Ub; Co; Su; Res.

Status: The house wren is a common summer resident almost throughout the state, including both national parks (Grand Teton and Yellowstone).

Habitats and ecology: This wren is generally most common in the lower elevation forests, but it occasionally reaches timberline. In Wyoming, the birds favor riparian woodlands, aspen groves, and the lower and more open coniferous forest zones as well as areas of human habitation. In the interior Pacific Northwest, this species breeds in mixed coniferous forests from the pole-sapling to the old-growth forest stages of succession (Sanderson, Bull, and Edgerton, 1980). The highest population densities found by Hutto and Young (1999) in 566 transects of 11 unaltered forest, grassland, shrub, and wetland habitats of Idaho and Montana were in cottonwood communities. Nesting occurs in natural tree cavities, old woodpecker holes, artificial cavities such as birdhouses, and the like. In Colorado, this species replaces the Bewick's wren above about 7,000 feet and has been reported breeding at elevations as high as 11,800 feet (Kingery, 1998).

Breeding biology: As house wrens arrive on their breeding grounds in the spring, adults tend to precede immature birds, and males arrive about nine days before females. An adult male that has nested previously normally returns to its old territory or establishes a new territory adjacent to it, and females also have a strong tendency to return to previous nesting areas. Males sing three kinds of songs, including a "territory song," "mating song," and "nesting song," and both sexes have a variety of call notes. Males typically have two or three possible nest sites within their territories and may have as many as seven, thus allowing the females considerable choice. When establishing nest sites, house wrens often destroy nearby eggs, nests, or young of their own or other species, and there is a good deal of territorial shifting owing to nest-site competition and the frequent changing of mates between broods. The clutch is usually six to eight eggs, incubation lasts 13 to 25 days, and the nestling period is approximately 15 days. In addition to mate-changing at this time, a second female may mate with a male and nest within his territory. In one study it was found that about 6 percent of the matings are polygynous and that about 40 percent of second matings are with the same mate. There is likewise about a 40 percent incidence of mating with the same individual in the following year, when both birds return to the same locality.

Suggested viewing locations: In Wyoming, and elsewhere in the region, this ubiquitous species can be found almost anywhere, from backyards where wren houses have been erected to brushy woods, wooded riparian edges, and even alpine timberline.

Suggested reading: BNA 380 (L. S. Johnson, 2014).

Family Polioptilidae (Gnatcatchers)

Blue-gray Gnatcatcher. *Polioptila caerulea.* Lo; Un; Su; Res.

Status: The blue-gray gnatcatcher is an uncommon and local summer resident in juniper woodlands of southern and central Wyoming as well as riparian deciduous woodlands of the eastern plains.

Habitats and ecology: Gnatcatcher breeding in the region occurs in pinyon-juniper and also adjacent oak woodland or sagebrush areas, up to at least 7,000 feet elevation. Arid park-like areas with scattered thickets are preferred for foraging, and nests are usually placed in low junipers. In Colorado, breeding birds selectively use pinyon-juniper woodlands but also occupy scrub oak thickets and mountain mahogany or serviceberry scrublands (Kingery, 1998).

Breeding biology: Shortly after they arrive on their breeding grounds, male gnatcatchers initiate territories; the time depends on the abundance of foliage-dwelling arthropods in the locality. All the breeding activities occur within the territory, which is defended by the male and sometimes also by the female. Pair-bonds may be established almost immediately after territoriality begins or may occur later. When a female appears on the territory of an unmated male, he accompanies her to various potential nest sites, frequently perching in an upright posture and singing an elaborate but whispered song sequence. Both members of the pair build the nest, and they frequently obtain materials by dismantling old nests. When only new materials are used, they need about two weeks to complete a nest, but when materials are already available they may finish in three to six days. The clutch is usually four or five eggs. The incubation period lasts 15 days. Both sexes incubate about equally, and both sexes brood the young, which remain in the nest for 12 to 13 days. They are at least occasionally fed by their parents for as long as 19 days after leaving the nest. In one California study, three of 12 pairs raised one brood successfully, and four managed to raise two broods. The length of the breeding season there suggests that three broods may sometimes be raised in favorable years.

Suggested viewing locations: In Wyoming, gnatcatchers are most common in the southwest but have expanded their range and now breed east to the Nebraska and South Dakota borders (Faulkner, 2010). Sweetwater County populations occur in the Flaming Gorge area and Powder Rim. Junipers around Alcova Reservoir in Natrona County may also support breeding birds, and they are known to nest north to Hot Springs County. Their status in the Wyoming Bighorns is unknown, but they are known to breed across the Montana line (Faulkner, 2010).

Suggested reading: BNA 23 (E. L. Kershner and W. G. Ellison, 2012).

Family Cinclidae (Dippers)

American Dipper. *Cinclus mexicanus*. SD; Un, Ch; Res.

Status: The American dipper is an uncommon resident in suitable habitats throughout western Wyoming and east and south to the Bighorn, Laramie, Medicine Bow, and Sierra Madre ranges. It is common in both national parks (Grand Teton and Yellowstone).

Habitats and ecology: Rapidly flowing mountain streams, often with waterfalls or cascades and ledges or crevices that provide safe nesting sites, are this species' prime habitat. Nesting is sometimes done on rock walls or overhangs near or even sometimes behind waterfalls, but more often the nests are constructed under bridges that cross creeks or rivers. The birds are highly territorial, and pairs tend to be well separated. In Colorado, breeding sites as high as 11,820 feet have been reported. At elevations below 7,600 feet, second nesting efforts may be attempted (Kingery, 1998).

Breeding biology: Dippers are very sedentary, and pairs tend to remain well separated from others. Nesting occurs in root tangles, rock crevices, or other cavities. Studies in Montana indicate that by November the birds begin to establish winter territories, which are held through February and may include from 50 yards to as much as a half-mile of stream. During winter the birds begin to sing, and singing increases in intensity to a peak in April. The songs of the two sexes are identical, and pair formation is accompanied by singing as well as wing-quivering and chasing. Both sexes participate in nest building, or in reconstructing old nests, which seems to be more common than building a new nest. The clutch is usually four eggs. Incubation lasts 16 days and is performed entirely by the female, but the male provides her with food. The young are hatched with a coating of down and grow relatively slowly, so the fledging period lasts 19 to 25 days. The female broods regularly for about a week after hatching, and the male apparently never enters the nest. Instead the young are fed by poking their heads out the nest entrance. When

the young leave the nest they are nearly as large as their parents and easily flutter to a safe landing below. After they have left the nest, one or both parents typically remove the nest lining, presumably to prepare the nest for use the following year. The young birds soon learn to clamber about on the wet rocks, and they remain in the vicinity of the nest for about 15 days after fledging. Adults likely return to the same nesting area each year, although it is known that one adult had different mates during two successive years.

Suggested viewing locations: In Wyoming, dippers can be found along streams in all the major mountains except for the Bear Lodge Mountains, but they do occur in several canyons in the adjacent South Dakota Black Hills. McEneaney (1988) judged the best observation areas in Yellowstone National Park to be along the Gibbon River, Firehole Canyon, and the Gardiner River.

Suggested reading: BNA 229 (M. F. Wilson and H. E. Kingery, 2011); Hahn, 1950.

Family Regulidae (Kinglets)

Ruby-crowned Kinglet. *Regulus calendula*. Lo; Un; Su; Res.

Status: The ruby-crowned kinglet is a common summer resident in coniferous forests throughout the state, more widespread and more numerous than the golden-crowned kinglet, which has an almost identical nesting distribution and very similar reproductive biology. It is common in both national parks (Grand Teton and Yellowstone).

Habitats and ecology: Breeding occurs in coniferous forests from the lower zones almost to timberline in the subalpine zone but is usually in taller and denser forests of medium altitude. In Colorado, the ruby-crowned kinglets breed exclusively in conifers but are not so closely associated with dense stands as is true of the golden-crowned kinglet. Like that species, they have been seen as high as 11,000 feet during the breeding season (Kingery, 1998). In the Pacific Northwest, this species breeds in mixed coniferous forests from the mature to the old growth forest stages of succession (Sanderson, Bull, and Edgerton, 1980). The highest population densities found by Hutto and Young (1999) in 566 transects of 11 unaltered forest, grassland, shrub, and wetland habitats of Idaho and Montana were in cottonwood-aspen communities, followed closely by spruce-fir, whereas for golden-crowned kinglets old-growth spruce-fir densities greatly outnumbered cottonwood-aspen numbers. During winter the birds often move toward lower elevations, including prairie riverbottom woodlands and sometimes cities.

Breeding biology: Few behavioral studies have been done on either the ruby-crowned or golden-crowned kinglet, perhaps because of their small size and often highly elevated nests. Males of both species utter high-pitched songs that consist of repeated *tew* or *tse* notes followed by a rapid trill (golden-crowned) or a thee-note trill of *liberty, liberty, liberty* (ruby-crowned). It is known that exhibition of their brilliant crown patches is an important part of aggressive and courtship display behavior. Both species build beautiful suspended nests from about 4 to 100 feet above ground. The nest, of mosses and lichens supported by spider webs and hair, is built over a period of five to nine days. The clutch size is usually 7 to 9 eggs, which are incubated by the female alone. The incubation period is about 14 days, and the young remain in the nest for about 16 to 17 days. Observations in Colorado suggest that the female broods the young and passes on food to them that is brought in by the male.

Suggested viewing locations: In Wyoming, this species occurs in all the major mountain ranges at middle to higher elevations, and sometimes also is found in conifers at lower elevations (Dorn and Dorn, 1990; Faulkner, 2010).

Suggested reading: BNA 119 (D. L. Swanson, J. L. Ingold, and G. E. Wallace, 2008).

Family Turdidae (Thrushes)

Mountain Bluebird. *Sialia currucoides*. Ub; Col; Su; Res.

Status: The mountain bluebird is a common and widespread summer resident throughout the state, especially in rather open woodlands. It is common in both national parks (Grand Teton and Yellowstone).

Habitats and ecology: Mountain bluebird breeding occurs in open woodlands and forest-edge habitats from mountain meadows downward through the ponderosa pine zone, the aspen zone, and into the pinyon-juniper zone. Typically the birds favor nesting where either dead trees are available for nest cavities or rock crevices or other suitable sites are present. In Colorado, pinyon-juniper woodlands are the most favored breeding habitat, followed closely by aspen forest and mountain grasslands. Nests there have been observed as high as 13,500 feet elevation (Kingery, 1998). In the Pacific Northwest, this species breeds in mixed coniferous forests from the young to old-growth forest stages of succession (Sanderson, Bull, and Edgerton, 1980). The highest population densities found by Hutto and Young (1999) in 566 transects of 11 unaltered forest, grassland, shrub, and wetland habitats of Idaho and Montana were in sagebrush communities, obviously a foraging habitat, whereas post-fire forest stands were heavily favored for nesting.

Breeding biology: Mountain bluebirds arrive relatively early in the central and northern plains and immediately begin searching for nesting sites. Paired birds often displace unmated males, which defend their territories only weakly, and when a nesting pair has established a territory they both defend it vigorously, the male defending the periphery and the female the actual nest site. One instance has been described of a male having two mates within his territory, nesting about 50 yards apart. Only the female builds the nest, which requires four to six days, and only the female incubates. The clutch is usually five to six eggs. The incubation period lasts 12 to 15 days.

Females brood their nestlings for about six days after hatching, and both parents actively feed the young. They fledge in 22 to 23 days, after which the female usually begins a second clutch and the male remains with the fledglings for about ten days. Young of the first brood of mountain bluebirds have been observed feeding the second brood, but only rarely.

Suggested viewing locations: Mature forests at all elevations support mountain bluebirds throughout Wyoming and elsewhere in the region. McEneaney (1988) judged the best observation areas in Yellowstone National Park to be around Roosevelt Lodge and the Upper Terrace Drive.

Suggested reading: BNA 222 (H. W. Power and M. P. Lombardo, 1996); Power, 1966.

Townsend's Solitaire. *Myadestes townsendi*. Wi; Co; Res.

Status: Townsend's solitaire is a common and widespread resident in wooded mountainous areas of the state during the breeding season. It extends into the lower pinyon-juniper woodlands during winter. Solitaires are present in both national parks, varying in abundance from common in Grand Teton to uncommon in Yellowstone.

Habitats and ecology: Forested mountain slopes that provide snow-free areas for nesting on or near the ground, and which also offer sources of berries for food, are favored for nesting. In Colorado, closed canopy forests, especially upland coniferous forests, are favored by breeding birds, but deciduous forests, pinyon-juniper communities, and other habitats including brushlands are also used, at elevations of 7,000 to 11,000 feet (Kingery, 1998). The highest population densities found by Hutto and Young (1999) in 566 transects of 11 unaltered forest, grassland, shrub, and wetland habitats of Idaho and Montana were in lodgepole pine and Douglas-fir communities. In the winter, the birds feed almost entirely on junipers or other kinds of berries, but while breeding the usual thrush diet of insects is the most important source of food.

Breeding biology: This little-studied montane species in some respects acts more like a flycatcher than a thrush. In the winter, the birds move from the mountains to lower elevations. Thus, in California and Arizona it has been reported that the birds establish winter territories that they hold from late September until April and that are associated with the distribution of juniper berries, their main winter food in those areas. Although they compete with jays and robins for these berries, they do not attempt to defend their territories against these larger birds, though they do attack juncos, bluebirds, and nuthatches. During the breeding season, the remarkably beautiful song of the territorial male is the species' most attractive feature, and it sings not only while perched but also while hovering high in the air, after which it makes a spectacular plunging flight back toward earth. Nesting occurs in root tangles, rock crevices, or other cavities. Relatively little is known of its nest building or incubation, but the clutch is usually four eggs, the incubation period is about 11 days, and the fledging period about two weeks. Both sexes help care for the young, and the range of available egg dates suggests that two broods may be raised in a season, but this is not yet established.

Suggested viewing locations: In summer, the loud songs of this species make it easy to find in all the mountain ranges of the entire region. In Wyoming, it breeds statewide in montane forests (Faulkner, 2010). The densest regional breeding populations probably occur in the Black Hills of Wyoming and South Dakota. They are also common throughout the Greater Yellowstone ecosystem and the mountains of southeastern Wyoming.

Suggested reading: BNA 269 (R. V. Bowen, 1997); Lederer, 1977.

Veery. *Catharus fuscescens*. Di; Co; Su; Res.

Status: The veery is a common and widespread summer resident in wooded areas of the state, especially near water. It is variably common in the national parks, ranging in abundance from occasional in Grand Teton to rare in Yellowstone.

Habitats and ecology: In this region, the favored habitats consist of wooded river valleys and canyons that range from deciduous gallery forests along prairie areas of Alberta, through aspen forests of the foothills, and willow-lined mountain streams up to about 8,000 feet at the southern end of the region. Areas with heavy and thickety undergrowth that are difficult for humans to penetrate are this species' favorite habitats, and most of its foraging is done on the ground. In Colorado, the birds favor moist and dense riparian thickets along mountain streams at elevations up to 10,000 feet (Kingery, 1998).

Breeding biology: Like the other North American *Hylocichla* and *Catharus* thrushes, males arrive in breeding areas before females and establish territories that they advertise by singing. When females arrive, they intrude on these territories and are initially chased by the resident males, presumably because there are no plumage differences between the sexes in these species. When females are chased, they tend to remain in the male's territory, flying in circles. Ultimately the male accepts the female's presence and a pair-bond is formed. Nest building requires six to ten days, and a clutch is begun soon thereafter. The clutch is usually four to five eggs. The incubation period lasts about 12 days. Although the male is strongly defensive of the nest, only the female incubates. The fledging period is 11 to 12 days, but the adults feed the young for some time thereafter. Veerys are frequently parasitized by cowbirds, in spite of their well-concealed nests, and they typically accept the cowbird eggs.

Suggested viewing locations: In Wyoming, the veery breeds in all the higher mountain ranges. Some good locations for finding veerys include the lower canyons on the east slope of the Bighorns, the western Black Hills (Crook County), and overgrown aspen groves in the southern Laramie Range of Albany County (Dorn and Dorn, 1990).

Suggested reading: BNA 142 (C. M. Heckscher et al., 2017).

Swainson's Thrush. *Catharus ustulatus*. Di; Co; Su; Res.

Status: The Swainson's thrush is common and widespread during summer throughout the forested montane areas of the state. It is common in both national parks (Grand Teton and Yellowstone).

Habitats and ecology: On migration these birds are likely to be found in almost any fairly dense woodlands, but during the breeding season the birds are found at higher and cooler elevations. There they use shaded canyons where fairly large areas of tangled brushy undergrowth permit ground-level foraging. Riparian thickets, often of willows or alders, and moist mountain slopes that support aspens, are usually used for nesting in this region. In the interior Pacific Northwest, this species breeds in mixed coniferous forests from the shrub-seedling to the old-growth forest stages of succession (Sanderson, Bull, and Edgerton, 1980). The highest population densities found by Hutto and Young (1999) in 566 transects of 11 unaltered forest, grassland, shrub, and wetland habitats of Idaho and Montana that also occur in Wyoming were in mixed conifer and young forest communities.

Breeding biology: This species is the most arboreal of the North American *Catharus* thrushes, and the birds spend much of their time foraging in the foliage and catching flies. Territorial males sing a distinctive song that consists of almost continuous melodic phrases that seem to spiral upward. Nests are built over a period of about four days, and eggs are laid daily. The clutch is usually three or four eggs. The incubation period lasts 10 to 13 days. Apparently only the female incubates, and only she broods the young. They are fed by both parents and leave the nest 10 to 12 days after hatching. Unlike hermit thrushes, males of this species frequently begin to sing while still migrating, and on a territory they may average nearly ten songs a minute, or more than 4,000 songs a day, from about 3:15 a.m. to 7:30 p.m.

Suggested viewing locations: In Wyoming, where it breeds in all the major mountain ranges, these birds might be found in streamside thickets and aspen groves with dense understories, such as occur at Fossil Butte National Monument (Lincoln County) (Dorn and Dorn, 1990). In Colorado, the birds prefer moist mountain valleys, typically breeding in riparian thickets of willow and alder but also nesting in aspen woodlands, mountain shrublands, and upper-level conifer forests (Kingery, 1998).

Suggested reading: BNA 540 (D. E. Mack and W. Yong, 2000).

Hermit Thrush. *Catharus guttatus*. Pan; Co; Su; Res.

Status: The hermit thrush is a common summer resident in wooded areas almost throughout the state. It is common in both national parks (Grand Teton and Yellowstone).

Habitats and ecology: Moist woodlands, especially of coniferous or mixed hardwoods and conifers, are preferred for breeding. Spruces, ponderosa pines, and higher zones of coniferous forests almost all the way to timberline are sometimes

Overleaf: Mountain bluebird, adult

used. Shady and leaf-littered forest floors are favored for foraging, and the altitudinal range of breeding often spans several thousand feet. In Colorado, they have been found to nest from 6,200 to 11,100 feet and perhaps to 12,000 feet. There they prefer coniferous forests to deciduous forests, where they mostly nest in spruce-fir communities—and have been strongly affected by clear-cut logging (Kingery, 1998). In the interior Pacific Northwest, this species breeds in mixed coniferous forests from the mature to the old-growth forest stages of succession (Sanderson, Bull, and Edgerton, 1980). The highest population densities found by Hutto and Young (1999) in 566 transects of 11 unaltered forest, grassland, shrub, and wetland habitats of Idaho and Montana were in spruce-fir communities.

Breeding biology: This species is the first of the thrushes to arrive in northern areas in the spring and the last to depart in fall, a reflection of its adaptation to boreal nesting. It also is perhaps the most famous songster; sometimes its beautiful and complex song can even be heard in wintering areas, although it does not sing during migration. So far as is known, only the female incubates the eggs, but the male regularly feeds her while she is sitting. The clutch is usually of four eggs. The incubation period lasts about 13 days. The male spends a good deal of time guarding the nesting territory, often standing on a sentinel post about 40 feet away from the nest itself. Both parents actively feed the young, which spend an average of 12 days in the nest. In the eastern states, the range of egg dates—which extend over about three months—is suggestive of double-brooding, but in Colorado and other areas of the region there seems to be no evidence of double-brooding.

Suggested viewing locations: In Wyoming, this species may be found during summer in all the major mountain ranges except the Bear Lodge Mountains and Black Hills (Dorn and Dorn, 1990; Faulkner, 2010).

Suggested reading: BNA 261 (R. Dellinger et al., 2012).

American Robin. *Turdus migratorius*. Ub; Co; Res.

Status: The American robin is a common and widespread resident throughout the entire state, especially in open woodland areas. It is abundant in both national parks; this is probably the most abundant of all North American songbirds.

Habitats and ecology: Open woodlands, whether natural or artificial, such as suburbs, city parks, and farmsteads, are typical habitats of the robin. The birds tend to occur almost anywhere there are at least scattered trees, soft ground suitable for probing for insects and worms, and mud that can be gathered for the nest. Nesting on human-made structures seems to be preferred over natural nest sites such as trees, at least in protected areas. In Colorado, robins nest from farmsteads though forested vegetation up to timberline in nearly every habitat type but most often in mountain conifers and aspen woodlands (Kingery, 1998). In the interior Pacific Northwest,

this species breeds in mixed coniferous forests from the shrub-seedling to the old-growth forest stages of succession (Sanderson, Bull, and Edgerton, 1980). The highest population densities found by Hutto and Young (1999) in 566 transects of 11 unaltered forest, grassland, shrub, and wetland habitats of Idaho and Montana were in wetland and cottonwood-aspen communities.

Breeding biology: A very early spring migrant, male robins tend to arrive on the breeding grounds slightly before females, and both sexes tend to return to the area where they were hatched. Males often establish essentially the same territory they held the previous year; the size of the territory seems to vary greatly with habitat and population density. The time that clutches are begun is closely associated with latitude, and both sexes apparently help select the nest site. The nest is sometimes completed in as little as 24 hours, with the male carrying much of the material and the female doing the shaping. However, most nests are built much more slowly, especially early ones, which often require five to six days. The clutch of three to five eggs are laid at daily intervals, and incubation is done almost exclusively by the female. The incubation period lasts 11 to 14 days. The fledging period is usually about 13 days but varies from 9 to 16 days, and the young are cared for until they are about a month old. Even at the northern edge of its range, the robin typically raises two broods, and pairs normally remain intact for the second brood. Sometimes the same nest is used for the second clutch, but often a new one is constructed nearby.

Suggested viewing locations: Robins breed commonly and ubiquitously throughout the entire region. Notably dense regional breeding populations occur in the Greater Yellowstone ecosystem.

Suggested reading: BNA 462 (N. Vanderhoff et al., 2016).

Family Mimidae (Thrashers and Catbirds)

Gray Catbird. *Dumetella carolinensis*. Ub; Un; Su; Res.

Status: The gray catbird is an uncommon summer resident in wooded habitats nearly throughout the entire state below 8,000 feet but rarer in the drier interior basins. It is present in both national parks, ranging in abundance from occasional in Grand Teton to rare in Yellowstone.

Habitats and ecology: Dense thickets, ranging from riverine forests or prairie coulees, city parks and suburbs, orchards, woodland edges, shrubby marsh borders, and similar overgrown areas that provide a combination of dense vegetation and transitional "edge" situations are the ideal habitats of this species. Coniferous forests are avoided, although aspen groves are used, as are other natural vegetational habitats that offer rich sources of insects and berries. In Colorado, catbirds usually nest in dense shrubbery along streamsides at fairly low elevations, but they also sometimes nest in mountain shrublands and willow thickets (Kingery, 1998).

Breeding biology: Although catbirds are distinctly territorial, active defense seems to be largely limited to the vicinity of the nest site, and much of the territorial proclamation is achieved by singing from within the dense vegetation the birds frequent. In one Michigan study, most nests were within two feet of the side or top of shrub cover, in sites providing good visibility for the sitting bird. Males frequently "point out" possible nest sites by sitting on branches with their wings spread and manipulating twigs or other objects as if nest building. However, once a nest is begun the female does most of the actual building, although the male may bring her materials. The first egg is usually laid two days after the nest is completed, and thereafter eggs are laid daily until the clutch has been completed. The clutch size is usually four eggs. The incubation period lasts 12 to 13 days. Incubation is by the female alone, and the male apparently feeds her very little during this time. The young remain in the nest for an average of 11 days, and they are cared for by their parents for approximately two more weeks. In many cases, the pair raises a second brood but rarely, if ever, is a third brood successfully reared in central or northern states.

Suggested viewing locations: In Wyoming, where breeding occurs statewide below 8,000 feet (Faulkner, 2010), good locations for finding catbirds include brushy North Platte River woodlands between Torrington and Casper, Fort Steele (Carbon County), and woodlands along the Bighorn River (Dorn and Dorn, 1990).

Suggested reading: BNA 167 (R. J. Smith et al., 2011).

Brown Thrasher. *Toxostoma rufum*. SD; Un; Su; Res.

Status: The brown thrasher is a local summer resident in the eastern plains along riparian deciduous corridors; it is a vagrant in Grand Teton National Park.

Habitats and ecology: Brown thrashers are associated with open brushy woodlands, scattered clumps of woodland in open environments, shelterbelts, woodlots, and shrubby residential areas. At the western edge of their range in Colorado, breeding birds are most often found in rural plantings of trees and shrubs (Kingery, 1998). In grassland areas, the birds are mostly confined to shrubby coulees or to riparian forests that provide sources of berries, but they forage in open grasslands as well.

Breeding biology: Males of this migratory species usually arrive on their breeding areas a few days ahead of females and apparently establish nesting territories almost immediately, although territorial singing may not begin for ten days or more. Once a territory has been established, the males become very sedentary, and all the nests of the season are built within this territory. Brown thrashers and catbirds have very similar territorial requirements, and at times thrashers will evict catbirds from their territory. The clutch size is usually four to six eggs. The incubation period lasts 11 to 14 days. Incubation is primarily by the female, and both birds also help brood the young, although males seem to be less efficient than females. The

average nestling period is 11 days, but in some cases the female leaves the care of the young to the male soon after hatching and begins a second nest. In other cases, the two parents may each take part of the brood after they fledge, later joining to begin a second nesting effort. Studies of banded birds have indicated that birds sometimes change mates between broods, even when the original mate is still available.

Suggested viewing locations: In Wyoming, breeding generally occurs east of a line from Sheridan south to Cheyenne, and also in the Bighorn Basin south to Lander (Faulkner, 2010). The species is most common in brushy areas along eastern rivers, such as the North Platte River from Torrington to Casper, around Sheridan, and at LAK Reservoir (Weston County) (Dorn and Dorn, 1990).

Suggested reading: BNA 557 (J. F. Cavitt and C. A. Hass, 2014).

Sage Thrasher. *Oreoscoptes montanus*. Ub; Un; Su; Res.

Status: The sage thrasher is a local summer resident in sagebrush habitats almost throughout the state. It is largely limited to lower elevations except during migration. In the national parks, it is an uncommon breeder in Yellowstone but a vagrant in Grand Teton.

Habitats and ecology: This species is closely associated with sage-dominated grasslands and to a much lesser extent other shrublands dominated by vegetation of similar growth forms to sage, such as rabbitbrush and greasewood. In Colorado, more than 80 percent of breeding season observations were in shrubland habitats, with sagebrush accounting for more than half (Kingery, 1998). The birds do most of their foraging on the ground, but nests are placed in shrubs. A greater array of shrublands is used in other seasons.

Breeding biology: Relatively little has been written on the breeding biology of this arid-adapted thrasher. Some early descriptions suggested a territorial song flight, with the bird zigzagging low over the ground, uttering a warbling song, and landing with upraised and fluttering wings. Apparently both sexes incubate, and incubation probably begins the day before the last egg is laid. The clutch size is usually four to five eggs. The incubation period lasts 15 days, and the nestling period is 11 to 13 days. When bringing food to the young, the adults are highly secretive, landing on a sagebrush about ten feet away, and then approaching the nest while hidden from view. Pairs often remain mated during successive years, and the birds are sometimes rather long-lived, with one banded individual known to have reached 13 years.

Suggested viewing locations: This species is common in sage-dominated places such as Hutton Lake National Wildlife Refuge (Albany County), below Fontenelle Dam in Lincoln County, the north side of Pathfinder Reservoir (Natrona County), and in valley bottoms south of Fossil Butte in Lincoln County (Dorn and Dorn, 1990).

Suggested reading: BNA 463 (T. D. Reynolds, T. D. Rich, and D. A. Stephens, 1999); Buseck, Keinath, and McGee, 2004.

Family Bombycillidae (Waxwings)

Bohemian Waxwing. *Bombycilla garrulus*. Pan; Un; Win; Mig.

Status: The Bohemian waxwing is an uncommon winter migrant over most of the state, mainly east of the Wind River range and in the northern half of the state. It is occasional at Grand Teton National Park and uncommon at Yellowstone.

Habitats and ecology: During the breeding season this species is associated with coniferous and mixed forests, often nesting as loosely associated groups in conifer groves. Outside the breeding season the birds move about opportunistically, seeking out sources of berries and small fruits in trees and hedges, such as mountain ash, crab apples, pyracantha, and other fruiting trees.

Suggested viewing locations: Bohemian waxwings are most likely to be seen among flocks of cedar waxwings, often during migration or winter when the birds gather at fruit- or berry-bearing trees.

Suggested reading: BNA 714 (M. C. Witmer, 2002).

Cedar Waxwing. *Bombycilla cedrorum*. Pan; Un; Res.

Status: The cedar waxwing is an uncommon resident that is widespread over the wooded areas of the state up to about 8,000 feet. It is present in both national parks, varying in abundance from occasional in Grand Teton to uncommon in Yellowstone.

Habitats and ecology: Somewhat open woodlands, primarily of broad-leaved species, are used for nesting, including riparian forests, farmsteads, parks, cedar groves, shelterbelts, and brushy edges of forests. Areas that have abundant growths of berry-bearing bushes are especially favored, although insects, buds, and other food sources are also consumed. In Colorado, cedar waxwings nest rather sparingly in deciduous riparian forests below 7,500 feet (Kingery, 1998). The highest population densities found by Hutto and Young (1999) in 566 transects of 11 unaltered forest, grassland, shrub, and wetland habitats of Idaho and Montana were in riparian shrub, wetland, and cottonwood-aspen communities.

Breeding biology: Not much is known of the courtship behavior of this rather common and highly gregarious species, which remains in flocks for much of the year. Adult birds often may be seen passing berries back and forth, but whether this is courtship feeding is questionable. Mutual breast-preening and bill-clicking are probable courtship activities. During the period of nest building, the female does perform begging behavior and is fed by her mate. Territoriality is virtually absent in cedar waxwings. The nests are frequently situated colonially, and breeding seems to correspond with the period when berries and fruit ripen. Both sexes build the nest, which requires two to six days. The clutch size is usually three to five eggs. The incubation period lasts 12 to 14 days. The female does the incubating, but her mate frequently feeds her, and she also broods the young for several days after hatching. The young birds leave the nest when about 16 days old and may remain in the nest vicinity for about a month.

Suggested viewing locations: In Wyoming, cedar waxwings are throughout the state at elevations up to about 8,000 feet, especially in riparian deciduous forests. They can often be found during summer in the northwestern mountains or at places such as Battle Creek Campground of the Sierra Madre Mountains (Carbon County) (Dorn and Dorn, 1990).

Suggested reading: BNA 309 (M. C. Witmer, D. J. Mountjoy, and L. Elliot, 2014).

Family Motacillidae (Pipits)

American Pipit. *Anthus rubescens*. Lo; Un; Su; Res.

Status: The American pipit is an uncommon summer resident in alpine areas of northwestern Wyoming as well as in the Bighorns, Medicine Bow, and Sierra Madre ranges. It is present and common in both national parks. In Colorado, these birds only rarely breed in subalpine areas or montane grasslands, but they nest on nearly every patch of alpine tundra in the state, including both wet and dry tundra (Kingery, 1998).

Habitats and ecology: During the breeding season this species is found on alpine tundra and high meadows, while during other seasons it occurs on similarly very open terrain, usually with only sparse vegetation and often a moist substrate. Migrants and wintering birds commonly use shorelines, flooded fields, river edges, and similar habitats.

Breeding biology: Like its grassland relative (Sprague's pipit), the American pipit performs a spectacular aerial display while on its breeding territory. After flying to a height of up to 250 feet, the male begins a slow descent while singing loudly, with legs dangling and tail cocked, exposing the white outer tail feathers. The song is a rapid, repetitive series of clear notes, *chee-chee-chee*, or *cheedal-chedal-cheedal*. The birds are monogamous with large territories. The pair constructs a grass- or hair-lined ground nest in a recess or shallow depression, or on a flat ground surface, where it is wholly concealed from above by vegetation. The clutch size is four to six eggs, which are incubated for 14 days by the female. The hatchlings are covered with fairly long and thick down, an adaptation to alpine or Arctic climates, and are tended by both parents. They leave the nest at 14 to 15 days but cannot fly well until a few days later.

Suggested viewing locations: In Wyoming, this species is widespread during summer in the alpine zone of the Greater Yellowstone ecosystem, and it also occurs in the alpine zones of the Beartooth and Medicine Bow–Sierra Madre ranges (Faulkner, 2010). Roads that travel across alpine tundra offer easy summer viewing, such as along the Beartooth Plateau in Park County, Libby Flats and Brooklyn Ridge in the Medicine Bow range, and near Medicine Wheel in Big Horn County (Dorn and Dorn, 1990).

Suggested reading: BNA 95 (P. Hendricks and N. A. Verbeek, 2012).

Cedar waxwing, adult

Family Fringillidae (Finches)

Pine Grosbeak. *Pinicola enucleator*. SD; Un; Res.

Status: The pine grosbeak is an uncommon resident in coniferous forests of northwestern Wyoming and the Bighorn, Medicine Bow, and Sierra Madres ranges. It is of uncertain status in the Black Hills and the Laramie Mountains. The species is present in both national parks, varying in abundance from common in Grand Teton to uncommon in Yellowstone.

Habitats and ecology: Breeding occurs in the subalpine levels of the coniferous forest, primarily the alpine fir–Engelmann spruce zone. Nesting usually occurs in such conifers, especially in open or scattered woods near meadows or streams. In Colorado, spruce-fir forests are easily the most common breeding habitat (Kingery, 1998). In the interior Pacific Northwest, this species breeds in mixed coniferous forests in the mature and old-growth forest stages of succession (Sanderson, Bull, and Edgerton, 1980). The highest population densities found by Hutto and Young (1999) in 566 transects of 11 unaltered forest, grassland, shrub, and wetland habitats of Idaho and Montana were in spruce-fir communities. Outside the breeding season the birds descend to lower conifer zones, especially the juniper zone. They eat primarily conifer seeds but also berries, grains, and other food sources.

Breeding biology: Like other finches, pine grosbeaks rely on seeds rather than insects to feed their young, and adults develop gular pouches during the breeding season to carry food to their nestlings. Males also feed females during courtship. Their territorial song is a clear musical whistle in a series of continuous but rather weak warbles lasting up to about two seconds. In one study, the nesting territory was judged to have a diameter about 1,200 feet, and the resident birds were observed to drive off gray jays and a red squirrel. The rather bulky nest is built in a bush or tree from 2 to 25 feet above ground, apparently by the female alone. The clutch varies from two to five eggs, and the incubation of 11 to 15 days is also undertaken by the female, who is fed regularly by the male. Both parents feed the nestlings and continue to do so for a time after they leave the nest at 13 to 20 days of age. Variations in the annual production of pine or other conifer seeds may bring about periodic irruptions of birds moving out of the mountains and onto the plains during winter.

Suggested viewing locations: Breeding in Wyoming occurs at high elevations in the Greater Yellowstone ecosystem, the Bighorn Mountains, and the Medicine Bow–Sierra Madre ranges. Breeding in the Bear Lodge Mountains and Laramie range is less certain (Faulkner, 2010). During summer, Wyoming sites that might have pine grosbeaks include the west side of Blackhall Mountain in the Sierra Madre Mountains (Carbon County), Brooklyn Lake in the Medicine Bow Mountains (Albany County), and the Brooks Lake area (Fremont County) (Dorn and Dorn, 1990). They can also be found at subalpine elevations in the Tetons, Bighorns, and other ranges, and in Yellowstone National Park (Scott, 1993).

Suggested reading: BNA 456 (C. S. Adkisson, 1999).

Gray-crowned Rosy-Finch. *Leucosticte tephrocotis*. Pan; Co; Win; Mig.

Black Rosy-Finch. *Leucosticte atrata*. Lo; Un; Res.

Brown-capped Rosy-Finch. *Leucosticte australis*. HL; Ra, CC4; Res.

Status: All three rosy-finch species (previously regarded as subspecies) are limited during breeding to alpine areas of high mountains. In the mountains of northwestern Wyoming, the nape color of breeding birds is gray, but the back and breast are dusky brown (black rosy-finch). However, in southeastern Wyoming (Medicine Bow range), the crown and nape are dark brownish, and the breast and back are grayish brown, becoming reddish on the belly (brown-capped rosy-finch). In winter most Wyoming birds are likely have gray crowns and reddish brown back and breast colors (gray-crowned rosy-finch). The brown-capped rosy-finch is classified as a species of Greatest Conservation Need by the Wyoming Game and Fish Department.

Habitats and ecology: During the breeding season, these birds inhabit cirques, talus slopes, alpine meadows with nearby cliffs, and adjacent snow and glacial surfaces, where foraging for frozen insects is common. Nesting is done in cliff crevices or among talus rocks. During fall and winter, the birds move to lower elevations and to habitats that include mountain meadows as well as to lower-altitude grasslands, sagebrush, and agricultural lands.

Breeding biology: Because of their high-alpine breeding, little information on the breeding biology of rosy-finches in Wyoming is available. Proven breeding for the black rosy-finch is limited in Wyoming to the Greater Yellowstone ecosystem, and proof of Wyoming nesting by the brown-capped rosy-finch did not occur until 2004, on Medicine Bow Peak (Faulkner, 2010). All three rosy-finches place their nests in well-hidden rock crevices, caves, or similar locations. The nest is made of moss and grass, with hair, feathers, and fine grass as lining. The clutch is four to five eggs, and the 11- or 12-day period of incubation is performed by the female. The female also broods the nestlings, which are fed by both parents over the 16- to 22-day nestling period.

Suggested viewing locations: In Wyoming, the black rosy-finch breeds in the alpine zone of the Greater Yellowstone ecosystem. It has also been reported during summer, but not proven to breed, in the Snowy Range, where the brown-capped rosy-finch is known to be the local breeder (Faulkner, 2010). Black rosy-finches nest in the alpine zone of the Absaroka, Teton, and Wind River ranges and the Beartooth Plateau. From Jackson they can be easily seen in summer by taking the tramway at Teton Village to the top of Rendezvous

Pine grosbeak, adult male

Mountain. They can also be seen along the foothills of western Wyoming during winter (Scott, 1993). McEneaney (1988) judged the Yellowstone population to be in the hundreds during the late 1980s, with the best observation area during summer months for the black rosy-finch being around Mount Washburn. These rosy-finches can be seen in Wyoming during summer at Libby Flats, Medicine Bow peak, and along Brooklyn Ridge in the Medicine Bow Mountains of Albany County (Dorn and Dorn, 1990). Between 2015 and 2017, there were at last nine August eBird sightings of brown-capped rosy-finches in the Medicine Bow range. During 2017 they were also reported several times during August on Medicine Bow peak in groups of up to five birds, and other recent eBird reports have been from Mirror Lake and the Snowy range pass. Gray-crowned rosy-finches are widespread during winter months at elevations that vary with the weather, moving lower as the intensity of winter conditions increase.

Suggested reading: Gray-crowned rosy finch: BNA 559 (S. A. Macdougall-Shackleton, R. E. Johnson, and T. P. Hahn, 2000. *Black rosy-finch:* BNA 678 (R. E. Johnson, 2002); French, 1959. *Brown-capped rosy-finch:* BNA 536 (R. E. Johnson et al., 2000).

House Finch. *Haemorhous mexicanus*. Ub; Ab; Res.

Status: The house finch is an abundant resident at lower elevations across Wyoming, especially on plains and foothills up to 8,000 feet. It has been present in Grand Teton National Park since the 1990s, but its occurrence in Yellowstone is apparently marginal.

Habitats and ecology: Now generally associated with human habitations over most of its range, the house finch nests on buildings in such areas. Otherwise, it nests in open woods, riverbottom woodlands, scrubby desert or semidesert vegetation such as sagebrush, and tree plantings. In the interior Pacific Northwest, this species breeds in mixed coniferous forests from the shrub-seedling to the mature forest stages of succession (Sanderson, Bull, and Edgerton, 1980). Deciduous underbrush, preferably close to water, is favored over dense coniferous woods, and sources of seeds, berries, or fruits are also needed throughout the year.

Breeding biology: This species has a courtship display similar to that of the purple finch, with the male approaching the female with his tail spread and cocked and his wings lowered, uttering chirps and trills. Courtship feeding of the female also occurs during pair formation and frequently during incubation. Both sexes help in nest building, which requires from 2 to 11 days, with males helping mainly in the early stages. The clutch size is usually four to five eggs. The female incubates and broods alone, but both sexes bring in food. The incubation period lasts 12 to 14 days. The young remained in the nests for an average of 15 days in a California study and nearly 18 days in a Hawaiian study. In Colorado, it has been reported that females frequently begin to gather nesting materials for their second brood while being followed by begging young that are still partially covered with down.

Suggested viewing locations: House finches occur statewide at lower elevations and probably can be found in any village or town, especially around bird feeders.

Suggested reading: BNA 46 (A. V. Badyaev, V. Belloni, and G. E. Hill, 2012).

Red Crossbill. *Loxia curvirostra*. Wi; Co, Ch; Res.

Status: The red crossbill is a common resident in coniferous forest areas throughout the state, including all the mountain ranges. It is present in both of the national parks, varying in abundance from occasional in Grand Teton to uncommon in Yellowstone.

Habitats and ecology: Breeding is associated with coniferous forest habitats, especially those of pines including ponderosa, lodgepole, and pinyon, but nesting in the region has also been observed in Engelmann spruces and subalpine firs at elevations from 4,000 to 10,000 feet or more. Breeding in the Rocky Mountain region is associated with the higher levels of coniferous forests, but nonbreeding birds often frequent the pinyon zone. In Wyoming, five call types (out of the nine known in North America) have been identified; some of these different call-type populations may act as biologically distinct species (Smith and Benkman, 2007; Faulkner, 2010). In Colorado, crossbills with different bill shapes and call types specialize on eating different conifer seeds (Douglas-fir, ponderosa pine, and lodgepole pine) during late winter and spring, but during nesting spruce-fir forests had the highest frequency of habitat use during Breeding Bird Atlas surveys (Kingery, 1998). In the interior Pacific Northwest, crossbills breed in mixed coniferous forests in mature and old-growth forest stages of succession (Sanderson, Bull, and Edgerton, 1980). Although usually found in conifers, crossbills also feed on ripe box elder seeds in late summer (Scott, 1993).

Breeding biology: Observations of this species west of Denver indicate colonial nesting; about 24 pairs were found within a square mile of forest, but few were found elsewhere. The clutch is usually three to four eggs. Hatched young were found there as early as January 16, along with nests in progress. The females did the building, which required about five days, and another four days elapsed before the first egg was laid. Four more days were spent in egg laying, 14 more in incubation, and 20 days were needed for fledging. Fledging times of 16 to 25 days have been reported; these variations probably depend on local food supplies. Some pairs raise two broods in rapid succession. The nesting cycles of these birds are highly irregular; not only may two widely spaced breedings occur in a single year but young birds may breed in the same year they are hatched.

Suggested viewing locations: In Wyoming, breeding occurs in all the major mountain ranges (Faulkner, 2010), including the Black Hills.

Suggested reading: BNA 256 (C. W. Benkman and M. A. Young,

2019); Bailey, Niedrach, and Bailey, 1953; Nethersole-Thompson, 1975; Benkman, 1993; Benkman, Parchman, and Mezquida, 2010.

Pine Siskin. *Spinus pinus*. Wi; Co; Res.

Status: The pine siskin is a common resident in coniferous forests throughout the state. It is present and common in both national parks (Grand Teton and Yellowstone).

Habitats and ecology: Breeding occurs in coniferous or mixed forests and rarely in deciduous woodlands. Nesting preferentially occurs in conifers of almost any type but has also been observed in cottonwoods, lilacs, and willows in the Rocky Mountain region. In Colorado, breeding siskins use spruce-fir forests more often than pines; ponderosa, lodgepole, and pinyon pines made up only about 20 percent of observed breeding habitats, although conifers collectively accounted for 70 percent of the total (Kingery, 1998). In the interior Pacific Northwest, this species breeds in mixed coniferous forests from the pole-sapling to the old-growth forest stages of succession (Sanderson, Bull, and Edgerton, 1980). Their foods are mainly conifer seeds but also may include those of alders, birches, or various weeds, and they seasonally feed on flower buds and insects.

Breeding biology: During the nonbreeding season, siskins are highly gregarious; in late winter courtship begins in large flocks, evidenced by singing and chasing. Courtship feeding of the female is frequent, as are song flights by the male around a particular female. Frequently, nesting occurs in rather loose colonies, with the birds alternating between nest building and social flocking. The female chooses the nest site and carries in the necessary materials; the male accompanies her and performs courtship feeding during this period as well as during incubation. The clutch size is usually three to four eggs, and the incubation period lasts 13 days. Gregarious tendencies persist through the incubation period, and thus there is little territorial exclusion. Only the female incubates, but both sexes feed the young, which leave the nest in 14 to 15 days.

Suggested viewing locations: In Wyoming, this is the most common breeding finch, nesting in coniferous forests of all the major mountain ranges during summer (Faulkner, 2010), and siskins are common at bird feeders in towns and cities during winter (Dorn and Dorn, 1990).

Suggested reading: BNA 280 (W. R. Dawson, 2014).

American Goldfinch. *Spinus tristis*. Ub; Co; Su; Res.

Status: The American goldfinch is a common summer resident almost throughout the state. It is present in both of the national parks, varying in abundance from common in Grand Teton National Park to rare in Yellowstone.

Habitats and ecology: Breeding occurs in open grazing country, especially where thistles are abundant, or where cattails are to be found. In Colorado, surveys produced a total of 26 different habitat codes used by breeding goldfinches, about two-thirds of which were from riparian forests and rural woodlots or shelterbelts (Kingery, 1998). The seeds of thistles and other composites are used for feeding the young, and the "down" of thistles or cattails is used in nest construction. Riparian woodlands near weed-infested fields provide an ideal nesting situation. During winter the birds range widely over weedy fields and farmlands, and often visit urban bird feeders.

Breeding biology: These gregarious birds remain in flocks well into late spring, and pair formation begins among flocked birds by May, or possibly earlier. It is achieved by courtship flights by a female and varying numbers of males, male-to-female courtship singing, a hovering song flight by the male, and an extended male song resembling that of a canary. Pair-bonds are maintained by courtship feeding of the female, which occurs from egg-laying through the nestling period. The clutch is usually five eggs. Nesting is delayed until there is an abundant supply of small composite seeds, such as thistle, to feed the young. Nest building and incubation are performed by the female alone, but both sexes feed the young by seed regurgitation. The incubation period lasts 12 to 14 days. The nestlings fledge in 10 to 16 days, at which time the male takes over most of the feeding. This frees the female to begin a new nest, which sometimes happens within three days after the brood's fledging.

Suggested viewing locations: In Wyoming, the goldfinch is a common breeder throughout the state up to 8,000 feet (Faulkner, 2010). It is most common eastwardly and least common in the northwest. Some of the better locations to find these birds in summer include LAK Reservoir (Weston County), the North Platte River near Guernsey, the Green River, Tongue River Canyon (Sheridan County), and Yellowtail Reservoir (Big Horn County) (Dorn and Dorn, 1990).

Suggested reading: BNA 80 (K. J. McGraw and A. L. Middleton, 2017).

Evening Grosbeak. *Coccothraustes vespertinus*. SD; Un; Res.

Status: The evening grosbeak is an uncommon resident in coniferous forests almost throughout the state, including most of northwestern Wyoming; the Bighorns; Black Hills; and the Laramie, Medicine Bow, and Sierra Madre ranges. It is present in both national parks, varying in abundance from common in Grand Teton to uncommon in Yellowstone.

Habitats and ecology: During the breeding season this species is primarily associated with mature coniferous forests, although nesting has been observed in riparian willow thickets and even in city parks and orchards. In Colorado, surveys indicated that coniferous forests, especially ponderosa pine forests, are the favorite breeding habitat, where the long needles of this pine help camouflage nests (Kingery, 1998). In the interior Pacific Northwest, this species breeds in mixed coniferous forests in

Overleaf: Red crossbill, adult male

mature and old-growth forest stages of succession (Sanderson, Bull, and Edgerton, 1980). Nesting in elms, maples, and box elders has also been reported. During the breeding season evening grosbeaks feed on insects, including spruce budworm larvae, which often reach high populations during outbreaks. During fall and winter, the birds often occur in flocks that feed on such large and nutritious seeds as maples, ashes, and sunflowers. They might appear anywhere in the region during winter, when they regularly visit bird feeders.

Breeding biology: Surprisingly little is known of the breeding biology of this handsome species. In winter the birds are gregarious and strongly attracted to box elder trees, where they feed on the hanging seeds, as well as to various other species of maples. Courtship displays seem to consist of the male's crouching, then spreading and quivering his wings while fluffing his plumage. Females solicit courtship feeding by bobbing their heads and swaying their bodies in front of males while fluttering their wings; this or a similar display with tail-raising precedes copulation. Males apparently accompany their mates while the females gather nesting materials, but presumably the female does all the nest building. The clutch is usually three or four eggs, and incubation lasts 12 to 14 days. The female incubates, but the male feeds her both off and on the nest. The fledging period is 13 to 14 days; in one case the first egg of a second clutch appeared when the single nestling of the first cycle was only 11 days old and still in the nest.

Suggested viewing locations: Evening grosbeaks breed in Wyoming's northwestern mountain ranges, including Wyoming's national parks. Nesting records are few, but most summer reports are from coniferous and conifer-aspen forests at middle elevations of Wyoming's major mountain ranges, where breeding is presumed to occur (Faulkner, 2010).

Suggested reading: BNA 599 (S. W. Gillihan and B. E. Byers, 2001).

Family Calcariidae (Longspurs and Snow Buntings)

Chestnut-collared Longspur. *Calcarius ornatus*. Wi; Un; Mig.

Status: The chestnut-collared longspur is an uncommon summer resident in the counties bordering the eastern edge of the state and a local migrant somewhat farther west of the breeding range.

Habitats and ecology: Primary breeding habitats consist of grazed or hayed mixed-grass prairies, short-grass plains, the meadow zones of salt grass around alkaline ponds or lakes, mowed hayfields, heavily grazed pastures, and the like. In Colorado, this longspur breeds in taller and damper grasslands than does the McGown's longspur, on sites having less bare ground and more singing posts provided by tall forbs (Kingery, 1998). Outside the breeding season the birds are often found in weedy cultivated fields, especially those that have such seed-rich species as amaranth.

Breeding biology: Males establish territories shortly after their spring arrival; they prefer grassy plains that have sparse vegetation and at least one large rock or fencepost to serve as a singing post. Such singing points are often a central part of the territory; the nest is usually within 25 feet, and the total territory is about 100 feet in diameter. In some marginal areas, however, territories of up to ten acres have been estimated. Although flight songs are used in territorial advertisement, they are not as frequent or as formalized as in the McCown's longspur. Both species gradually gain altitude with rapid wingbeats, but the McCown's longspur sails downward quickly with wings upstretched, whereas the chestnut-collared longspur circles and undulates while singing before gradually descending. The female builds the nest alone and also does the incubation. The clutch size is usually three to five eggs. The incubation period lasts 11 to 13 days. Both sexes feed the young, which leave the nest in 9 to 11 days or rarely as late as the fourteenth day. By the fourteenth day, the young can fly very well, and by about 26 days they are independent of their parents.

Suggested viewing locations: In Wyoming, breeding occurs almost entirely along the easternmost tier of counties but extends west to Albany County at the southern end of the state (Faulkner, 2010). In common with McCown's longspurs, migrating chestnut-collared longspurs are most likely to to be supported by the grasslands of the eastern plains. During breeding season they may be found on grasslands south of Van Tassel, but these birds favor mixed-grass prairies rather than the short-stature grasslands that are used by McCown's longspurs.

Suggested reading: BNA 288 (B. Bleho et al., 2015).

McCown's Longspur. *Rhynchophanes mccownii*. Di; Co, CC1, CC2; Su; Res.

Status: McCown's longspur is a common summer resident on the shortgrass plains in the eastern half of Wyoming. It is rare in Yellowstone National Park and a vagrant in Grand Teton.

Habitats and ecology: During the breeding season this species is mostly limited to short-grass prairies and grazed mixed-grass prairies, but it also breeds to some degree on stubble fields or newly sprouting grain fields. In Colorado, the birds breed almost exclusively in very sparse grasses, with a large amount of exposed bare soil and a low diversity of other plants, including prickly pear cactus, lupine, and locoweed (Kingery, 1998). While on migration and during the winter period, the birds occur on open grasslands, low sage prairies, mountain meadows, and similar open habitats.

Breeding biology: Male longspurs arrive on their breeding grounds of eastern Wyoming in late April and soon begin to select territories. These are marked by flight songs as well as by singing from shrubs or rocks. As competition increases, territories gradually decrease in size to an area about 250 feet in diameter. The courtship display is remarkable; the male moves around the female in a narrow circle, holding the nearer wing erect and thus exposing the white lining. The female gathers the nesting material and makes any nest excavation that may be necessary. The clutch size is usually three or four eggs, and the incubation period lasts 12 to 13 days. The female also performs all the incubation and does most of the brooding, although the male occasionally relieves her during the later brooding stages. The young leave the nest at ten days, and two days later are able to fly for short distances.

Suggested viewing locations: In Wyoming, breeding mostly occurs east of a line from Sheridan to the western edge of the Laramie plains in Albany County (Faulkner, 2010). Good numbers of McCown's longspurs may be seen in places such as the native grasslands south of Van Tassel, the high grasslands around Cheyenne, and in and around Hutton Lake National Wildlife Refuge (Albany County) (Dorn and Dorn, 1990; Scott, 1993). The densest regional breeding populations probably occur in southeastern Wyoming.

Suggested reading: BNA 96 (K. A. With, 2010); Greer and Anderson, 1989.

Family Passerellidae (Sparrows, Towhees, and Juncos)

Green-tailed Towhee. *Pipilo chlorurus*. Ub; Co; Su; Res.

Status: The green-tailed towhee is a common summer resident in western and central Wyoming, and east to the Bighorn, Laramie, and Medicine Bow ranges. It is present in both national parks, varying in abundance from common in Grand Teton to uncommon in Yellowstone.

Habitats and ecology: During the breeding season, this species occurs in brushy foothill areas dominated by sagebrush, scrub oaks, saltbush and greasewood flats, scrubby riparian woodlands, and similar open and semiarid habitats. Forested areas are avoided, but scattered trees in brushlands are used as singing posts. Spreading shrubs that allow for easy movement and foraging on the ground surface below are favored vegetation types. In Colorado, the birds breed at an average altitude of 7,300 feet, most often in montane shrublands such as snowberry, serviceberry, chokecherry, mountain mahogany, scrub oaks, and sagebrush (Kingery, 1998). The highest population densities found by Hutto and Young (1999) in 566 transects of 11 unaltered forest, grassland, shrub, and wetland habitats of Idaho and Montana were in sagebrush communities. Like the other towhees and the fox sparrow, this species is a ground forager; it is the most arid-adapted of the North American towhees. Like the spotted towhee it often utters catlike mewing call notes. In spite of its desert adaptations, however, it is unable to drink salt water.

Breeding biology: In spite of this bird's widespread occurrence throughout the arid parts of the western United States, very little is known of its breeding biology. Probably it is much like the brown thrasher in its basic biology, but the birds nest primarily in sagebrush, often placing their nests two to three feet above ground in the fork of a sage bush. The nest is probably built by both sexes. The clutch size ranges from three to five eggs, usually four, and rarely as many as seven. Both sexes incubate, and the incubation period is 15 days. Fledging occurs at 11 to 12 days. Probably the birds are double-brooded or are at least prone to renest.

Suggested viewing locations: In Wyoming, this towhee breeds mostly west of a line from Sheridan to Laramie, especially in mountain-foothills shrublands and juniper-woodland shrub-steppe under 8,000 feet (Faulkner, 2010). Grand Teton National Park is a prime location for seeing green-tailed towhees. They are uncommon in Yellowstone National Park, where there is less sagebrush. They are common in nearly all sage-dominated areas of Wyoming but are absent from the Black Hills.

Suggested reading: BNA 368 (R. C. Dobbs, P. R. Martin and T. E. Martin, 2012).

Spotted Towhee. *Pipilo maculatus*. Di; Co; Su; Res.

Status: The spotted towhee is a common summer resident over the eastern three-fourths of the state at lower elevations. It is present but rare in both national parks (Grand Teton and Yellowstone).

Habitats and ecology: Breeding occurs in brushy fields, thickets, woodland openings or edges, second-growth forests, city parks, and well-planted suburbs. Habitats that have a good accumulation of litter and humus as well as a protective screen of shrubby foliage above the ground are highly favored by these birds. The highest population densities found by Hutto and Young (1999) in 566 transects of 11 unaltered forest, grassland, shrub, and wetland habitats of Idaho and Montana were in riparian shrub communities.

Breeding biology: Pair formation in towhees is achieved by males singing persistently from a variety of locations in their territory. As pair-bonds form the rate of singing drops off, and the two birds forage within the male's territory. The female builds the nest with little or no help from the male, although he sometimes carries about small twigs. The usual clutch size is three to four eggs. The female incubates during the 12- to 14-day incubation period, but both sexes feed the young. The young leave the nest in 9 to 11 days. Double-brooding sometimes occurs.

Suggested viewing locations: Wyoming is at the western edge of the hybrid zone with eastern towhees, and spotted towhees breed widely over the eastern three-fourths of the state (Faulkner, 2010). Brushy areas almost statewide are good birding sites for this species, such as around LAK Reservoir (Weston County), along Birdseye Creek (Fremont County), and the North Platte River bottomlands from Torrington to Casper (Dorn and Dorn, 1990).

Suggested reading: BNA 263 (S. B. Smith and J. S. Greenlaw, 2015); Sibley and West, 1959.

American Tree Sparrow. *Spizelloides arborea*. Co; Win; Mig.

Status: The American tree sparrow is a common wintering migrant throughout the state, including both montane areas and plains.

Habitats and ecology: While in Wyoming, this species occupies brushy prairie areas, roadside thickets, farmsteads, old orchards, overgrown and weedy pastures, and similar relatively open habitats. The birds often occur in company with juncos and other gregarious and hardy sparrows, and feed about on the ground or snow surface, industriously searching out small seeds. During the breeding season they are associated with Arctic timberline habitats.

Suggested viewing locations: During winter this species is likely to appear in open to somewhat brushy areas at lower elevations throughout the entire region.

Suggested reading: BNA 37 (C. T. Naugler, P. Pyle, and M. A. Patten, 2017).

Chipping Sparrow. *Spizella passerina*. Ub; Co; Su; Res.

Status: The chipping sparrow is a common summer resident throughout the state in all wooded areas. It is present and common in both of the national parks (Grand Teton and Yellowstone).

Habitats and ecology: Breeding in this species is done in open deciduous or mixed forests, the margins of forest clearings, the edges of muskegs, in timberline scrub, riparian woodlands, pinyon-juniper or oak–mountain mahogany woodlands, and similar diverse habitats. Generally scattered trees, an unshaded forest floor, and a sparse ground covering of herbaceous plants seem to be the kinds of habitat considerations that are important. In Colorado, the birds nest in diverse habitats but most often in coniferous woodlands, especially those dominated by ponderosa pine or pinyon-juniper (Kingery, 1998). In the interior Pacific Northwest, this species breeds in mixed coniferous forests from the shrub-seedling to the old-growth forest stages of succession (Sanderson, Bull, and Edgerton, 1980). The highest population densities found by Hutto and Young (1999) in 566 transects of 11 unaltered forest, grassland, shrub, and wetland habitats of Idaho and Montana were in ponderosa pine communities.

Breeding biology: Territorial establishment begins almost immediately after the males return to their breeding grounds in spring, and the males spend a good deal of time each day in singing and chasing intruders. Territories average about an acre in area but sometimes are as small as half an acre. The female gathers all the nesting material and constructs the nest. Usually she also does all the incubating, but an exceptional instance of male incubation has been reported. The usual clutch size is four eggs and the incubation period lasts 11 to 14 days. The female broods the young, but both sexes feed them, and they fledge in about ten days, with an observed range of 8 to 12 days. By the time they are 14 days old, the young are able to fly several feet.

Suggested viewing locations: This species can be found throughout the region in most wooded habitats, at low to moderate elevations. In Wyoming, birds breed statewide, mostly in open coniferous forests at middle elevations but with some use of junipers, aspens, and other wooded habitats (Faulkner, 2010). The densest regional breeding populations probably occur in the Black Hills and the Greater Yellowstone ecosystem.

Suggested reading: BNA 334 (A. L. Middleton, 1998).

Brewer's Sparrow. *Spizella breweri*. Ub; Co, CC1, CC2; Su; Res.

Status: The Brewer's sparrow is a common summer resident over nearly the entire state. It is usually found in semidesert scrub habitats, but it also breeds at alpine timberline in Colorado, Montana, and possibly also Wyoming. It is common in both national parks (Grand Teton and Yellowstone).

Habitats and ecology: In the Rocky Mountain region, this species breeds in two very different habitats. The first is in short-grass prairies with sage or other semiarid shrubs present in varying densities. In Wyoming, it breeds in sage shrubsteppe statewide but is not yet known to have an alpine breeding population (Faulkner, 2010). In Colorado, sage shrubland accounts for most breeding habitats, but mountain mahogany or currants growing on brushy hillsides or mesa edges are sometimes used (Kingery, 1998). In Idaho, the birds have been found breeding on both sagebrush flats as well as in serviceberry-covered slopes of mountain ridges. The highest population densities found by Hutto and Young (1999) in 566 transects of 11 unaltered forest, grassland, shrub, and wetland habitats of Idaho and Montana were in sagebrush communities. In southern Canada, the birds also breed on short-grass plains with scattered sage and cacti, and as well along timberline in Banff and Jasper national parks in stunted spruces, firs, willows, and alders. This alpine-adapted population was once proposed as constituting a new species, the timberline sparrow (*S. taverneri*), but it is now considered to be a subspecies.

Breeding biology: This species occurs in two widely different climatic zones: the arid sage-dominated western states and the Arctic timberline of northern Canada. In both it is dependent upon open shrub-dominated habitat. There is little information on its breeding behavior, although in eastern Washington as many as 47 pairs may occur in 100 acres of favorable habitat. In a Montana study, spray-killing all the sagebrush on a study area reduced the Brewer's sparrow population by about half. The sparrows will nest in dead sagebrush, but it provides considerably less concealment than do live plants. The usual clutch is three or four eggs, incubation lasts 11 to 13 days, and fledging occurs at eight to nine days. Double-brooding has been reported.

Suggested viewing locations: Sagebrush is an ideal habitat for Brewer's sparrows, such as occurs in abundance around Rock Springs, along the Green River northwest of Daniel, the

Upper Birdseye Pass Road at Boysen Reservoir, and along the Laramie range east of Pole Mountain in Albany County (Dorn and Dorn, 1990).

Suggested reading: BNA 390 (J. T. Rotenberry, M. A. Patten, and K. L. Preston, 1999).

Vesper Sparrow. *Pooecetes gramineus*. Ub; Co; Su; Res.

Status: The vesper sparrow is a common summer resident throughout the state in grassland areas. It is present and common in both national parks (Grand Teton and Yellowstone).

Habitats and ecology: During the breeding season this species is found in overgrown fields, prairie edges, grasslands with scattered shrubs and small trees, sagebrush areas where the plants are scattered and stunted, and similar open habitats but not mountain meadows or tundra zones. In Colorado, nesting occurs widely but most commonly in middle- to higher-elevation sagebrush, extending to lower sagebrush stands locally, especially where the sage is interspersed with a good grass cover (Kingery, 1998). The highest population densities found by Hutto and Young (1999) in 566 transects of 11 unaltered forest, grassland, shrub, and wetland habitats of Idaho and Montana were in grassland and sagebrush communities.

Breeding biology: Vesper sparrows occupy a considerably larger home range than do many prairie-adapted sparrows and frequently defend territories of about two acres. Most singing is done from fairly high perches, and, rarely, song flights are also performed. The usual clutch is three to five eggs, and the incubation period lasts 11 to 13 days. It is believed that the female does most of the incubating, but males have been seen covering eggs, and they sometimes also brood the young. On average, the young remain in the nest for nine days, but they remain semidependent on their parents until they are 30 to 35 days old. In one Michigan study, a pair hatched a second brood 29 days after the hatching of the first, and among another group of 29 pairs, 15 pairs raised a single brood, 13 raised two broods, and 1 raised three broods in a single season.

Suggested viewing locations: In Wyoming, this species breeds statewide, from low-elevation grasslands to mountain meadows. Open grasslands, especially where some shrubs and bare ground are present, are good places to look for this large sparrow.

Suggested reading: BNA 624 (S. L. Jones and J. E. Cornely, 2002).

Lark Sparrow. *Chondestes grammacus*. Wi; Un; Su; Res.

Status: The lark sparrow is an uncommon summer resident over most of the state in grassland habitats but rarer at high elevations. It is present but rare in both national parks (Grand Teton and Yellowstone).

Habitats and ecology: This species favors grasslands that have scattered trees, shrubs, large forbs, or adjoin such vegetation; thus, weedy fencerows near grasslands, open brushland on slopes, sagebrush flats, scrubby and open oak woodlands, orchards, and similar habitats are all suitable. Generally, open views and a variety of plants, including scattered woody vegetation, some grasses, and herbs are preferred. In Colorado, grasslands with junipers, greasewood, or yucca are commonly used, as are plains grasslands within cottonwood stands, and shortgrass prairie with cholla cacti present (Kingery, 1998).

Breeding biology: Lark sparrow nests are made on the ground by the female, with a base of thin twigs, walls of thick grasses, and a lining of finer grasses, rootlets, and sometimes hair. The usual clutch is four to five eggs, and the incubation period lasts 11 to 13 days. The female does most of the incubating, but males help tend the young. On average, the young remain in the nest for 9 to 11 days.

Suggested viewing locations: Lark sparrows are widespread regionally, and in Wyoming they occur statewide except for the higher elevations in the Greater Yellowstone ecosystem, the Bighorns, and other major ranges (Faulkner, 2010).

Suggested reading: BNA 488 (J. W. Martin and J. R. Parrish, 2000).

Sagebrush Sparrow. *Artemisiospiza nevadensis*. Di; Un; Su; Res.

Status: The sagebrush sparrow (previously known as the sage sparrow) is among the sage obligates of the American West, with a distribution nearly contiguous with that of sagebrush. It is an uncommon summer resident in sagebrush areas throughout the state, especially in the southwest. It is a vagrant in Grand Teton National Park.

Habitats and ecology: The species is closely associated with fairly dense to sparse and scrubby sagebrush vegetation during the breeding season, but it also sometimes breeds in similar semidesert vegetation types, such as in saltbush. Foraging is done on rather bare-ground areas of gravel or alkali soil around the bushes, and escapes are made by fleeing into the shrubbery.

Breeding biology: Male sagebrush sparrows establish their territories in stands of big sagebrush. Their rather weak songs are a high-pitched tinkling series of notes, *tsi, tsi, tsi', you*, with the third note accented, or a jumbled series of notes with a see-saw rhythm, *twee, si, tity, slip*. The birds are elusive and hard to observe, as they are prone to running invisibly through the brushy cover. The species is monogamous and territorial. The nest is sometimes a simple depression in the ground under a sage shrub but more often is located in the bush at heights of from 3 to 40 inches. Elevated nests often include sage twigs, and they are lined with wool, cow hair, feathers, or grass. The clutch varies from three to five eggs but usually is three or four. The incubation period is 13 to 16 days, and the fledging period is 9 to 10 days, but few details are available on parental and juvenile behavior. Apparently two broods are typically raised.

Suggested viewing locations: In Wyoming, these birds are sagebrush obligates and are most common in tall, dense stands of sage. They are most abundant in the southwestern parts of the state, especially in Carbon, Lincoln, Sweetwater, and Uinta Counties, but they also occur north in central Wyoming at least as far as Hot Springs and Washakie Counties (Faulkner, 2010).

Suggested reading: BNA 326 (J. W. Martin and B. A. Carlson, 1998) ("Sage sparrow")

Lark Bunting. *Calamospiza melanocorys*. Wi; Co, CC2, CC3; Su; Res.

Status: The lark bunting is a common summer resident throughout the state, mainly on plains and foothill grasslands. It is present but rare in both national parks (Grand Teton and Yellowstone).

Habitats and ecology: This species favors mixed-grass prairies for nesting but can also be found in short-grass and tall-grass prairies as well as sage grasslands, retired croplands, alfalfa fields, and stubble fields. Areas with abundant shrubs are avoided, but fence posts or scattered trees may be used as song posts. In Colorado, more than 40 percent of breeding birds were found in shortgrass prairie, and taller prairies plus croplands compose an additional 42 percent of the total (Kingery, 1998).

Breeding biology: Lark buntings arrive on their breeding areas in flocks, within which courtship begins, and thus dispersal occurs gradually. There seems to be relatively little territorial development since nests are often placed only 10 to 15 yards apart, and males sometimes sing from adjacent fence posts. Both sexes incubate, but females evidently do the most. The usual clutch is of four or five eggs, and the incubation period lasts an average of 12 days. At least through the incubation period, to about the middle of July, the males continue to sing and perform song-flight displays. The abundance and local distribution of nesting birds seem to vary considerably from year to year; areas with dense populations one year may be virtually deserted the next. By late August the males lose their distinctive nuptial plumage, and the fall migration begins soon afterward.

Suggested viewing locations: In Wyoming, breeding occurs nearly statewide, except for the Greater Yellowstone ecosystem, the Bighorns, and other major mountain ranges (Faulkner, 2010). Lark buntings are most likely to be found on drier grasslands of the eastern plains, but their local distribution varies greatly from year to year, depending on local rainfall or, in drier areas, irrigation activities. Irrigated alfalfa fields seem to be especially favored. They are common at Hutton Lake National Wildlife Refuge. The densest regional breeding populations probably occur in the eastern parts of Wyoming.

Suggested reading: BNA 542 (T. G. Shane, 2000); Butterfield, 1969; Taylor and Ashe, 1976.

Song Sparrow. *Melospiza melodia*. Ub; Co; Res.

Status: The song sparrow is a common resident in suitable habitats throughout the state. It is present in both national parks, varying in abundance from common in Grand Teton to uncommon in Yellowstone.

Habitats and ecology: Breeding habitats include such woodland edge types as the brushy margins of forest openings, the edges of ponds or lakes, shelterbelts, farmsteads, coulees on prairies, aspen groves, and the like. Foraging occurs mostly on the ground, both in open areas and leaf-covered spots, where the birds scratch to expose foods. The highest population densities found by Hutto and Young (1999) in 566 transects of 11 unaltered forest, grassland, shrub, and wetland habitats of Idaho and Montana were in wetland communities, followed closely by riparian shrub habitats.

Breeding biology: The song sparrow is one of America's best-studied songbirds, thanks to the pioneer banding and behavior studies of Margaret M. Nice. Males are highly territorial and often maintain the same territories year after year. Females return to their old territories about half the time but only infrequently mate again with their previous partners (8 of 30 cases in one study). They often settle into adjacent territories if their males have found new mates, or may move as far as a mile from their place of hatching. Usually the female builds the nest alone, but there are cases of unmated males building nests and of helping mates in nest construction. About three to four days are needed for nest building, and the female incubates alone. The usual clutch is three to five eggs, and the incubation period lasts 12 to 14 days. Both sexes feed the young, which remain in the nest for about ten days and become independent when 28 to 30 days old. Periods between the fledging of two successful broods range from 30 to 41 days.

Suggested viewing locations: Song sparrows are common to abundant summer residents throughout the lower elevations of the Wyoming mountains. They breed in virtually all the state's riparian areas and in brushlands near marshes or beaver ponds as high as 10,000 feet (Faulkner, 2010).

Suggested reading: BNA 704 (P. Arcese et al., 2002).

White-crowned Sparrow. *Zonotrichia leucophrys*. Wi; Co; Su; Res.

Status: The white-crowned sparrow is a summer or permanent resident in suitable habitats almost throughout the state, including both national parks where it is a common (Yellowstone) to abundant (Grand Teton) breeder.

Habitats and ecology: During the breeding season this species occurs in riparian brush, in coniferous forests with well-developed wooded undergrowth, in aspen groves with a shrubby understory, in willow thickets around beaver ponds or marshes, and on mountain meadows with alders or similar low and thick shrubbery, often to timberline. Damp, grass-covered ground and nearby shrubbery seem to be important habitat components. In Colorado, these birds were most often found to breed in willow thickets at medium to high elevations, with a secondary use of krummholz (stunted, often wind-shaped, timberline conifers) (Kingery, 1998). On migration and during winter the birds are found in a variety of lower-elevation habitats that offer a combination of brushy cover and open ground for foraging. The highest population densities found by Hutto and Young (1999) in 566 transects of 11 unaltered forest, grassland, shrub, and wetland habitats of Idaho and Montana were in sagebrush and grassland communities.

Breeding biology: In Wyoming, this sparrow commonly breeds at higher elevations (from about 7,500 feet to more than 10,000 feet) in the major mountain ranges, but it is absent from the Black Hills (Faulkner, 2010). On their montane and subarctic breeding grounds, territorial males sing all day long from low shrubby vegetation or a rock. The song has some geographic variation but often sounds like *dear-dear buzz buzz buzz*. Although usually monogamous, the male are sometimes polygynous. The nest may be placed on the ground, at the base of a small bush, or in a bush or tree but usually no more than about ten feet above ground. The clutch size varies from three to five eggs, the latter number more typical of northern breeding regions. The incubation period lasts 11 to 14 days. The female alone incubates, although she might be fed by the male, and both sexes feed the young. The nestlings fledge in 9 to 11 days.

Suggested viewing locations: Roads over high passes during summer, such as over the Snowy Range pass or across Libby Flats in the Medicine Bow Range, are good places to find this species (Dorn and Dorn, 1990; Scott, 1993). The birds are common breeders in the Greater Yellowstone ecosystem, especially around Jackson Hole.

Suggested reading: BNA 183 (G. Chilton et al., 1995).

Dark-eyed Junco. *Junco hyemalis*. Wi; Co; Res.

Status: The dark-eyed junco is a common resident in wooded habitats throughout the state, including both national parks where the species is an abundant breeder. It winters at lower elevations over the entire state. Breeding populations of slate-colored juncos (*J. h. hyemalis*) winter widely southwardly through the Wyoming mountains and plains. The breeding juncos of this region include a "white-winged" race (*J. h. aikeni*) that is endemic to and semiresidential in the Black Hills region and surrounding areas. A complex of up to eight intergrading northwestern races collectively known as the "Oregon junco" breeds east from the Pacific Coast to the Pacific slope of Montana, of which one race (*J. h. shufeldti*) winters in Wyoming. The most interior "pink-sided" variant of the Oregon complex (*J. h. mearnsi*) breeds in northwestern Wyoming and the Bighorn range. The "gray-headed" form (*J. h. caniceps*) of the southern Rockies breeds in southern Wyoming north to the Wyoming and Wind River ranges, intergrading with the pink-sided form in southwestern Wyoming.

Habitats and ecology: Breeding habitats include open coniferous forests, especially pinyon-juniper woodlands, ponderosa pine forests, mixed forests, aspen woods, forest clearings, the edges of muskegs or jack pine–covered ridges, and similar habitats that offer ground-foraging and ground-nesting opportunities as well as tree or brush cover for escape. In Colorado, 70 percent of breeding birds were found in coniferous forests, with most of the remainder in aspens, and a few in other deciduous or shrubby habitats (Kingery, 1998). The highest population densities found by Hutto and Young (1999) in 566 transects of 11 unaltered forest, grassland, shrub, and wetland habitats of Idaho and Montana were in lodgepole pine, ponderosa pine, and young forest communities.

Breeding biology: Juncos are notable for their sociable winter flocking behavior, which persists until the birds return to their breeding areas. In the Black Hills, territorial singing sometimes can be heard in early March. When a female enters a male's territory, he follows her with tail lifted and fanned and wings drooping. Several days are spent in establishing and strengthening the pair-bond, during which the birds remain close together and the male continues to display frequently by wing-drooping and tail-fanning. The female builds the nest over a period of several days, and she apparently does all the incubation and brooding. The usual clutch is of three to five eggs, and the incubation period lasts 12 to 13 days. Both parents bring food to the young, which fledge in 10 to 13 days. The young continue to be semidependent on their parents for about three weeks after leaving the nest, and juveniles have been seen with their fathers as late as 46 days after fledging. Presumably by that time the females might be incubating a second clutch, if not feeding young.

Suggested viewing locations: Collectively, four races of juncos occur throughout Wyoming, and most are present throughout the year. Three of these subspecies breed in Wyoming, occupying all of the state's montane forests (Faulkner, 2010). All of the races are likely to appear commonly at bird feeders during winter.

Suggested reading: BNA 716 (V. Nolan, Jr., et al., 2002).

Family Icteriidae (Chats)

Yellow-breasted Chat. *Icteria virens*. SD; Un; Ch; Su; Res.

Status: The yellow-breasted chat is an uncommon summer resident at lower elevations over most of the drier portions of the state. It is a vagrant in Grand Teton National Park.

Habitats and ecology: During the breeding season this species occurs along the shrubby coulee areas of the plains, the oak and mountain mahogany woodlands of the foothills, along alder- and willow-lined creeks of the prairies, brushy forest edges, and in shrubby overgrown pasturelands. In Colorado, most nesting occurs below 7,000 feet but sometimes is as high as 8,000 feet, with nearly 95 percent in dense riparian thickets (Kingery, 1998).

Breeding biology: Males are unusual in their remarkable diversity of vocalizations, which often have six to ten different song phrases, almost randomly uttered and varying greatly in loudness, pitch, and duration. The usual clutch is three to five eggs, and the incubation period lasts 11 to 12 days. Only the female incubates the eggs and broods the young, which leave the nest in 8 to 11 days. This species is perhaps the most aberrant of all the traditional warbler assemblage, and in 2017 was removed from the warbler family, although its relationships are still obscure.

Suggested viewing locations: In Wyoming, this species occurs statewide at elevations up to about 7,500 feet (Faulkner, 2010). During the breeding season it can be easily found in the Laramie River bottoms, the Yellowtail Wildlife Habitat Management Area near Lovell, the LAK Reservoir (Weston County), and Richard's Gap (Sweetwater County) (Dorn and Dorn, 1990).

Suggested reading: BNA 575 (K. P. Eckerle and C. F. Thompson, 2001).

Family Icteridae (Blackbirds, Orioles, and Meadowlarks)

Bobolink. *Dolichonyx oryzivorus*. SD; Un; Su; Res.

Status: The bobolink is an uncommon summer resident in lower-altitude grasslands and wet meadows in northeastern Wyoming, the Bighorn range, and locally in western and northwestern Wyoming. It is occasional in Grand Teton National Park and declining nationally.

Habitats and ecology: Breeding occurs in tallgrass prairies, ungrazed or lightly grazed mid-grass prairies, wet meadows, hayfields, retired croplands, and similar habitats. Scattered bushes or other singing posts in the territory add to its attractiveness.

Breeding biology: Males arrive on their breeding areas about a week before females and quickly spread out, although specific territorial establishment and defense seems to be weak or lacking. Although the nests are well scattered, males tolerate other males surprisingly near the nest site. The usual clutch is five to six eggs, and the incubation period lasts 11 to 13 days. The female incubates alone; the male seldom visits her and apparently never feeds her. But males do help feed the young. Broods usually remain in the nest for about 10 to 14 days but have been reported to leave when only 7 to 9 days old, or well before they are able to fly. Males often acquire second mates after their first mate has begun nesting. These secondary mates tend to lay smaller clutches than the primary mates, perhaps because they often are young birds or are renesting. This smaller clutch size of secondary mates is adaptive, since males less frequently assist in feeding their second broods, and unassisted females are more likely to be able to tend smaller broods.

Suggested viewing locations: In Wyoming, bobolinks are only local breeders. They are most common along the eastern foothill slopes of the Bighorns, but small populations occur elsewhere, such as at the National Elk Refuge (Faulkner, 2010). They are often found in fields southwest of Sheridan and on the Wagon Box Road west of Fort Phil Kearny (Scott, 1993).

Suggested reading: BNA 176 (R. Renfrew et al., 2015).

Red-winged Blackbird. *Agelaius phoeniceus*. Ub; Co; Su; Res.

Status: The red-winged blackbird is a common summer resident in suitable habitats throughout the region, including both lowlands and montane areas. It is present and a common breeder in both national parks (Grand Teton and Yellowstone).

Habitats and ecology: Typical breeding habitats are wetlands ranging from deep marshes or the emergent vegetation zones of lakes and reservoirs through variably drier habitats including wet meadows, ditches, brushy patches in prairies, hayfields, and weedy croplands or roadsides. Wetlands with bulrushes or cattails are especially favored for nesting, but sometimes shrubs or other woody vegetation are used for nest sites. Outside the breeding season the birds often stray far from water and seek grain fields, city parks, pasturelands, and other habitats that offer food sources.

Breeding biology: This species is one of the commonest and most thoroughly studied of all North American songbirds, with more than 200 million birds probably present in North America. Adult males arrive on their breeding marshes well before females and begin to advertise their territories by flight song and "song-spread" displays, both of which prominently exhibit the red upper wing-coverts. Experiments with surgically muting males or painting these red markings black before they acquire mates result in the loss of territories by so-altered males. Pair-bonds last only during the breeding season, and most territorial males manage to acquire at least two females. In one Wisconsin study, it was found that experienced males tend to return to their old territories in successive years and that first-year males are usually unable to hold territories long enough to breed. In that study, no more than three females were mated to a single male, but a few instances of double-brooding (producing two broods in a single nesting season) were found. The usual clutch is four eggs, and the incubation period lasts 10 to 12 days. The young birds leave the nest at 10 to 11 days but are dependent for some time thereafter.

Suggested viewing locations: Perhaps the most abundant songbird in North America, this species can be found breeding in marshlands, ditches, and agricultural lands anywhere in the region.

Suggested reading: BNA 184 (K. Yasukawa and W. A. Searcy, 2019).

Western Meadowlark. *Sturnella neglecta*. Ub; Ab; Su; Res.

Status: The western meadowlark is an abundant summer resident throughout the state, overwintering locally in southern Wyoming. It is present and common in both national parks (Grand Teton and Yellowstone).

Habitats and ecology: During the breeding season, this species occupies mixed-grass to tallgrass prairies, wet meadows, hayfields, the weedy borders of croplands, and retired croplands, and to some extent short-grass and sage-steppe up to mountain meadows. In Colorado, shortgrass prairies are the most common breeding habitat, followed closely by croplands (Kingery, 1998).

Breeding biology: The breeding biology of this species is virtually identical to that of the eastern meadowlark, and these two species provide an interesting problem in terms of their ecology and evolutionary relationships. Where the two species occur together, they are sometimes intermediate in their primary songs, but this does not prove frequent hybridization; their call notes are more diagnostic and indicative of ancestry. One area of apparent hybridization is the Platte River valley of Nebraska, where intermediate birds are several times more frequent than elsewhere in the Great Plains. In areas of overlap, there has been no evolutionary shifting in song types, but apparently some of the aggressive displays of the two species do exhibit convergent elements. The usual clutch is five eggs, and the incubation period lasts 13 to 15 days. The young birds leave the nest at 9 to 12 days but are unable to fly well until they are nearly three weeks old.

Suggested viewing locations: Probably Wyoming's most numerous bird, this meadowlark is abundant in all grasslands at lower elevations. It is less common in the grasslands of western Wyoming, and it also occurs in the juniper scrublands of southwestern Wyoming (Scott, 1994). The densest regional breeding populations probably occur in eastern Wyoming.

Suggested reading: BNA 104 (S. K. Davis and W. E. Lanyon, 2008); Rohwer, 1971.

Yellow-headed Blackbird. *Xanthocephalus xanthocephalus*. Ub; Co; Su; Res.

Status: The yellow-headed blackbird is a common summer resident in wetland habitats throughout almost the entire state. It is rarer in montane areas but common in both national parks (Grand Teton and Yellowstone).

Habitats and ecology: This blackbird species is restricted during the breeding season to relatively permanent marshes, the marsh zones of lakes, and the shallows of river impoundments where there are good stands of cattails, bulrushes, or phragmites. Although sometimes breeding in the same areas as red-winged blackbirds, yellow-headed blackbirds occupy the deeper areas adjacent to open water. In Colorado, they breed at elevations up to 9,000 feet (Kingery, 1998).

Breeding biology: The displays of the yellow-headed blackbird are very similar to those of the red-winged blackbird, but the species differs ecologically in that the males normally participate in brood care and are more dependent on emerging aquatic insects such as damselflies. Thus, they are more dependent on marshes than are redwings. In both species, the males' conspicuous and prolonged displays seem to be related to the importance of territorial size and quality in attracting the maximum number of females. The usual clutch size is four eggs. As in the red-winged blackbird, only the female incubates during the 10- to 13-day incubation period, but males often help feed the young, particularly those of their first mate. The young leave the nest at 9 to 12 days.

Suggested viewing locations: Nearly all of Wyoming's cattail marshes that are deep enough to have some open water are likely to support this conspicuous blackbird. Large colonies occur in places such as Hutton Lake National Wildlife Refuge,

Table Mountain Wildlife Habitat Management Area, Ocean Lake, and Loch Katrine (Dorn and Dorn, 1990).
Suggested reading: BNA 192 (D. J. Twedt and R. D. Crawford, 1995).

Bullock's Oriole. *Icterus bullockii*. Ub; Co; Su; Res.

Status: The Bullock's oriole is a common summer resident in wooded plains and foothills areas throughout the state but is rare in the mountains, breeding up to about 8,000 feet. It is present in both national parks, where it is rare (Yellowstone) to occasional (Grand Teton). The Baltimore oriole (considered conspecific with Bullock's from 1983 to 1995, when they were collectively named the northern oriole) is limited to the eastern border of Wyoming. There it is in contact and sometimes hybridizes with the Bullock's oriole.

Habitats and ecology: During the breeding season, males of the Bullock's oriole especially favor riverbottom forests of willows and cottonwoods, but they also occur in city parks and on plains or foothill slopes and valleys with aspen, poplars, birches, and similar vegetation. In Colorado, mature native cottonwoods and exotic landscaping trees provide a major breeding habitat for both orioles, but all the Baltimore records came from only three habitat types, primarily cottonwoods plus some rural and urban habitats (Kingery, 1998). During summer and fall, orioles are attracted to trees and bushes that provide berries.

Breeding biology: The remarkable pendant nests of this species are built mostly by the female, sometimes in as little as four or five days though usually a week or so is needed. No true knots are tied in the process, but a loose tangle of fibrous materials is gradually pulled together and tightened, forming a woven structure. A new nest is made each year, but certain trees or territories from previous years seem to be favored, as the remains of old nests are often found near new ones. Incubation is by the female, who is fed on the nest by her mate. The usual clutch is four to five eggs, and the incubation period lasts about 14 days. The nestling period is approximately two weeks, and the young are dependent on the adults for another two weeks. Until a few decades ago, the western Bullock's oriole was regarded as a species distinct from the eastern Baltimore oriole (*Icterus galbula*), even though extensive hybridization in the Great Plains favored the view that they are biologically a single species. However, some recent evidence suggests that hybrids are becoming less frequent in the overlap zone, favoring the recognition of two species.

Suggested viewing locations: The densest regional breeding populations of the Bullock's oriole occur in eastern Montana, eastern Wyoming, and eastern Colorado. Some Baltimore orioles may also breed in the easternmost parts of these states, although Faulkner (2010) stated that no evidence exists of breeding by phenotypically pure Baltimore orioles in Wyoming.

Suggested reading: BNA 416 (N. J. Flood, 2016); Sibley and Short, 1964.

Brewer's Blackbird. *Euphagus cyanocephalus*. Ub; Co; Su; Res.

Status: The Brewer's blackbird is a summer resident virtually throughout the entire state, breeding in most habitats from plains to mountain meadows. It is present and a common breeder in both national parks (Grand Teton and Yellowstone).

Habitats and ecology: Low-stature grasslands are the primary breeding habitats of this species, including mowed or burned areas; farmsteads and residential areas; the edges of marshes, especially where scattered shrubs are present; aspen groves; the brushy banks of prairie creeks; and similar locations. Nesting occurs on the ground or in low shrubs such as sage, and shrubs or fence posts also serve as singing posts where they are available. Outside of the breeding season, a wider array of open habitats are used, including grain fields, orchards, berry farms, and similar agricultural lands. In Colorado, juxtaposed cropland and rural areas appear to be the ideal breeding habitats. The birds breed at elevations from the plains up to about 10,000 feet, but they are mostly in the foothills, intermountain valleys, and along the Western Slope (Kingery, 1998). They are especially common in the North Park region. In the interior Pacific Northwest, this species breeds in mixed coniferous forests from the shrub-seedling to the old-growth forest stages of succession (Sanderson, Bull, and Edgerton, 1980).

Breeding biology: Brewer's blackbirds are often colonial nesters, but unlike red-winged and yellow-headed blackbirds they are more frequently monogamous than polygynous. In contrast to these species, pair formation begins when the birds are still in winter flocks, and frequently mates of the previous season re-form pair-bonds. The female builds the nest, although the male may often accompany her as she gathers material. From 10 to 14 days are spent in building the nest and laying the clutch. The usual clutch size is five to six eggs. Males do not assist in incubation but do visit the nest, rarely, to feed the incubating female during the 12- to 13-day incubation period. Both sexes feed the young, which leave the nest after about 13 days, and fledglings may be cared for by their parents for at least three weeks. At least in some areas, double-brooding is fairly frequent, and as many as three nesting attempts may be made in a single season by an unsuccessful pair.

Suggested viewing locations: This species is a common to abundant summer resident across Wyoming, in sage-dominated scrub, ranchlands, and agricultural lands up to about 9,500 feet (Faulkner, 2010). The most extensive, dense, regional breeding populations are in Wyoming—for example, at Hutton Lake National Wildlife Refuge and across the Greater Yellowstone ecosystem.

Suggested reading: BNA 616 (S. G. Martin, 2002).

Common Grackle. *Quiscalus quiscula*. Wi; Co; Su; Res.

Status: The common grackle is a common summer resident in suitable plains or foothills habitats over most of the state at lower elevations. It is present but rare in the national parks (Grand Teton and Yellowstone).

Habitats and ecology: Breeding habitats consist of woodland edges, areas partially planted to trees such as residential areas, farmsteads, shelterbelts, coniferous or deciduous woodlands of an open nature, woody shorelines around lakes, and riparian woodlands. Junipers, spruces, and other small and dense conifers are preferred for nesting, although hardwoods, shrubs, buildings, birdhouses, and even cattails are sometimes also used. In Colorado, about 80 percent of breeding birds used rural habitats, and the remainder used wooded riparian habitats of various deciduous trees including cottonwoods (Kingery, 1998).

Breeding biology: Males of this colonial-nesting species usually arrive on their breeding areas well before females and remain in flocks until the females arrive. There is only a gradual breakup of migratory flocks as pairs are progressively formed. A major component of pair formation is a flight involving a single female and up to five males, which follow her while keeping their tails strongly keeled. After pairing occurs, the females begin to select nest sites, and their mates defend only a small area of the nesting tree. The female gathers most of the material and does all of the actual construction, which sometimes takes about a week and sometimes may be spread out over several weeks. The usual clutch size is four to five eggs, and the incubation period lasts 12 to 14 days. The female incubates alone and also does all the brooding. Both sexes feed their young, which leave the nest at 10 to 17 days.

Suggested viewing locations: Probably all of Wyoming's lower-elevation cities, towns, and villages have resident flocks of grackles, which can be almost as much of a nuisance as starlings. The densest regional breeding populations probably occur in eastern and central Wyoming.

Suggested reading: BNA 271 (B. D. Peer and E. K. Bollinger, 1997).

Brown-headed Cowbird. *Molothrus ater*. Ub; Co; Su; Res.

Status: The brown-headed cowbird is a common summer resident throughout the state, including both national parks (Grand Teton and Yellowstone).

Habitats and ecology: Breeding occurs in a variety of woodland edge habitats, including brushy thickets, forest clearings, brushy creek-bottoms in prairies, aspen groves, sagebrush, desert scrub, agricultural lands, and open coniferous forests at lower elevations (up to about 7,000 feet in southern parts of the region). In the interior Pacific Northwest, this species parasitizes birds breeding in mixed coniferous forests from the shrub-seedling to the mature forest stages of succession (Sanderson, Bull, and Edgerton, 1980).

Breeding biology: This cowbird is the only species of North American bird that is an obligatory nest parasite. Before colonization, the species was largely limited to the Great Plains, but recently it has come into contact with many new potential host species that have not had time to evolve defensive mechanisms. These include many forest-adapted songbirds, including the extremely rare Kirtland's warbler. There is no egg mimicry in this species, but apparently each cowbird normally lays only one egg in each host's nest, usually during the host's egg-laying period. The female sometimes but not always removes a host's egg from the nest as well, but only when at least two eggs are already present. The eggs do not always hatch at the same time as or before the host's eggs, and the nestling period is roughly ten days, about the same as for many of its hosts. However, the young grow much more rapidly than the host nestlings and thus the amount of food available for the young of the host is correspondingly reduced, causing their starvation.

Suggested viewing locations: Wyoming's ubiquitous cowbird population is centered at lower elevations where cattle grazing is common. Cowbirds can often be seen feeding on insects around the feet of cattle and other ungulates. The densest and most extensive regional breeding populations probably occur in eastern Montana, but the birds are nearly ubiquitous. In Colorado, breeding birds were reported from 46 different habitat types and have been implicated as parasitizing at least 59 host species (Kingery, 1998).

Suggested reading: BNA 47 (P. E. Lowther, 1993); Newman, 1970; Johnsgard, 1977b.

Family Parulidae (New World Warblers)

Orange-crowned Warbler. *Oreothlypis celata*. Di; Un; Su; Res.

Status: The orange-crowned warbler is an uncommon summer resident in most wooded areas of all the major mountain ranges except for the Black Hills. It is present in both national parks, where it is rare (Yellowstone) to occasional (Grand Teton).

Habitats and ecology: A variety of woodland and brushy habitats are used by this species for breeding, ranging from riparian woodlands, pinyon-juniper habitats, and aspen groves. In montane areas, they favor willow or alder thickets near streams or willow thickets at tree line. At lower elevations they tend to breed along riverine woods or in brushy vegetation surrounding beaver ponds in northern coniferous habitats. In Colorado, they breed from 6,500 to 9,500 feet, in lower montane up to subalpine vegetation, such as scrub oak, aspen, and montane shrubs, including mountain mahogany and willows. The highest population densities found by Hutto and Young (1999) in 566 transects of 11 unaltered forest, grassland, shrub, and wetland habitats of Idaho and Montana were in riparian shrub and young forest communities. On migration the birds are found in a wide variety of brushy or wooded habitats, but they favor the brushy areas of river bottoms.

Breeding biology: The territorial song of this drab-colored warbler consists of a fairly weak staccato series of trilled notes that become lower and slower toward the end. Their usual nesting habitat consists of shrubby growth along rivers, tangled growth at the edge of woodlands, and clearings with shrubs and vines. Nests are on the ground or in low bushes, up to about two feet above ground. They are built by the female over a period of two to four days. The usual clutch size is five eggs. Incubation is performed by the female and lasts 12 to 14 days. The young are fed by both parents, with fledging occurring at 12 or 13 days.

Suggested viewing locations: During migration this species is common across all of Wyoming. Breeding occurs locally at higher elevations, over all the major mountain ranges except the Black Hills (Faulkner, 2010). They often may be found in moist deciduous woodlands, such as the headwaters of Coral Creek on Muddy Mountain (south of Casper Mountain in the Laramie range) and near Battle Lake in the Sierra Madre range (Scott, 1993).

Suggested reading: BNA 101 (W. M. Gilbert, M. K. Sogge, and C. van Riper, 2010).

MacGillivray's Warbler. *Geothlypis tolmiei*. Wi; Co; Su; Res.

Status: MacGillivray's warbler is an uncommon and widespread summer resident in woodlands and brushy areas of all the mountain ranges throughout the state. It is present in both national parks (Grand Teton and Yellowstone) and is common in both.

Habitats and ecology: This warbler is generally associated with brushy thickets, especially riparian woodlands. Less often it occurs in dense deciduous woods or mixed woodland on upland slopes, or in mature riverbottom forests. In Colorado, 80 percent of breeding birds were found in montane willow and alder thickets, montane shrublands, riparian deciduous forests, and scrub oaks (Kingery, 1998). In the interior Pacific Northwest, this species breeds in mixed coniferous forests from the shrub to the mature forest stages of succession (Sanderson, Bull, and Edgerton, 1980). In Alberta, the birds are usually found close to water in thick brushy growth, in prairie coulees, mountain slopes with dense shrubbery, or along forest clearings. The highest population densities found by Hutto and Young (1999) in 566 transects of 11 unaltered forest, grassland, shrub, and wetland habitats of Idaho and Montana were in riparian shrub and wetland communities.

Breeding biology: This western counterpart of the mourning warbler has been much less studied than that species. It builds its nests on or near the ground in dense shrubs or in clumps of grass or ferns. Apparently the female undertakes virtually all the parental duties, including incubation and brooding. The usual clutch size is four eggs, and the eggs are laid at daily intervals. The incubation period is 11 to 13 days and is done by the female. The male participates in feeding the young, which leave the nest eight or nine days after hatching.

Suggested viewing locations: In Wyoming, this species breeds in lower to middle-level zones of all the major mountain ranges (Faulkner, 2010). It is a common species in brushy stream bottoms throughout the state, from Sand Creek in the Black Hills (Crook County) to Jenny Lake in Grand Teton National Park (Scott, 1993).

Suggested reading: BNA 159 (J. Pitocchelli, 2013).

Common Yellowthroat. *Geothlypis trichas*. Ub; Un; Su; Res.

Status: The common yellowthroat is an uncommon and widespread summer resident that inhabits low-elevation woodlands and brushy areas throughout the region. It is common in both national parks (Grand Teton and Yellowstone).

Habitats and ecology: The yellowthroat is generally associated with brushy thickets, especially riparian woodlands, below 8,000 feet. Less often it occurs in dense deciduous woods or mixed woodland on upland slopes, or in mature riverbottom forests. In Alberta, the birds are usually found close to water in thick brushy growth, in prairie coulees, on mountain slopes with dense shrubbery, or along forest clearings. In Colorado, 80 percent of breeding birds were found in montane willow and alder thickets, montane shrublands, riparian deciduous forests, and scrub oaks (Kingery, 1998). In the interior Pacific Northwest, this species breeds in mixed coniferous forests from the shrub to the mature forest stages of succession (Sanderson, Bull, and Edgerton, 1980). The highest population densities found by Hutto and Young (1999) in 566 transects of 11 unaltered forest, grassland, shrub, and wetland habitats of Idaho and Montana were in wetland communities.

Breeding biology: As soon as they arrive in spring, males establish territories that usually are less than two acres in area but in some instances may exceed three (one bigamous male was found to occupy an area of 3.4 acres). Nests are usually built in the drier and more open parts of the territory, but water is always nearby. The female builds the nest over a period of two to five days and also performs all the incubation. The usual clutch size is four eggs. Feeding of the young is done by both sexes, and the young may leave the nest when only 7 or 8 days old; however, they cannot fly until they are 11 or 12 days old, and they do not begin feeding on their own until about three weeks of age. One or both parents may tend them until they are four to five weeks old. Most or all females attempt to raise second broods, but few are successful. At least some females build at least three nests, and mate changing between broods apparently occurs infrequently.

Suggested viewing locations: In Wyoming, this species breeds in lower to middle-level zones of all the major mountain ranges (Faulkner, 2010). It is a common species in brushy stream bottoms throughout the state, from Sand Creek in the Black Hills (Crook County) to Jenny Lake in Grand Teton National Park (Scott, 1993).

Suggested reading: BNA 448 (M. J. Guzy and G. Ritchison, 1999).

American Redstart. *Setophaga ruticilla*. Di; Un; Su; Res.

Status: The American redstart is an uncommon summer resident in woods over the eastern half of the state, especially in the Black Hills and the eastern slope of the Bighorns. It is present but rare in both national parks (Grand Teton and Yellowstone).

Habitats and ecology: Breeding habitats of this species include moist bottomland woodlands, the margins or openings of mature forests, young or second-growth stands of various types of forests, and especially deciduous forests. Nearby water and a brush layer seem to be important habitat components. In Colorado, redstarts typically nest in streamside habitats with undergrowth shrubs, but they may also nest in either deciduous or mixed deciduous-coniferous forests (Kingery, 1998). The highest population densities found by Hutto and Young (1999) in 566 transects of 11 unaltered forest, grassland, shrub, and wetland habitats of Idaho and Montana were in cottonwood-aspen communities.

Breeding biology: The American redstart is one of the few warblers whose social behavior has been extensively analyzed, including both courtship and more general aggressive behavior. Males arrive on the breeding grounds before females and immediately become territorial. Females typically arrive at night and may obtain a mate by the following morning, presumably being attracted to the males by their singing. Up to 60 hours may elapse between the formation of the pair-bond and the start of nest building, during which the female investigates the territory and begins to restrict her activities to the eventual nest site. The female selects the site alone and immediately begins to build a nest there, which may require about three days. The usual clutch is four eggs, and the incubation period lasts 12 days. The female alone performs incubation and brooding, although males are typically highly attentive to the young during the nestling period, which usually lasts eight to nine days.

Suggested viewing locations: Breeding in Wyoming occurs in the Black Hills National Forest, the Bighorn Mountains, Laramie Mountains, and the Snowy Range (Faulkner, 2010). Good summer birding sites in Wyoming include the Black Hills, Tongue River Canyon in Sheridan County, Sand Creek in Crook County, and LAK Reservoir in Weston County (Dorn and Dorn, 1990). The densest regional breeding populations probably occur in the Black Hills.

Suggested reading: BNA 277 (T. W. Sherry et al., 2016); Hubbard, 1969.

Yellow Warbler. *Setophaga petechia*. Ub; Co; Su; Res.

Status: The yellow warbler is a common summer resident throughout the region in suitable habitats, including the national parks where it is abundant (Grand Teton) to common (Yellowstone).

Habitats and ecology: Generally moist habitats, such as riparian woodlands and brush, the brushy edges of marshes, swamps, or beaver ponds, and also drier areas including roadside thickets, hedgerows, orchards, and forest edges, are the yellow warbler's preference. A combination of open areas and dense shrubbery seem to be important for breeding, although migrant birds are rather more widely distributed. In Colorado studies, deciduous stream bottoms provided the most typical nesting habitat during breeding bird surveys, but the birds used a wide variety of other deciduous sites, with only 2 percent occurring in conifers (Kingery, 1998). The highest population densities found by Hutto and Young (1999) in 566 transects of 11 unaltered forest, grassland, shrub, and wetland habitats of Idaho and Montana were in riparian shrub, wetland, and cottonwood-aspen communities.

Breeding biology: This widespread and abundant warbler seems to have rather generalized territorial needs, including suitable nest sites, tall singing posts, concealing cover, and foraging areas in shrubs and trees, within an area of about two-fifths of an acre. Territorial behavior begins soon after the males arrive, and pairs may be formed in one to four days. Approximately four days are needed for nest construction, which is done mostly or entirely by the female. The female also does all the incubating, often beginning before the clutch has been completed. The usual clutch is four eggs, and the incubation period lasts 11 days. Both sexes feed the young, and occasionally males have been seen brooding them, but this seems to be atypical. The young fledge in 9 to 12 days and remain in the general vicinity for another 7 to 10 days.

Suggested viewing locations: This species breeds commonly throughout the entire region. It occurs throughout Wyoming below 8,000 feet, especially in river bottoms with mature cottonwoods and an understory of willows or Russian olives (Faulkner, 2010).

Suggested reading: BNA 454 (P. E. Lowther et al., 1999); Hubbard, 1969.

Yellow-rumped Warbler. *Setophaga coronata*. Ub; Co; Su; Res.

Status: The yellow-rumped warbler is a common and widespread summer resident in wooded areas throughout the state; it is perhaps the state's commonest breeding warbler. It's also present and abundant in both national parks (Grand Teton and Yellowstone).

Habitats and ecology: This species breeds in a wide array of coniferous forests, from the ponderosa pine zone upward, and also breeds in riparian forests with conifers present. Habitats range from open, park-like ponderosa pine communities through dense montane forests to timberline species, foraging from low branches to the highest crown levels. In the Pacific Northwest, this species breeds in mixed coniferous forests from the young to the old-growth forest stages of succession (Sanderson, Bull, and Edgerton, 1980). The highest population densities found by Hutto and Young (1999) in 566 transects of 11 unaltered forest, grassland, shrub, and wetland habitats of Idaho and Montana were in Douglas-fir

communities. In Colorado studies, these birds bred in conifer and aspen forests from 7,000 to 11,500 feet, with two-thirds of the observations in coniferous forests (Kingery, 1998). During winter the habitats used are more varied, and foraging may range from berry-eating and nectar-drinking to aerial fly-catching.

Breeding biology: This warbler is among the first of its family to move northward in spring, often becoming fairly abundant as the first tree leaves appear. Nesting typically occurs in coniferous woodlands, but sometimes nests are constructed in aspen groves or among scattered conifers. The nests are built by the female, on branches 4 to 50 feet above ground. The usual clutch size is four to five eggs. Only the female incubates over the 12 or 13 days of incubation, but both sexes tend the young. The young fledge at about 12 to 14 days.

Suggested viewing locations: This species breeds commonly throughout the entire region. In Wyoming, it occurs widely in coniferous forests of all of the state's mountain ranges and in some low-elevation habitats, such as cottonwood-lined riparian corridors (Faulkner, 2010). Extensive regional breeding populations probably occur in the Greater Yellowstone ecosystem.

Suggested reading: BNA 376 (P. D. Hunt and D. J. Flaspohler, 1998); Hubbard, 1969.

Wilson's Warbler. *Cardellina pusilla*. SD; Co; Su; Res.

Status: Wilson's warbler is a common summer resident in woodlands in western Wyoming and east to the Bighorns, Sierra Madre, and Medicine Bow ranges. It is present in both national parks, where it ranges from common (Grand Teton) to uncommon (Yellowstone).

Habitats and ecology: On their breeding grounds, these birds inhabit willow and alder thickets along rivers or beaver ponds, brushy edges of lakeshores, the edges of mountain meadows, timberline areas of low shrubby vegetation, and sometimes aspen thickets. In Colorado, they regularly breed at elevations of 9,000 to 10,500 feet, especially in willow thickets around high mountain lakes, beaver ponds, or other montane wetlands (Kingery, 1998). In the interior Pacific Northwest, this species breeds in mixed coniferous forests from the pole-sapling to the mature forest stages of succession (Sanderson, Bull, and Edgerton, 1980). The highest population densities found by Hutto and Young (1999) in 566 transects of 11 unaltered forest, grassland, shrub, and wetland habitats of Idaho and Montana were in spruce-fir communities.

Breeding biology: The male Wilson's warbler song consists of a series of staccato *tsee* or *chip* notes, which descend in pitch and slightly accelerate toward the end. Territories are established where there is a well-developed brushy understory in coniferous woodland, especially if there is a stream nearby. Sometimes several pairs nest in close proximity, and although the birds are usually monogamous, the males are sometimes polygynous. The birds build their nests on the ground, often in a grass hammock or at the base of a shrub. It is a bulky

cuplike structure of grasses, leaves, and mosses lined with fine grasses. The clutch size is usually five eggs but can vary from four to six. Incubation is performed by the female and lasts 11 to 13 days. Both sexes tend the young, which fledge in 10 to 11 days.

Suggested viewing locations: In Wyoming, this species breeds in most high mountain ranges, including those in the Greater Yellowstone ecosystem, the Bighorns, and the Medicine Bows (Faulkner, 2010). It most often nests in montane willow thickets, such as those in Grand Teton and Yellowstone National Parks, as well as in the Bighorn Mountains (Scott, 1993).

Suggested reading: BNA 478 (E. M. Ammon and W. H. Gilbert, 1999).

Family Cardinalidae (Tanagers, Grosbeaks, and Buntings)

Western Tanager. *Piranga ludoviciana.* Wi; Co; Su; Res.

Status: The western tanager is a common summer resident in coniferous forests throughout the state, including all the mountain ranges. It is present and common in both national parks (Grand Teton and Yellowstone).

Habitats and ecology: Breeding occurs in various habitats, including riparian woodlands, aspen groves, ponderosa pine forests, and occasionally in Douglas-fir forests and pinyon-juniper or oak–mountain mahogany woodlands. It is usually found in areas having a predominance of coniferous trees, preferably those that are fairly open, but it also occasionally extends into fairly dense forests. In Colorado, nesting occurs from 5,000 to 8,000 feet, especially in ponderosa pine and aspen woodlands, and in various coniferous forests ranging from foothills to mid-elevations (Kingery, 1998). In the Pacific Northwest, this tanager breeds in mixed coniferous forests from the young forest to the old-growth forest stages of succession (Sanderson, Bull, and Edgerton, 1980). The highest population densities found by Hutto and Young (1999) in 566 transects of 11 unaltered forest, grassland, shrub, and wetland habitats of Idaho and Montana were in ponderosa pine communities.

Breeding biology: Surprisingly little has been written on the breeding biology of this beautiful species, but it apparently closely resembles that of the better-studied scarlet tanager. In spite of the bright coloration of the males, breeding pairs are not conspicuous, since the olive-colored females remain high in the trees, and they tend to be very elusive during nesting. The nest is placed on the branch of a tree from 8 to 60 feet above ground. The female incubates the clutch of three to five eggs alone throughout the 13-day incubation period, and the male evidently rarely if ever approaches the nest during this time. After hatching, he helps feed the young, which leave the nest in 10 or 11 days and probably are fully fledged in about two weeks.

Suggested viewing locations: In Wyoming, this bird breeds from the western Black Hills to the Greater Yellowstone ecosystem,

including the Bighorn, Laramie, and Medicine Bow ranges (Faulkner, 2010). It is notably common in the Black Hills and around Moose and Jenny Lake in the Tetons.

Suggested reading: BNA 431 (J. Hudon, 1999).

Black-headed Grosbeak. *Pheucticus melanocephalus.* Co; Su, Wi; Res.

Status: The black-headed grosbeak is a common summer resident and variably common breeder over most of the state in wooded areas up to about 8,000 feet. It is present in both national parks, where it is common (Grand Teton) to rare (Yellowstone).

Habitats and ecology: During the breeding season this species is associated with open deciduous woodlands having fairly well-developed shrubby understories and usually on floodplains or upland areas. It extends into wooded coulees and riparian forests of cottonwoods and similar vegetation in the plains, and sometimes also nests in orchards, oak–mountain mahogany woodlands, and aspen groves. In Colorado, most nesting occurs from 5,000 to 8,000 feet, in a variety of habitats that especially include ponderosa pine, aspen, and riparian foothill forests, as well as pinyon-juniper and scrub oak woodlands. The presence of tick clover (*Desmodium*), either as a canopy or undergrowth cover plant, might be a positive habitat selection factor (Kingery, 1998). In the Pacific Northwest, this species breeds in mixed coniferous forests, from the young forest to the old-growth forest stages of succession (Sanderson, Bull, and Edgerton, 1980). The highest population densities found by Hutto and Young (1999) in 566 transects of 11 unaltered forest, grassland, shrub, and wetland habitats of Idaho and Montana were in ponderosa pine and cottonwood-aspen communities.

Breeding biology: So far as is known, the breeding biology of this species is essentially identical to that of the rose-breasted grosbeak. Studies of these two closely related forms in North Dakota indicate that the courtship behavior of the two is very similar, and thus the color differences among the males are likely to be important in avoiding more widespread hybridization than occurs. Males apparently do not discriminate between the songs of their own and the other species but do make visual discriminations when confronted with mounted males placed in their territories. The nests are built by the females and are placed at about 6 to 12 feet above ground in the fork of a twig. The usual clutch is three to five eggs and it is incubated by the female. When the chicks hatch after about 13 days, they are tended by both sexes over the 10- to 11-day fledging period.

Suggested viewing locations: In Wyoming, this grosbeak breeds nearly statewide at elevations below 8,000 feet (Faulkner, 2010). It is most likely to be found in deciduous riparian woodlands across the state but especially in the eastern half where these habitats are more common. Favorable birding areas include the oak woodlands of the western Black Hills

Western tanager, adult males

(Crook County), aspen groves in Fossil Butte National Monument (Lincoln County), and lower parts of Grand Teton National Park (Dorn and Dorn, 1993).

Suggested reading: BNA 143 (C. Ortega and G. E. Hill, 2010); West, 1962; Kroodsma, 1970.

Lazuli Bunting. *Passerina amoena*. Un; Su, Wi; Res.

Status: The lazuli bunting is an uncommon summer resident in suitable habitats nearly throughout the state, up to about 9,500 feet. Present in both national parks, it varies from occasional (Grand Teton) to common (Yellowstone).

Habitats and ecology: In the mountain areas, these birds breed along the edges of deciduous forests on gentle valley slopes, such as in aspen groves or thickets of willow or alder. On the foothills and plains, the birds are usually found in riparian woodlands that support a mixture of shrubs, low trees, and herbaceous vegetation. Plant diversity and discontinuity of cover seem to be important habitat characteristics for this species. In Colorado, the birds mostly breed between 5,500 and 7,000 feet, especially in riparian habitats with an abundance of shrubs. Mountain shrublands of mountain mahogany or serviceberry are also important breeding habitats (Kingery, 1998). The highest population densities found by Hutto and Young (1999) in 566 transects of 11 unaltered forest, grassland, shrub, and wetland habitats of Idaho and Montana were in riparian shrub communities.

Breeding biology: The breeding biology of this species can be considered nearly identical to that of the indigo bunting. In the western Great Plains, including extreme eastern Wyoming, these two species overlap appreciably, and about a third of the pairs there may be of mixed species (Baker and Boylan, 1999). Playback of songs indicates that in some areas males respond only to the song of their own species and ignore that of the other, but in one area of Nebraska sympatry, males responded to both song types. This suggests that learning may be involved in male song recognition. Mixed matings in areas of sympatry are infrequent, and such pairs seem to exhibit delayed breeding characteristics, as compared with nonmixed pairs. The clutch size is usually four eggs but varies from three to four. Incubation is performed by the female and lasts 12 days. Both sexes tend the young, which fledge in 10 to 15 days.

Suggested viewing locations: Lazuli buntings breed statewide in Wyoming, in shrub-dominated habitats up to about 9,500 feet (Faulkner, 2010). Streamside thickets provide ideal habitats, such as along the Tongue River at Dayton and in brushlands such as above LAK Reservoir (Weston County), Richard's Gap south of Rock Springs, and near Battle Creek campground in the Sierra Madre Mountains (Carbon County) (Dorn and Dorn, 1990).

Suggested reading: BNA 232 (E. Greene, V. R. Muehter, and W. Davison, 2014); Sibley and Short, 1959; Emlen, Rising, and Thompson, 1975.

Dickcissel. *Spiza americana*. HL; Ra, CC1, CC2; Su; Res.

Status: The dickcissel is a rare summer resident in the Great Plains east of the Bighorn Mountains, breeding locally in northeastern Wyoming. It is declining nationally and is a species of conservation concern.

Habitats and ecology: This is a prairie-adapted bird that breeds in grasslands that have a combination of tall forbs, grasses, and shrubs, or in grassy meadows that have nearby hedges or brushy fencerows.

Breeding biology: This species is one of the rather few North American passerines that regularly practices polygyny; a Kansas study indicated that 18 percent of the males had more than one mate, 40 percent were monogamous, and 42 percent were unmated. The variable success of males in attracting females seems to be related to the nest sites available in their individual territories. The females build the nests and perform all the incubation; the male helps feed the young only a little if at all. Males usually obtain second mates during the laying or incubation phases of the first nesting cycle. The usual clutch size is five eggs, and the incubation period is 11 to 13 days. Young birds remain in the nest 8 to 10 days.

Suggested viewing locations: In Wyoming, breeding occurs mostly east of a line from Sheridan to Torrington, and the amount of breeding may vary from year to year (Faulkner, 2010). The best populations occur in the western Black Hills (Crook County) and in the lower North Platte valley from Fort Laramie eastward (Scott, 1993). Other localized regional breeding populations probably occur in eastern Wyoming grasslands.

Suggested reading: BNA 703 (S. A. Temple, 2002).

Chapter 4 • Reptiles and Amphibians

The following profile summaries include 22 of Wyoming's total reptiles and amphibians. Wyoming supports 32 species of "herpetiles," including 12 species of amphibians (1 salamander and 11 frogs and toads) and 22 species of reptiles (4 turtles, 7 lizards, and 11 snakes). Koch and Peterson (1995) reported seven amphibians and nine reptiles from Yellowstone and Grand Teton National Parks, reflecting that region's markedly colder environments than are typical of most of Wyoming. By comparison, the more southerly latitudes of Nebraska support 13 amphibians (2 salamanders and 11 frogs and toads) and 48 reptiles (9 turtles, 10 lizards, and 29 snakes) (Ballinger, Lynch, and Smith, 2010). The taxonomic nomenclature used here is based on the Center for North American Herpetology database. A field guide with current taxonomic treatment and excellent range maps was recently published by Robert Powell, Roger Conant, and Joseph T. Collins (2016). Taxa are arranged alphabetically, initially by genus and secondly by species. Measurements below refer to body plus tail length, except for turtles, which instead are the length of the carapace (dorsal bony shell). Some alternate English and older Latin names are shown in parentheses.

Amphibians (Salamanders, Toads, and Frogs)

Order Caudata—Salamanders

Salamanders are unique among amphibians in that they retain their long tails into adulthood and develop four legs of equal length. As aquatic larvae they also have external gills, which in most species are lost by adulthood, after which lung breathing occurs. In a few permanently aquatic species (and rarely in tiger salamanders), the external gills may be retained throughout the animal's lifetime, and their moist skin might also allow for some oxygen exchange.

Family Ambystomatidae: Mole Salamanders

Mole salamanders (locally often called hellbenders) remain below ground (like moles) for most of the year but move into pools and ponds for breeding. During their spring breeding period, they sometimes can be found in water tanks used by livestock.

Barred Tiger Salamander. *Ambystoma malvortium*.
Ubiquitous.

Identification: This species is the only salamander that occurs in Wyoming. Its color pattern is highly variable, but the ground color is usually dark gray or very dark brown, on which there is a pattern of large black spots. The Wyoming race *Ambystoma tigrinum nebulosum* is described as being dark gray, with some blotches of black or brown (Baxter and Stone, 1985). In southeastern Wyoming and parts of western Nebraska, the usual black color of the local population (*A. t. malvortium*) is reduced, and the resulting pattern is likewise more one of dark blotches and spots over a yellow to greenish body. Thirdly, in northeastern Wyoming the local population (*A. t. melanostictum*) is said to have a more mottled or reticulated body pattern, although the taxonomic situation would seem to be much confused and not worth further discussion in a popular text. *Length:* Males average 6.6 inches (168 mm), females 5.5 inches (140 mm).

Status: Tiger salamanders are common and widespread in Wyoming and have been reported from nearly all counties. However, they are nocturnal, and when living on land they emerge from their burrows to forage only at night. In western Nebraska, they sometimes live in the burrows of prairie dogs. Sandy areas or loose soils with nearby water are favored habitats.

Habitats and ecology: Tiger salamanders begin breeding with the onset of early spring rains. At that time small ponds and stock tanks might become full of salamanders, but they avoid ponds with large fish; predatory fish sometimes eat salamanders. Breeding occurs in early spring, usually during March and April in Nebraska. Mating occurs in water, and the eggs are also laid in water, either singly or in clusters. The larvae hatch within a few weeks and soon develop into voracious predators, even eating other salamander larvae. The larvae are apparently distasteful to some predatory fish, allowing the salamanders to survive in fish-rich waters. Most of them metamorphose into the summer months, gradually losing their gills and tail fins, and then move onto land. Other individuals retain their larval structures and continue an aquatic existence, overwintering in that form (so-called "neotenous" form). Sexual maturity is reached the following spring, and mating begins again.

Suggested reading: Stebbens and Cohen, 1995; Baxter and Stone, 1985.

Order Anura—Frogs and Toads

Frogs and toads are easily recognized on the basis of their lack of a tail as adults (Anura means "without a tail"); moist, glandular skin; and four legs, the rear pair being much larger than the front pair. Except for spadefoot frogs, all the frogs and toads of Nebraska have horizontal pupils. Like other amphibians, frogs and toads produce gelatinous eggs that develop in water, resulting in tailed, legless larvae that have external gills and mouthparts adapted for foraging on algae.

Family Pelabatidae: Spadefoots

Spadefoot toads are unique among Wyoming's frogs and toads in having vertical pupils as well as black nail-like tubercles ("spades") on both of their hind feet. Spadefoots also lack obvious parotoid glands (glands that are typically visible behind the eyes in toads and exude distasteful or poisonous substances). Some people have a strong allergic reaction after handling them. Spadefoots are found in sandy soils, where they use the horny "spades" on their rear legs to dig into the sand and rapidly become hidden. When threatened, spadefoots can also inflate their bodies, making them more difficult to be swallowed by predators. Males produce rasping or snoring calls in spring, usually while in temporary pools.

Plains Spadefoot. *Spea bombifrons*. Local.
Great Basin Spadefoot. *Spea intermontanus*. Highly
   local.
Identification: Wyoming has two species of spadefoot toads, which have complementary ranges and are very similar in appearance. The plains spadefoot is found in north-central, northeastern, and eastern Wyoming, west to Park and Fremont Counties, and southeast to the Colorado border. The Great Basin spadefoot occurs in southwestern Wyoming, northeast to Natrona County. The plains spadefoot's horny "spades" on its hind feet are short and broad, whereas in the Great Basin spadefoot they are long and narrow (Baxter and Stone, 1985). *Length:* Snout–vent length is about 2 inches (50 mm).
Voice: The plains spadefoot has a call that is a short, loud trill that lasts about one-half second, whereas the Great Basin spadefoot has a similar call that is shorter and lasts about one-third second. A nighttime chorus of spadefoots may be heard as far as a mile away (Baxter and Stone, 1985).
Status: The plains spadefoot is associated with grasslands and sagebrush communities below 6,000 feet elevation. The Great Basin spadefoot is found in similar communities and comparable elevations, mostly in the Wyoming Basin and Green River valley. The species are in limited contact in Natrona and Fremont Counties (Baxter and Stone, 1985).
Habitats and ecology: Both species feed on insects and spiders, especially nocturnal species, while small larvae forage on plankton and organic matter. Breeding occurs opportunistically between May and July, when temporary ponds, playas, and perhaps other water sources are fed by summer rains (Baxter and Stone, 1985).
Breeding biology: During breeding, males typically call in chorus while floating on the water. Eggs are laid the night following mating and are elliptical masses containing 250 or more eggs. Except when breeding, spadefoots are rarely seen, as they are nocturnal and spend daytime hours under sand or mud.
Suggested reading: Stebbens and Cohen, 1995; Baxter and Stone, 1985.

Tiger salamander, adult and larvae (*upper*); adults, eggs, and larvae of spadefoot, (*middle*), northern leopard frog (*lower left*), and plains leopard frog (*lower right*)

Family Bufonidae: True Toads

Toads, as distinguished from frogs, are semiterrestrial amphibians that have warty skin and a large parotoid gland on each side of the neck, either above the tympanum (ear drum) or on it, which secretes a highly poisonous mucus. The parotoid glands sometimes extend upward to the bony midpoint of the cranium (the cranial crest). Males vocalize in spring while inflating their throats, producing loud musical or metallic-sounding calls that serve to attract females.

Woodhouse's Toad. *Anaxyrus woodhousei*. Local.
Great Plains Toad. *Anaxyrus cognatus*. Highly local.
Identification: Easily identified as a toad by its stocky shape and highly warty skin, the Woodhouse's toad is very similar to the Great Plains toad, but the Woodhouse's is more likely to be found in urban yards and gardens. The Woodhouse's has a pale line down the middle of the back, an unspotted

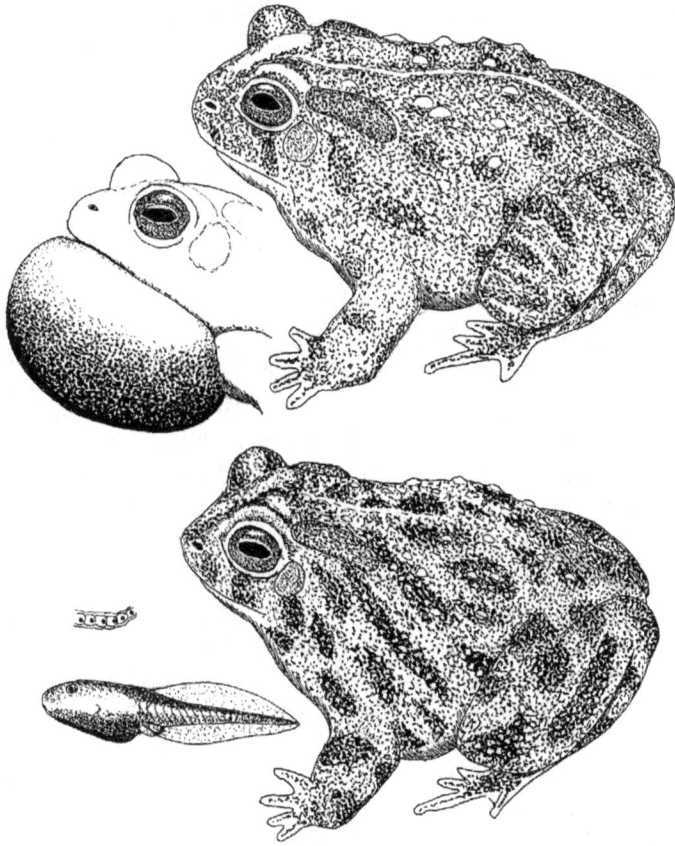

Woodhouse's (*upper*) and Great Plains (*lower*) toads, showing throat inflation during male calling, eggs, and larvae

abdomen, and a parotoid gland behind the eye that extends to the rear end of the bony cranial ridge that is present along the top of the head. The tympanum is inconspicuous but about as large as the eye. When calling, the throat area (the vocal sac) is strongly inflated. Krupa (1994) has described the breeding biology of the Great Plains toad. *Length:* Snout–vent length of both species average about 3.3 inches (83 mm); females are slightly larger than males.

Voice: The Woodhouse's toad utters a nasal trill that is similar to the bleating of a calf or sheep and lasts one to three seconds, whereas the Great Plains toad produces a very loud, trilled scream that lasts up to 50 seconds and resembles the sound of riveting.

Status: The Woodhouse's toad has been reported from all of Wyoming's eastern tier of counties and extends locally west to Natrona and Big Horn Counties, whereas the Great Plains toad inhabits northeastern Wyoming, east of the Powder River and north of the North Platte River.

Habitats and ecology: The Woodhouse's toad favors sandy areas but can be found anywhere that is fairly close to water. It is nocturnal, remaining in its burrow throughout the day. It emerges at night to feed on insects, especially beetles and

ants, as well as snails, spiders, and other arthropods. Breeding begins in late spring, with male choruses lasting until mid-summer. Eggs are laid in long strings of at least 25,000 eggs. The tadpoles hatch in less than a week, and metamorphosis occurs 50 to 60 days after hatching. The biology of the Great Plains toad is little studied in Wyoming but is apparently much like that of the Woodhouse's.

Suggested reading: Stebbens and Cohen, 1995; Baxter and Stone, 1985.

Family Ranidae: True Frogs

Unlike tree frogs, true frogs lack toe pads and have a large rounded tympanum (external ear drum) behind each eye. The North American species all have long legs, pointed toes that lack terminal pads, and extensive toe webbing on the muscular hind legs. They are powerful jumpers. They are also excellent swimmers, and adults are carnivorous, eating anything they can capture and swallow whole. Males typically call in chorus during spring, and during that season they have swollen forearms and thumbs that enable them to firmly grasp females during mating. Eggs are laid in water, in long strings that might contain up to 20,000 eggs. From 6 to 24 months are needed for the hatched larvae to metamorphose and reach adulthood.

American Bullfrog. *Lithobates* (*Rana*) *catesbiana*.
Local.

Identification: The American bullfrog is the largest frog in Wyoming, with adults reaching at least five inches long. Adults vary from lime green to olive, with warty backs, and often reddish brown to blackish dorsal markings. The tympanum is large (especially in males), unmarked, and conspicuous.

Length: Snout–vent length averages about 6 inches (150 mm).

Voice: The breeding call of the male has been described as a low snore, sounding like *jug-o-rum*. It can be heard from May into July and serves as both a territorial and sexual advertisement signal.

Status: This frog is abundant only in eastern Wyoming, where there is more available surface water, but it has also been introduced into Fremont County near Dubois.

Habitats and ecology: Bullfrogs usually can be found along the banks of almost any wetland, from ponds and marshes to slow rivers, where they patiently wait for prey to come within reach of their huge mouths. They try to capture anything they are able to swallow, from insects to other amphibians, turtles, small snakes, mammals, and birds. Breeding begins in late spring, when males advertise and defend their territories. After mating, females deposit clusters of 25,000 to 45,000 eggs as surface films on water. Following hatching, the larvae disperse and spend two years growing and completing their metamorphosis into adults, becoming torpid over the winters.

Suggested reading: Stebbens and Cohen, 1995; Baxter and Stone, 1985.

Northern Leopard Frog. *Lithobates* (*Rana*) *pipiens.*
Widespread.

Plains Leopard Frog. *Lithobates blairi* (= *Rana pipiens*
in Baxter and Stone, 1985)

Identification: Recognizing leopard frogs is easy; they are the commonest amphibians in Wyoming. However, separating the Plains leopard frog from the northern leopard frog is another matter; they were long considered to be a single species. Visual distinction between the two is difficult; both are smaller than bullfrogs, ranging in length from 2.5 to 4 inches, and their overall color is green to brown, with dark brown spots and blotches that are often edged with white. Two pale dorsolateral folds that extend from the back of the head are discontinuous toward the rear in the Plains leopard frog (upper right in drawing), but these folds are continuous in the northern species. The Plains leopard frog also usually has a white spot in the middle of the tympanum, a color pattern lacking in the northern leopard frog, and the former species also has a generally more stocky body conformation. *Length:* Adult snout–vent length about 2.8 inches (70 mm) in both species.

Voice: Males of the two leopard frog species differ in their mating calls. In the Plains species the calls are a long (35- to 40-second) series of chuck notes that end in a longer *cu-u-u-uck*, the series sounding like a finger being rubbed over a balloon. In the northern species the call is a long snore that is followed by two or three chuck notes, each series lasting only two to three seconds. Females of *L. pipiens* show differential responses to the calls of *L. blairi*, a presumed hybrid, and conspecifics (Kruse, 1981).

Status: Plains leopard frogs are common in southeastern Wyoming, west to about Natrona and Carbon Counties and north to about Niobrara County, beyond which northern leopard frogs replace them, although exact boundaries of the two appear to remain undetermined.

Habitats and ecology: Leopard frogs begin their mating activities very early, usually by the end of March. Males call while floating at the water surface, beginning after sunset. The newly fertilized female deposits a globular cluster of 4,000 to 6,000 eggs that hatch within a few days. Metamorphosis into the adult frog occurs in 50 to 60 days. As adults, the frogs eat a variety of insects and other invertebrates as well as vertebrates up to the maximum size that their mouths can accommodate. During summer the adults often wander about on land, and prior to winter they bury themselves in mud and leaves at the bottom of ponds.

Suggested reading: Stebbens and Cohen, 1995; Baxter and Stone, 1985.

Family Hylidae: Chorus Frogs

Tree frogs are small frogs that are often found in trees, adapted for climbing by the presence of adhesive pads on their toes. One group of frogs in this family (*Acris*) doesn't climb trees; these species are variously known as chorus frogs, cricket frogs, and spring peepers. These species have reduced toe webbing and smaller toepads. Males of this group have a round, inflatable vocal sac that expands during calling, the different species producing buzzing (*Hyla*), cricket-like clicking (*Acris*), or sounds that resemble a comb's teeth being stroked (*Pseudacris*). Some tree frogs are able to closely match their background by adjusting their skin color from gray, tan, or brown to bright green.

Boreal Chorus Frog. *Pseudacris* (*triseriata*)
maculata. Widespread.

Identification: This tiny frog barely exceeds a inch in length, and is the only Wyoming frog having a white streak along the upper lip, and three dark streaks along the back, plus wide dark stripes extending from the nose through the eyes and back along the flanks. *Length:* Snout–vent length about 1.5 inches (35–40 mm); females are larger than males.

Voice: Males begin advertisement calling as soon as the snow melts, typically during March and early April. Their calls are uttered with the vocal sac greatly enlarged and while clinging to vegetation. Calling continues both throughout the night and during the day. The call is a mechanical trill that is similar to the sound produced by running one's fingers over the teeth of a comb. Breeding may continue until early June.

Status: This frog is widespread throughout Wyoming, having been reported from most counties. Water bodies of almost any size are used, including vernal ponds, flooded fields, sewage lagoons, and lakes. During the summer the frogs also may be found on grasslands and woods far from water.

Habitats and ecology: Immediately after mating, females release their eggs, which are adhesive and cling to vegetation. Individual clutches may contain up to nearly 200 eggs, and the total seasonal ovarian production may be 500 to 800 eggs. Metamorphosis occurs after about six weeks, and thereafter both juvenile and adult frogs feed on insects and other small arthropods. In one study some individuals were found to survive for as long as six years.

Suggested reading: Stebbens and Cohen, 1995; Baxter and Stone, 1985.

REPTILES (TURTLES, LIZARDS, AND SNAKES)

Order Chelonia—Turtles

All turtles are easily recognized by the presence of a bony dorsal "shell" (the carapace) and a corresponding but variably smaller ventral supporting bony structure (the plastron). In most turtles these protective structures are covered by thick, horny scutes. Turtles lack true teeth but have sharp rims along their upper jaws that serve to cut and tear their food. All turtles are egg-layers, typically depositing their eggs in sandy soil and then abandoning them, requiring the hatchlings to independently dig their

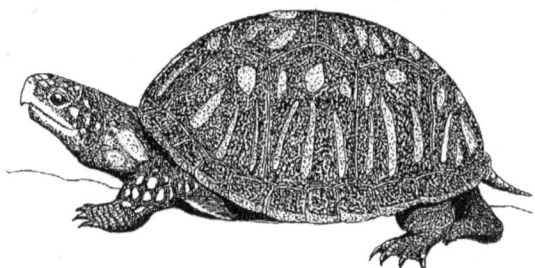

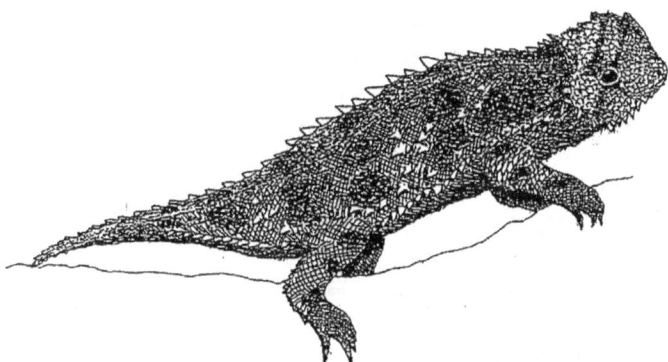

Striped chorus frog, male calling (*upper*), adult ornate box turtle (*middle*), and adult short-horned lizard (*below*)

way out. All the North American turtles can defensively retract their fairly long necks back under their carapace.

Family Chelydridae: Snapping Turtles

Snapping turtles are notable for their hard, bony, rough carapace, which is raised into three keel-like enlargements that extend from front to rear at its center, but the plastron is relatively small and cross-shaped. The tail is at least as long as the carapace, and the upper scales also have saw-toothed keels. Snapping turtles have large heads and powerful jaws and can weigh up to 50 pounds, making them the largest (and much the most dangerous) of Wyoming's turtles. Like all turtles, they are oviparous, and females may lay clutches of up to more than 100 eggs, starting at about 10 to 12 years of age.

Common Snapping Turtle. *Chelydra serpentina*.
Localized.

Identification: Snapping turtles are unmistakable; the large head, long tail, and powerful front and hind legs set them apart from other Wyoming turtles. The triple-keeled rather than smoothly curved carapace is unique, and its posterior portion has uniquely serrated edges. Snapping turtles are the largest turtles in the state, with some individuals weighing up to about 50 pounds. *Length:* Adult carapace length more than 12 inches (300 mm); females are larger than males.

Status: Snapping turtles are mostly limited to eastern Wyoming but extend west to the Bighorn Basin. Snapping turtles are most often found in shallow bodies of water, especially where aquatic vegetation or debris is present.

Habitats and ecology: Snapping turtles move to land to deposit their eggs, the females digging out holes in areas of sandy soil and open vegetation. The sandy substrate evidently provides the best incubation conditions, and the short vegetation makes it easier for the hatchlings to disperse from the nest site. In Wyoming, nesting begins in May and mostly occurs over a three-week period. Clutches of up to as many as 109 eggs have been found, but they average about 50 eggs. The female does not protect them, and raccoons, skunks, and minks are all major egg predators. Hatching occurs from as early as August until as late as October. The sex of the hatchlings depends on incubation temperature, with males being produced at intermediate temperatures and females at both high and low temperatures. From 10 to 12 years are required to reach sexual maturity. Snapping turtles must be treated with great care when closely approaching them or lifting them by their carapace; their long necks can extend surprisingly far forward or backward, and their jaws can easily cut off a finger. I once found an incapacitated western grebe with one missing leg that had obviously been cleanly sheared off by a snapping turtle. Snapping turtles are sometimes hunted for food and are often killed simply because of their threat potential.

Suggested reading: Ernst, Lovich, and Barbour, 1994; Baxter and Stone, 1985.

Family Embydidae: Pond and Box Turtles

All of Wyoming's turtles other than the snapping turtle and spiny softshell turtle belong to the pond (or basking) and box turtle family, which have bony carapaces that are covered with smooth, horny scutes. They have plastrons that are almost as large as their carapaces and often distinctively colored or patterned. Their tails are not as long as their carapaces. One Wyoming species, the ornate box turtle, is hinged on the plastron so to allow the front section to be elevated and permit the animal's head to be fully retracted when threatened. Most species are semiaquatic, and the aquatic species in this family tend to rest on floating logs or at the water's edge (thus the common name "basking" turtles). The highly terrestrial box turtle rarely, if ever, enters water and is often found far from water.

Western Painted Turtle. ***Chrysemys picta belli***. Highly localized.

Identification: The painted turtle is well named: its plastron is red-dish, with an irregularly shaped black to brown central figure, and some red color is often present along marginal scutes of the carapace. *Length:* Carapace length is usually less than 8 inches (200 mm); females are larger than males.

Status: This aquatic turtle occurs locally in north-central, northeast-ern, and southeastern Wyoming. It tends to favor large ponds and lakes rather than small ponds and shallow marshes.

Habitats and ecology: Painted turtles can be found in larger per-manent ponds, lakes, and streams and larger slow-flowing rivers. Basking sites are an important part of the species' hab-itat, and individuals sometimes fight over preferred locations. The turtles are basically diurnal, with most foraging occur-ring in late morning and late afternoon, but nighttime activi-ties have also been reported. They perform limited move-ments between ponds and have been found able to return home after displacements of up to 100 meters, evidently us-ing a sun-compass guide. During winter they might remain active under the ice or burrow into mud at the bottom of the wetland. Mating occurs during early spring and fall, and fe-males lay clutches of about 13 to 14 eggs, with as many as three clutches produced per year in the Nebraska Sandhills. Warmer nests produce young that are all or mostly females, colder nests all males, and intermediate nests produce a mix-ture of the sexes. Apparently the temperature present dur-ing the middle part of the incubation period determines sex. Adult painted turtles eat larger aquatic invertebrates such as crayfish, insects, and insect larvae. They also consume small vertebrates, including salamanders, frogs, and fish, as well as plant materials. Iverson and Smith (1993) have described this species' natural history in Nebraska.

Suggested reading: Ernst, Lovich, and Barbour, 1994; Iverson and Smith, 1993; Baxter and Stone, 1985.

Ornate Box Turtle. ***Terrapene ornata ornata***. Highly localized.

Identification: Box turtles are easily identified. They are Wyoming's only entirely terrestrial turtle, and they have a hinge on the plastron that allows its front end to be lifted and the turtle's head to be withdrawn protectively behind the high-domed carapace. The box turtle has a mixture of yellow lines and spots radiating outward from the center of the carapace; the plastron is dark brown with yellow spots. The iris color is red in adult males and varies from green to yellow-brown or ma-roon in females. *Length:* Carapace length about 4–5 inches (100–125 mm).

Status: The box turtle has been reported from only Goshen County. Box turtles tend to wander during spring and summer, prob-ably in search of females, and have been proven to be able to find their way back to their home range after being displaced over distances of up to nine kilometers. During their travels the turtles often cross roads and frequently are accidentally (or purposefully) run over by motorists. They have few natural predators, otherwise, and Nebraska Sandhills box turtles have been found to survive for up to at least 29 years.

Habitats and ecology: Legler (1960) has thoroughly monographed and described this species' natural history. In Nebraska, box turtles become active as early as mid-April, when males be-gin to seek out females. Mating extends from late April to early June. Clutch sizes are small in box turtles, averaging only about four to five eggs, but females probably produce more than one clutch per year in some populations. As for many other turtles, sex determination apparently depends on temperatures during the incubation period. Hatchlings overwinter in burrows below their nests in western Nebraska and perhaps Wyoming, and during hot weather adults tend to be inactive and remain in burrows (Ballinger, Lynch, and Smith, 2010).

Suggested reading: Ernst, Lovich, and Barbour, 1994; Dodd, 2001; Baxter and Stone, 1985.

Family Trionychida: Softshell Turtles

Softshell turtles are easily defined by a carapace covered not in scutes but rather a leathery skin (sandpaperlike in one spe-cies, smooth in the other), which overlays the bony support-ing skeleton. Softshell turtles also have relatively long tails and large, webbed hind legs that make them powerful swimmers. They have flattened, pancake-shaped carapaces, long necks, and heads with long, tapering noses, which allow them to breathe by lifting only the tip of the nose above water. They eat crayfish, snails, insects, and other animal materials.

Western Spiny Softshell Turtle. ***Apalone spinifera***. Slightly dispersed.

Identification: This distinctive turtle is unique in its flattened, oval body shape; very long neck; and narrow head with a nose that tapers to a point. The carapace is olive-colored with small black spots (more evident in females), and its anterior margin has many spiny tubercles. The tail in males is very long and thick, whereas females have very short tails. *Length:* The dorsal carapace in adult females may be up to 18 inches (400 mm) in length, and in males to 9 inches (200 mm); adult females can be twice as heavy as adult males.

Status: This species occurs in ponds, lakes, and large streams un-der 6,000 feet elevation from northeastern Wyoming west to the Bighorn Mountains and south to Goshen County.

Habitats and ecology: This highly aquatic turtle is rarely found on land except when it is basking on sand bars or laying eggs. It is a powerful swimmer; its forelimbs are adapted to swim in an effective rowing manner that increases its swimming ef-ficiency. Females have large home ranges and may move as far as seven kilometers to find a nesting site. They evidently can use the sun as a compass and have an associated internal clock. Softshells eat a variety of foods, including fish, frogs, tadpoles, crayfish, and aquatic insects. They are sometimes killed and eaten by larger turtles, such as snapping turtles, and like horned lizards, they have been reported to be able to squirt blood from their eyes defensively (Ballinger, Lynch, and Smith, 2010).

Breeding biology: Little is known of the breeding of this species in Wyoming, but in Nebraska breeding occurs from early to mid-June (first clutch) and from late June to early July (second clutch). Clutch sizes range from 16 to 42 eggs, averaging about 30. Eggs incubated under moister conditions are larger than those in drier situations. Egg predation often affects hatching success rates (Ballinger, Lynch, and Smith, 2010).

Suggested reading: Ernst, Lovich, and Barbour, 1994; Baxter and Stone, 1985.

Order Lacertilia—Lizards

Family Scincidae: Skinks

Skinks comprise a family of lizards that have eyelids, small legs, smooth scales, and long, cylindrical bodies and tails. Many species have brightly colored tails with fracture plates that allow the tail to broken off when grasped, thereby permitting the animal to escape with minimal damage and to gradually regenerate its tail. All skinks are egg-layers, and females of the locally occurring species tend their clutches of up to about a dozen eggs during the three- to four-week incubation periods. Skinks are diurnal, eating mostly insects and other arthropods, and are strongly patterned with multiple dark and white body stripes that extend into the tail.

Many-lined Skink. *Eumeces multivirgatus.* Local.
Identification: Skinks are notable among lizards for their very long and often colorful tails that easily detach and are left behind when grabbed by a predator, letting the skink escape and later regrow a new tail. Wyoming has only one skink species, making identification relatively easy. As its common name, the many-lined skink, indicates, it has many alternating dark stripes (8–12) and light dorsal stripes (7–9) that extend from head to tail. Breeding males have red lips. In western Nebraska, a patternless gray to tan variant body pattern is also present, and this morph possibly also occurs in Wyoming. The legs are short, but the tail is very long (1.5 times the snout to vent length); it is bright blue in young individuals. Otherwise, the body is mostly dark brown to greenish and longer in females than in males. *Length:* 5–7.6 inches (125–184 mm).
Status: This skink is limited to southeastern Wyoming in Laramie, Platte, and Goshen Counties.
Habitats and ecology: This lizard is closely associated with sandy soils, such as upland sandsage grasslands and sandy riverside habitats. In Wyoming, it occurs in eastern plains grasslands

and scarp woodlands; it can often be found under loose boards and metal scraps in junkyards. It is often found in Nebraska under cow pies, and in Colorado it has been found in rocky habitats. Insects and insect larvae are typical foods, and spiders are also eaten.
Breeding biology: Breeding occurs in spring and summer; in Wyoming, a female with internal eggs was found in May, and in Nebraska a female with enlarged follicles was found in late June (Ballinger, Lynch, and Smith, 2010). Females brood their eggs, and clutch sizes vary from three to seven eggs. In Colorado, the breeding season extends from late March or April to September or October.
Suggested reading: Baxter and Stone, 1985; Ballinger, Lynch, and Smith, 2010.

Family Phrynosomatidae: Spiny, Earless, Tree, and Horned Lizards

The family of lizards contains more than 130 species in North and Central America. They all have small teeth, and most of the US species have keeled scales. Most of them are diurnal and active insectivores up to about 12 inches in length. Five species are in Wyoming, two of which are in the widespread and species-rich genus *Sceloporus*, and the other three species are members of three different genera. None of them exceed seven inches in total length, and most are insectivorous. Baxter and Stone (1985) included all of them in the very large family Iguanidae. The older terminology of Baxter and Stone (1985) is parenthetically included here for the genera *Sceloporus* and *Phrynosoma* to provide conformity with that widely used state reference, and in the knowledge that it now is outdated as a result of new molecular taxonomic work.

Northern Sagebrush Lizard. *Sceloporus graciosus graciosus.* Slightly dispersed.
Northern Plateau Lizard. *Sceloporus consobrinus (undulatus) elongates.* Highly localized.
Red-lipped Prairie (Fence) Lizard. *Sceloporus consobrinus (undulatus) erythrocheilus.* Highly localized.
Northern Prairie (Plateau) Lizard. *Sceloporus consobrinus (undulatus) garmani.* Highly localized.
Identification: The prairie lizard and sagebrush lizard can be separated by the scales on their posterior thighs, which are small and granular-like in the generally paler and widespread sagebrush lizard and larger, keeled, and overlapping in the variably darker prairie lizard. The three subspecies of the prairie lizard differ in range: the northern plateau lizard is confined to the lower Green River valley, the red-lipped prairie lizard (named for the red or orange-red lips of breeding males) has a scattered distribution in the eastern third of the state, and the northern prairie lizard (also widely known as the eastern

fence lizard) is limited to a few southeastern Wyoming counties. Adult males of the sagebrush, northern plateau, and red-lipped prairie lizards have bright blue throat and belly patches, while the northern prairie lizard differs in lacking blue throat markings. It should also be noted that what Baxter and Stone (1985) termed *S. undulatus* is named *S. consobrinus* by recent authorities (Powell, Conant, and Collins, 2016), although the application of *undulatus* as a valid species category seems to be still unsettled (Ballinger, Lynch, and Smith, 2010). *Length:* Snout–vent length of sagebrush lizard (*graciosus*) is about 2.2 inches (55 mm); of *consobrinus* about 2.2–2.8 inches (55–70 mm). Females are larger than males in both species.

Status: The statuses of these several lizards are varied, but the sagebrush and northern prairie lizards are common, and the other two forms are of peripheral or undetermined status (Baxter and Stone, 1985).

Habitats and ecology: The northern sagebrush lizard is closely associated with sagebrush, while the others are often found among rocky outcrops, canyons, or, in the case of the northern prairie lizard, sandy grasslands. They are all mostly found below 6,000 feet of elevation.

Breeding biology: All of these species are egg-layers, often producing two or three laying cycles in a breeding season. The northern prairie lizard is probably the best studied of these. The breeding season is extended in Nebraska, with the animals emerging from hibernation in late March or early April. The males quickly establish territories, which they maintain until the onset of winter. Mating occurs as early as late April, and a succession of clutches is produced until mid-July. Yearling females produce up to three clutches of four to six eggs; older females have clutch sizes that vary with body size. Hatching continues until mid-September. Mortality is highest during the overwintering period, and the maximum longevity in the wild is about four years (Ballinger, Lynch, and Smith, 2010).

Suggested reading: Baxter and Stone, 1985; Ballinger, Lynch, and Smith, 2010.

Greater (Eastern) Short-horned Lizard.
Phrynosoma hernandesi (*douglassi*). Dispersed.

Identification: The distinctive greater short-horned lizard is unique in having a flattened body covered by granular to spine-like scales and fringed along its widest margins with soft spines. The head has a crown of blunt spines that point posteriorly. The entire animal is gray to brown with large, dark, and irregular spots on each side of the midline and a series of fainter and smaller spots more laterally. The undersides are white, except for a spotted or mottled throat. Length: Snout–vent length 2.2–3.0 inches (57–77 mm); females average slightly larger than males.

Status: This lizard occurs statewide at lower elevations and is especially associated with sagebrush and sage-grasslands

on sandy soils. It can quickly burrow and hide under sandy soil, and its grayish mottled coloration closely resembles sandy or pebbly soil, making it hard to see even when exposed.

Habitats and ecology: Short-horned lizards primarily consume ants but also eat grasshoppers, beetles, other insects, and spiders. Unlike the other lizards in this family, they are slow moving and rely on camouflage to catch their prey. This is the only Wyoming lizard that bears its young alive (vivipary) rather than laying a clutch of eggs that undergoes an incubation period outside the female's body (ovipary). It is also the only Wyoming lizard that squirts blood out of it eyes, which apparently serves as a deterrent when threatened by a canid predator such as a coyote.

Breeding biology: Few regional studies have been done on this inconspicuous species. It is active in Nebraska from May to September and mating occurs in early spring. Litter sizes vary geographically, with reports of 13 to 24 young in Colorado, 10 in Alberta, and 5 to 6 in North Dakota, the number trending smaller in northern populations than in the southern (Ballinger, Lynch, and Smith, 2010).

Suggested reading: Baxter and Stone, 1985; Ballinger, Lynch, and Smith, 2010.

Order Serpentes—Snakes

Snakes are in general easily distinguished from other reptiles, their scaled bodies and legless condition instantly separating them from all other reptiles except for a few legless lizards (glass lizards), which unlike snakes have movable eyelids and external ear openings. Snakes lack any auditory structures, so sounds can be perceived only by vibrations received through the body proper. Both jaws possess teeth, which are slanted backward, and in some species specialized teeth (fangs) exist that serve to inject venom into prey, which is then swallowed whole. The skull bones and jaws are connected loosely, so as to allow for the swallowing of very large prey. Most snakes have platelike ventral scales that can be tilted and by means of muscular action and sinuous body movements used to facilitate locomotion.

Family Colubridae

The snake family Colubridae is the largest of all snake families, numbering nearly 800 species worldwide. Many genera of this family occur in North America. These snakes are all nonvenomous, lack heat-sensitive (infrared-detecting) sensory pits, and have round pupils (most of them). None of the snakes in this family pose a serious threat to humans, although all of them eat whole animals, ranging in size from insects to large rodents and other snakes. Most colubrids are egg-layers (oviparous), producing clutches of up to about 25 eggs. Most of them kill by constriction, although some "rear-fanged" snakes are mildly poisonous, having grooved teeth near the rear of the jaw that transmit toxic saliva to their prey while the snake is holding or swallowing it. Some species of this group, including hog-nosed snakes, are sometimes recognized as a separate family, Dipsadidae.

Plains Hog-nosed Snake. *Heterodon nasicus nasicus*. Local.

Identification: Hog-nosed snakes are unique among American snakes in that they have an upturned and pointed snout (an enlarged and strongly keeled rostral scale). This provides a digging tool that helps in burrowing and in uncovering toads, their primary prey. They also have enlarged teeth at the back of the upper jaw that are used to deflate toads and frogs that have inflated their bodies with air to avoid being swallowed. There is only one Wyoming species of hog-nosed snake, but in western Nebraska the very similar eastern hog-nosed snake also occurs. The two can be distinguished by the fact that the snout of the eastern species is only slightly up-turned and also by differing underpart coloration (Ballinger, Lynch, and Smith, 2010). Adult total length: 21–24.5 inches (535–622 mm).

Status: The hog-nosed snake occurs in the grasslands of eastern Wyoming, east of a line extending from the eastern edge of the Bighorn Mountains southeast to Laramie County.

Sound production: Although mute, a hog-nosed snake will produce a hissing sound when threatened, while spreading its head and thereby somewhat resembling a rattlesnake. It may even strike, often with its mouth closed. On further provocation, the snake will writhe convulsively, turn over on its back, apparently gasp, and become "lifeless." If left alone, the snake will soon turn over and crawl away, or if placed in water will swim away.

Habitats and ecology: Hog-nosed snakes are found at lower elevations in Wyoming, where toads (their favorite prey) are abundant, and thus they are often found in areas of sandy soil. They also eat frogs, salamanders, lizards, turtles, turtle eggs, birds, and mammals. The snakes are active for about six months of the year, from early May to late October in Kansas (Ballinger, Lynch, and Smith, 2010).

Breeding biology: The hog-nosed snake is oviparous, laying clutches of 3 to 16 eggs, averaging about 13 in Kansas and Nebraska. Frequently females will skip reproducing for a year and develop a biennial reproductive cycle (Ballinger, Lynch, and Smith, 2010).

Suggested reading: Baxter and Stone, 1985; Ballinger, Lynch, and Smith, 2010; Shaw and Campbell, 1974.

Bullsnake (Gophersnake). *Pituophis catenifer.*

Identification: The bullsnake, or gophersnake, is Wyoming's most common and largest snake, with older adults often reaching 60 to 70 inches in length, and rarely approaching 100 inches,

at least historically. Its head is large and somewhat triangular, with alternating dark and yellow bands extending from the eye to the jaw, and a brown band extending diagonally from behind the eye to the angle of the jaw. More than 40 black to reddish brown blotches are scattered along the back, sides, and tail, and the yellow belly is also spotted with black. The dorsal scales are strongly keeled. *Length:* Average of 15 adults, 51.7 inches (1.313 m).

Sound production: Like other snakes, bullsnakes are voiceless, but when threatened they can make a hissing sound by exhaling and passing air over a unique laryngeal structure. They also often shake their tail, which might resemble the rattle of a rattlesnake as it brushes through shrubby vegetation. A threatened bullsnake can even change the shape of its head to become more triangular, something like a rattlesnake's. Such features suggest that bullsnakes may be mimicking rattlesnakes as a protective adaptation (Kardong, 1980).

Status: Bullsnakes have been recorded over most of Wyoming's northeastern counties. They are also among the state's largest snakes.

Habitats and ecology: This species is primarily a grassland snake that especially favors taller prairies but also often enters farmlands and urbanized areas; it rarely occupies woodlands. Bullsnakes are diurnal and active from about April to October. They breed in spring, shortly after emerging from their winter burrows, and females lay eggs until as late as July. Clutches average 12 to 13 eggs but range from 8 to 17. The incubation period is about 70 days. At night they return to their burrow, which is often a gopher burrow (thus the alternative common name of gophersnake) or prairie dog burrow. Bullsnakes are also able to dig soil by using head movements. They eat a variety of mammals (especially rodents), lizards, and ground-nesting and cavity-nesting birds as well as their eggs and nestlings. They are remarkably adept at climbing trees and can even climb vertical concrete walls to reach the nests of culvert-nesting cliff swallows.

Suggested reading: Baxter and Stone, 1985; Kardong, 1980; Ballinger, Lynch, and Smith, 2010; Shaw and Campbell, 1974.

Family Natricidae: Live-bearing Snakes

Like the closely related colubrid snakes (and sometimes included as part of that family), members of the Natricidae assemblage have round pupils and lack facial sensory pits. However, they are livebearers, producing litters of rarely up to more than 80 offspring. Gartersnakes overwinter socially in collective underground retreats (hibernacula), which in northern latitudes might support hundreds of individuals. Mating may occur among the adults either before entering a hibernaculum or after spring emergence. Male gartersnakes typically emerge in spring before the females and patrol the vicinity for weeks while waiting for the females to emerge.

Wandering (Western Terrestrial) Gartersnake.
 Thamnophis elegans. Widespread.

Common Gartersnake. *Thamnophis sirtalis*. Local.

Western Plains Gartersnake. *Thamnophis radix*.
 Highly local.

Identification: The wandering gartersnake can be easily separated from the other two species by the absence of red or bright yellow on its body. Instead it has a brown or sometimes greenish body with three longitudinal stripes, one dorsal and two lateral, between which are two alternating rows of darker brown spots. The head is also brown, with pale yellow spots, and the underparts are gray to pale blue. The common gartersnake closely resembles the plains gartersnake. Both species are strongly striped dorsally and along their sides with dark and light stripes. The common gartersnake alternates brick red and black vertical bars and wedges along its entire body, except for an unmarked white to bluish underside, whereas most populations of the plains gartersnake alternate yellow and black patterning. Both species have a conspicuous pale stripe down the middle of the back, which is yellow in the common gartersnake but usually bright orange in the plains gartersnake. Adult total length, for both species, is about 24 to 27.5 inches (606–653 mm).

Status: The wandering gartersnake has been recorded in nearly all the state's counties, the common gartersnake in at least six, and the plains gartersnake in at least four eastern counties. All three species tend to favor locations close to water, but the common gartersnake seems to also have a preference for grassland-woodland transitional habitats where rocks and other escape cover are present. The wandering gartersnake is the state's most widespread and common snake, extending in elevation up to 10,000 feet.

Habitats and ecology: Gartersnakes are diurnal and seasonally active in Wyoming, emerging in March from underground hibernacula that are shared by large numbers of others, sometimes including other species of snakes. They probably return to their winter dens in late October but may briefly emerge during warm winter days. With spring emergence, mating begins (Joy and Crews, 1985). Males emerge from the hibernacula well before the females, and then patrol the area while waiting for the females to emerge. From pheromone (cloacal secretion) trails left by females, the males can recognize those from the male's own den and may prefer to court them. Males may also use visual clues in their choices of females to be courted. Writhing "knots" of several simultaneously mating snakes can sometimes be seen. Female common gartersnakes produce litters that vary from about 13 to 40 young. Plains gartersnake litters average about 17 to 18 young. The young of all species hatch from July to September, and most adults and young probably move into hibernacula by October. During summer, common gartersnakes are surprisingly mobile, with males

having home range of up to 35 acres, and females up to nearly 23 acres. Displaced common gartersnakes are able to orient themselves in the correct homeward direction, apparently using the sun as a compass. Adult common gartersnakes prey on a variety of animals, especially leopard frogs, but also small rodents, earthworms, and others, evidently using chemical clues to identify potential prey. Plains gartersnakes similarly mostly feed on amphibians, fish, earthworms, and slugs. Wandering gartersnakes are most common near water, where they feed on frogs and small fish but also small mammals and some invertebrates. Various studies indicate annual survival rates of gartersnakes to range from 34 to 50 percent, meaning that most individuals don't live beyond three to four years.

Suggested reading: Baxter and Stone, 1985; Joy and Crews, 1985; Ballinger, Lynch, and Smith, 2010; Shaw and Campbell, 1974.

Family Viperidae: Pitvipers

All the members of this uniformly dangerous Viperidae snake family have large, erectile, and hollow upper teeth (fangs) that can inject venom, which is used to subdue their prey as well as defensively. Pitvipers have large triangular heads, eyes with vertical pupils, and paired infrared (heat-sensitive) pits located just behind their nostrils. All of these snakes specialize in eating warm-blooded vertebrates, although other snakes, lizards, frogs, and a variety of invertebrates may also be consumed. All of the most dangerous North American snakes belong to this family.

Prairie Rattlesnake. *Crotalus viridis*. Widespread.

Identification: The rattle of a rattlesnake is enough to identify it, although rattlesnakes also have the family traits of large triangular heads, eyes with vertical pupils, and sensory pits behind their nostrils. Most Wyoming individuals are pale brown to dark brown with about 40 dark dorsal spots that become bands posteriorly and rings on the tail. Two light stripes extend from behind and below the eye back to the angle of the jaw. The "midget faded" snake population from the Green River valley (*C. v. concolor*) is paler brown to tan throughout with 43 dorsal blotches and a faint white line extending from the angle of the jaw forward to the front of the eye (Baxter and Stone, 1985). *Length:* Rattlesnakes are one of Wyoming's larger and longer snake species (second to the bullsnake), often reaching 45 inches (114 cm), and elsewhere reported as reaching nearly 59 inches (123.6 cm). Among 117 Wyoming specimens, males of *C. v. viridis* averaged 35.6 inches (905 mm) and females 31 inches (787 mm). The "midget" race (*C. v. concolor*) that occurs in the Green River valley has an adult length of usually less than 25.6 inches (650 mm) (Baxter and Stone, 1985).

Sound production: The defensive rattle of a rattlesnake is easily recognized—and is sometimes mimicked by other species, such as bullsnakes, which shake their rattleless tails in a similar manner.

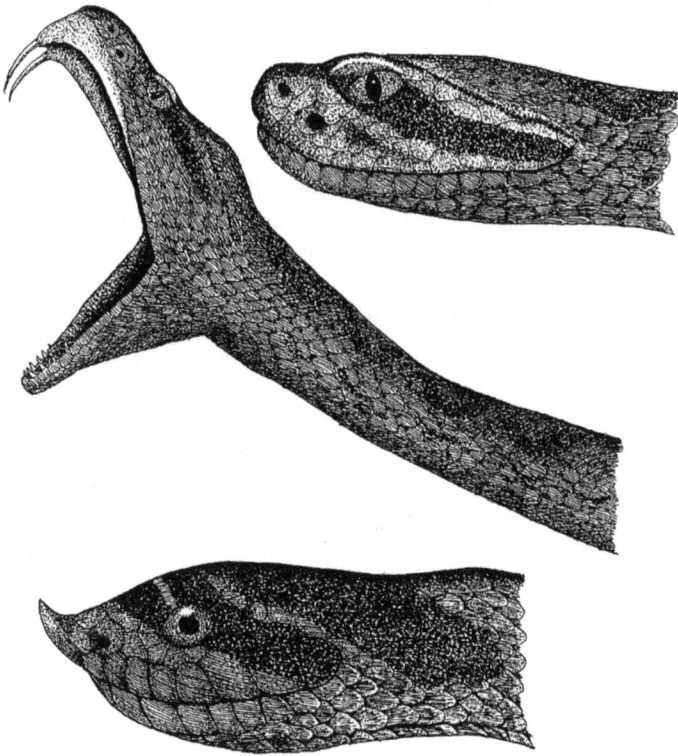

Prairie rattlesnake, head (*upper*) and striking posture (*middle*); western hog-nosed snake, head (*below*)

Status: This species is widespread on the plains and foothills up to about 7,000 feet of elevation east of the Continental Divide, and it occurs locally west of the Divide in Carbon County.

It is most common in scarp woodlands and foothills, where rimrock and outcrops of limestone occur. On the plains it is often found among prairie dog colonies, using the burrows for shelter and the prairie dogs as a food source (Baxter and Stone, 1985).

Habitats and ecology: The primary food of rattlesnakes consists of rodents, making them a useful addition to the fauna, but they also eat rabbits, lizards, birds, and other snakes. They are mainly diurnal in activity and hibernate in large aggregations, often in deep rocky crevices of limestone outcrops, or in prairie dog burrows. Females may begin to gather as early as August, followed by hatchlings only nine to ten inches long that were born in August and September. Yearling snakes then average about 24 inches (610 mm) in length, and two-year-olds about 30 inches (760 mm). Snakes of the latter age usually have six to seven buttons on the tail, reflecting the minimum number of their previous molts. The average amount of venom injected during a bite even in adult snakes is only about one-eighth the lethal dose for an adult human (Baxter and Stone, 1985).

Breeding biology: Mating in rattlesnakes may occur in the fall, and courtship in Nebraska has been seen as late as autumn (Ballinger, Lynch, and Smith, 2010), but it is much more likely to occur in the spring, soon after the snakes emerge from their dormancy. The females give birth to their young in August and September. Mean litter size in a Kansas study averaged 10.3 young, increasing with female body size, whereas a population of the midget race in Wyoming averaged only 4.7 young (Baxter and Stone, 1985).

Suggested reading: Baxter and Stone, 1985; Klauber, 1972; Ballinger, Lynch, and Smith, 2010; Shaw and Campbell, 1974.

Chapter 5 • Taxonomic Checklist and Species Codes

Explanation of Abbreviations and Symbols Used in the Taxonomic Synopses

Latilongs and Biological Mapping
A very useful application of latitudes and longitudes is to allow the concept of a latilong, which represents a rectangular area encompassing one degree of latitude and one degree of longitude. Wyoming's latilongs provide a shorthand means of describing the state's distribution of individual vertebrate species. Latilongs were first used biologically to plot bird distributions in Montana in 1975 and later were applied in Colorado to define the state's distributions of birds, mammals, reptiles, and amphibians (Bissell and Graul, 1981). They were similarly first used in Wyoming in 1997 (Luce et al., 1997) and have been periodically updated (Cerovski et al., 2004; Orabona et al., 2012).

Encompassing five lines of latitude, the state of Wyoming extends 275 miles in north-south distance. It also extends 365 miles in east-west distance at Wyoming's southern boundary with Colorado, and 342 miles at its northern boundary with Montana, encompassing eight lines of longitude. Wyoming thus has 28 latilongs, which occupy the state's 97,914 square miles for an average of about 3,500 square miles per latitlong (Map 7). All latilongs extend 68.75 miles north-south but because lines of longitude converge northwardly, the latilongs in Wyoming vary in width from 51.7 miles at the state's southern border to 48.8 miles at the northern border, making the southernmost latilongs at most about 4 percent larger than the northernmost. Fom a biological mapping standpoint, though, they can be considered as essentially the same.

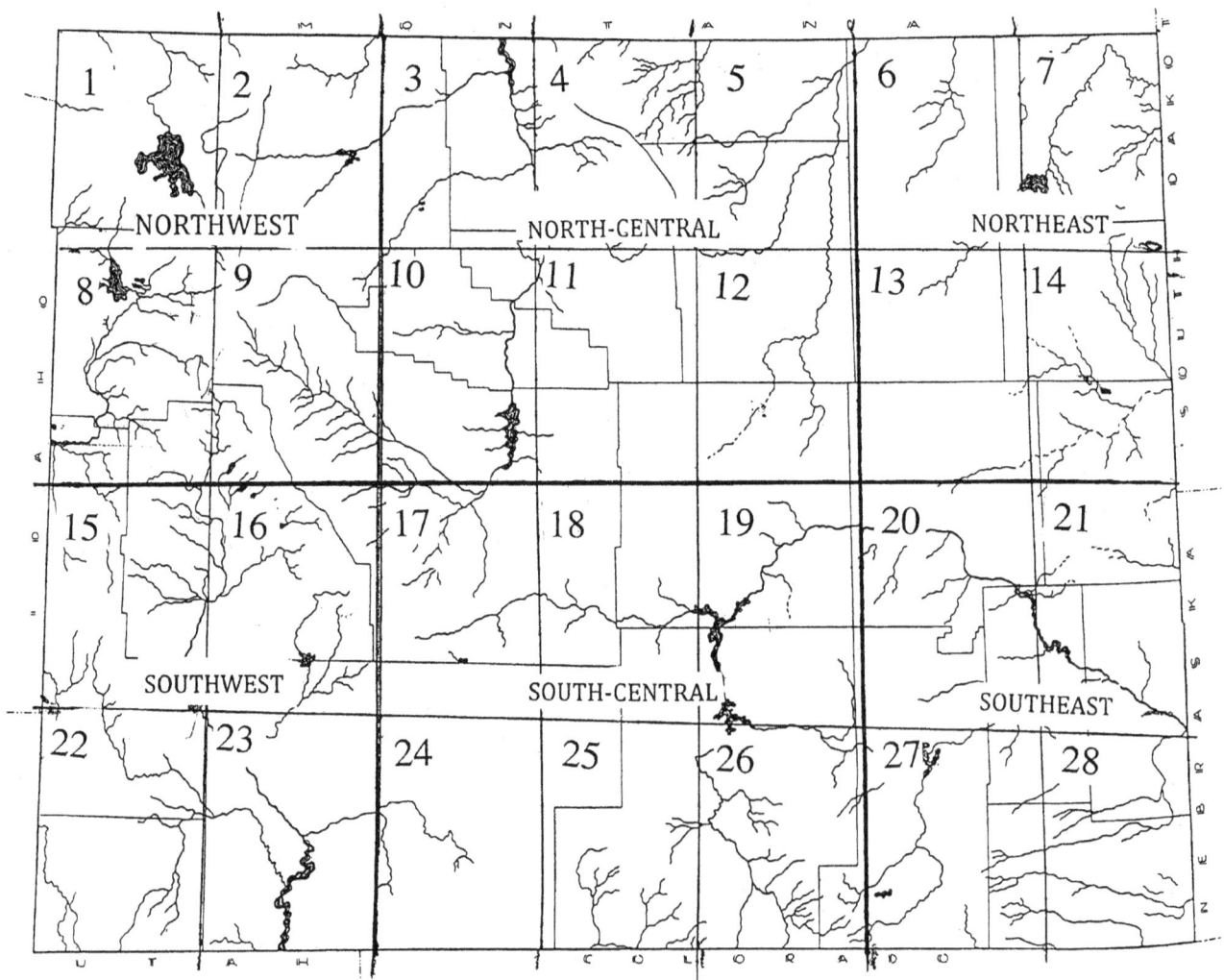

Map 7. Wyoming latilongs and subregion names. After Orabana et al. (2012).

Breeding/Distribution Symbols
(Mammals, Birds, and Herpetiles)*

Ub   Ubiquitous; resident or breeding in 26–28 latilongs
Wi   Widespread; resident or breeding in 21–25 latilongs
All   Breeding reported within all six regional quadrants (birds) or sightings/specimens reported within all six quadrants (mammals)
Di   Dispersed; resident or breeding in 16–20 latilongs (regional distribution quadrants are listed parenthetically)
SD   Slightly dispersed; resident or breeding in 11–15 latilongs (regional distribution quadrants are listed parenthetically)
Lo   Localized; resident or breeding in 6–10 latilongs (latilong numbers are listed parenthetically)
HL   Highly localized; resident or breeding in 1–5 latilongs (latilong numbers are listed parenthetically)

*Based on number of latilongs reported by Orabona et al. (2012) for confirmed (bold) and circumstantial post-1949 breeding bird records with some additions or updates based on more recent information. All post-1964 latitlong records for mammals and herpetiles were counted; latilongs where good evidence of breeding was obtained are shown in boldface, and latilongs where breeding was assumed but not established are shown in *italics*. Where regional distribution symbols are listed parenthetically, that regional quadrant symbol is shown in bold if all latilongs within a region have reported presumptive or established breeding by the species.

Regional Distribution Symbols
(All Vertebrate Groups)

Pan   Pandemic, reported from at least 24 latilongs
Spr   Spread over at least 20 latilongs
NW   Northwestern WY quadrant record(s), latilongs 1, 2, 8, 9
NC   North-central WY quadrant record(s) latilongs 3–5, 10–12
NE   Northeastern WY quadrant record(s), latilongs 6, 7, 13, 14
SE   Southeastern WY quadrant record(s), latilongs 15, 16, 22, 23
SC   South-central WY quadrant record(s), latilongs 17–19, 24–26
SW   Southwestern WY quadrant record(s), latilongs 20, 21, 27, 28
Ea   Mainly distributed east of Continental Divide
We   Mainly distributed west of Continental Divide
No   Mainly distributed north of Lat. 43°N
So   Mainly distributed south of Lat. 43°N
?   Data inadequate for regional attribution

Abundance/Conservation Symbols
(Birds and Herpetiles)*

Ab   Abundant
Co   Common
Un   Uncommon
Sca   Scattered records only
Ra   Rare
VR   Very rare
Vagrant   Extralimital species out of normal range (birds)
Pe   Peripheral (edge-of-range) species (herpetiles)
?   Abundance uncertain
CC   Conservation Concern species**
Ch   Charismatic species***

* Bird abundance estimates are nearly all from Faulkner (2010); herpetile abundance estimates are from Baxter and Stone (1985).
** CC1 Watch List bird species of national conservation concern (Rich et al., 2004). CC2 Bird species with surveywide annual breeding population declines (1966–2015) exceeding 1.0% (Sauer et al., 2017). CC3 Threatened, endangered, and vulnerable vertebrate species in the Rocky Mountains (Finch, 1991). CC4 Wyoming Conservation Priority Species.
*** Ch is considered (by author) to be a charismatic species high in tourist-viewing interest.

Seasonal Occurrence Symbols
(Birds)

Sp   Spring (March through May)
Su   Summer (June and July)
Fa   Fall (August through November)
Win   Winter (December through February)
?   Seasonal occurrence uncertain

Temporal Status Symbols
(Birds)

Mig   Migrant (spring and fall)
WM   Wintering migrant
SM   Spring migrant
Res   Resident (breeding assumed or proven)
Vis   Visitor (nonbreeding)
?   Current temporal status uncertain

Habitat Symbols
(Mammals, Birds, and Herpetiles)*

AlSh	Alkaline/Desert shrubland
AlWe	Alkaline wetland
FoAs	Forest; aspen; aspen-coniferous
FoLP	Forest; lodgepole pine
FoMC	Forest; mixed coniferous montane
FoDC	Forest; deciduous-coniferous
FoDe	Forest; deciduous
FoOG	Forest; old-growth coniferous
FoPP	Forest; ponderosa pine foothills
FoSF	Forest; spruce-fir subalpine
FoSG	Forest; second-growth (burned or logged)
GrMG	Mixed-grass prairie
GrSG	Short-grass prairie (steppe)
GrTG	Tallgrass prairie
GrVa	Grassland; variably weedy, brushy
HuCL	Human-modified; cultivated land
HuFR	Human-modified; farmsteads, ranches
HuPG	Human-modified; parks, gardens
HuTC	Human-modified; towns, cities
MoAl	Montane; alpine (tundra, rock fields, etc.)
MoMe	Montane; meadow
MoRo	Montane/Canyon; exposed rock (cliff, talus, etc.)
MoTi	Montane; timberline (stunted trees)
RiDe	Riverine; deciduous woods shoreline
RiHe	Riverine; herbaceous shoreline
RiLe	Riverine; lentic (flowing-water)
RiSS	Riverine; shrub shoreline
SaSG	Sage-shortgrass (steppe)
SaSh	Sage scrubland
WeDW	Wetlands; deep water (lakes, reservoirs)
WeSW	Wetlands; shallow-water (marshes, playas, ponds)
WoJu	Woodland; juniper, juniper-deciduous

* Personally chosen symbols

Elevation/Community Symbols
(Herpetiles)*

Alp	Alpine, above 10,500 ft; alpine tundra and bare rock
Sub	Subalpine, 9,600–10,500 ft; high coniferous forest and timberline
Mon	Montane, 8,000–9,500 ft; mixed or coniferous montane forests
Foo	Foothills, 6,000–8,000 ft; woodlands, savanna, and forest transition
Pla	Plains, below 6,000 ft; grasslands, scrublands, and semidesert

* Based on Baxter and Stone (1985)

Habitat and Elevation Symbols
(Mammals)*

1. Grassland (plains, foothills, and montane parks, 3,100–ca. 9,000 ft)
2. Semidesert shrubland (plains to subalpine zone, 3,100–10,000 ft)
3. Pinyon-juniper woodland (plains and foothills zone, 3,100–8,000 ft)
4. Montane shrubland (foothills and lower montane zones, 6,000–ca. 8,500 ft)
5. Montane woodland and forest (foothills to upper montane zone, 6,000–9,500 ft)
6. Subalpine forest (subalpine to timberline zone, 9,500–10,500 ft)
7. Alpine tundra (above timberline, 10,500–13,800 ft)
8. Riparian and wetland habitats (plains to subalpine zone, 3,100–9,500 ft)

* Habitats are based on Armstrong, Fitzgerald, and Meaney (2011). Elevations are based mainly on Reed (1969) and Hoffman and Alexander (1976).

MAMMALS

Family and generic sequences generally follow Clark and Stromberg (1987); congeneric species are listed alphabetically by specific names. English names and binomial nomenclature mostly follow Kays and Wilson (2002) except where more recent taxonomic research has required name changes. Older or alternate nomenclature is shown parenthetically.

Order Marsupialia (Marsupials)
Family Didelphidae (Opossums)
Virginia Opossum. *Didelphis v. virginiana.* Lo (4, 11, 14, 20, *21, 22,* 23, 27, 28); NE, SE; RiDe, RiSS; WeSW, HuFR, HuTC, HuCL.

Order Soricomorpha (Shrews and Moles)
Family Soricidae (Shrews)
Cinereus (Masked) Shrew. *Sorex cinereus.* Pan; Habs. 5–8, also diverse moist habitats.
Prairie (Hayden's) Shrew. *Sorex haydeni.* HL (*5, 7*); NE; Habs. montane bogs, willow flats.
Pygmy Shrew. *Sorex hoyi.* HL (26); Habs. 6, 8.
Merriam's Shrew. *Sorex merriami.* Wi (all quadrants); Habs. 1–5.
Montane (Dusky) Shrew. *Sorex monticolus.* Wi; NW, NC, SW, SC, SE; Habs. 5–8, also moist montane, alpine tundra.
Dwarf Shrew. *Sorex nanus.* Lo (2, 8, 15, *16,* 17, *19, 20,* 24, 26, 27); GrSG, HuFR, WoJu; Habs. 4–8, also alpine rock fields.
Western Water Shrew. *Sorex (palustris) navigator.* Di; NW, NC, SW, SC, SE; Hab. 8, cold, clear streams and wet meadows.
Preble's Shrew. *Sorex preblei.* HL (1, *2,* 8); Hab. 4, also diverse habitats from sage to tundra.

Vagrant Shrew. *Sorex (monticulus) vagrans*. Di; SW, NW, NC, NE; Habs. 5–8, also riparian habitats.

Family Talpidae (Moles)
Eastern Mole. *Scalopus aquaticus caryi*. HL (5, 21, 28); NC, SE; Hab. 1.

Order Chiroptera (Bats)
Family Vespertilionidae (Vesper Bats)
Pallid Bat. *Antrozous pallidus*. Lo (3, 4, 10, 11, 17, 19, 21, 22, 25, 27); Habs. 2–4, 8.
Big Brown Bat. *Eptesicus fuscus*. Wi (all); Habs. 2–6, 8.
Eastern Red Bat. *Lasiurus borealis*. HL (7, 10, 11, 14, 22, 25, 26, 27); Habs. 8.
Hoary Bat. *Lasiurus cinereus*. Wi (all); Habs. 2, 3, 5, 6, 8.
California Myotis. *Myotis californicus*. HL (3, 23); Habs. 2, 3.
Western Small-footed Myotis. *Myotis ciliolabrum*. Wi (all); Habs. 1–5, 8.
Long-Eared Bat. *Myotis evotis*. Wi (all); Habs. 3–5.
Little Brown Myotis. *Myotis lucifugus*. Ub; Habs. 5, 6, 8.
Northern Myotis. *Myotis septentrionalis*. HL (7, 8, 21); FoPP, FoMC, also dense forest.
Fringed Myotis. *Myotis thysanodes*. Di; NW, NC, NE, SE, SC; Habs. 3–5.
Long-legged Bat. *Myotis volans*. Wi (all); Habs. 3–6.
Yuma Myotis. *Myotis yumanensis*. HL (3, 5); Habs. 2–5, 8.
Silver-haired Bat. *Lasionycteris noctivagans*. Wi (all); Habs. 2, 5, 6, 8.
Spotted Bat. *Euderma maculatum*. Lo (2, 3, 10, 11, 12, 23); Habs. 2–4, 8.
Townsend's Big-eared Bat. *Corynorhinus townsendii*. Wi (all); Habs. 1–5.
American Perimyotis (Eastern Pipistrelle). *Perimyotis (Pipistrellus) subflavus*. HL (21); Hab. 8.
Brazilian Free-tailed Bat. *Tadarida brasiliensis*. HL (3, 16, 28); Habs. 1–4.
Big Free-tailed Bat. *Nyctinomops macrotis*. HL (8); Hab. 1.

Order Lagomorpha (Rabbits, Hares, and Pikas)
Family Ochotonidae (Pikas)
American Pika. *Ochotona princeps*. Di (all); Habs. 6, 7.

Family Leporidae (Hares and Rabbits)
Snowshoe Hare. *Lepus americanus*. Di (all); Habs. 6, 8.
Black-tailed Jackrabbit. *Lepus californicus*. SD; NC, NE, SW, SC, SE; Habs. 1, 2.
White-tailed Jackrabbit. *Lepus townsendii*. Ub; Habs. 1–5, 7.
Desert Cottontail. *Sylvilagus audubonii*. Ub; Habs. 1–4, 8.
Eastern Cottontail. *Sylvilagus floridanus*. HL (21); Habs. 8.
Mountain (Nuttall's) Cottontail. *Sylvilagus nuttallii*. Wi; Habs. 4–6, 8.
Pygmy Rabbit. *Brachylagus idahoensis*. HL (15, 16, 17, 22, 23, 24, 25); Hab. 2, obligate in sage, sage-grassland.

Order Rodentia (Rodents)
Family Sciuridae (Squirrels, Marmots, and Prairie Dogs)
Yellow-bellied Marmot. *Marmota flaviventris*. Wi (all); Habs. 4–8.
Yellow-pine Chipmunk. *Tamias amoenus*. Lo (1, 2, 8, 9, 10, 15, 16, 22); FoAs, FODC, FoLP, FOMC, WOJU.
Cliff Chipmunk. *Tamias dorsalis*. HL (23, 24); MoRo; Habs. 2, 3.
Least Chipmunk. *Tamias minimus*. Ub; Habs. 2–7.
Uinta Chipmunk. *Tamias umbrinus*. SD (all); Habs. 2–8.
Northern Flying Squirrel. *Glaucomys sabrinus*. SD (all); FoAs, FoPP, FoOG, FoPP.
Eastern Fox Squirrel. *Sciurus niger*. Di (all); Hab. 8.
Red (Pine) Squirrel (Chickaree). *Tamiasciurus hudsonicus*. Wi; NW, NC, NE, SW, SC, SE; Habs. 5, 6.
Abert's Squirrel. *Sciurus aberti*. HL (26, 27); Hab. 5.
Thirteen-lined Ground Squirrel. *Ictidomys (Spermophilus) tridecemlineatus*. Wi; NC, NE, SW, SC, SE; Habs. 1, 2.
Golden-mantled Ground Squirrel. *Callospermophilus (Spermophilus) lateralis*. Di; NW, SW, SC, SE; Habs. 2–8.
Uinta Ground Squirrel. *Urocittellus armatus*. SD; NW, NE, SE, SC, SW; Habs. 2–8.
Wyoming Ground Squirrel. *Urocittellus (Spermophilus) elegans*. Wi; NW, NC, SW, SC, SE; Habs. 1, 2, 4, 5, 7.
Spotted Ground Squirrel. *Xerospermophilus (Spermophilus) spilosoma*. HL (20, 21, 28); Habs. 1, 2.
White-tailed Prairie Dog. *Cynomys leucurus*. Wi (all); Habs. 1, 2, 4.
Black-tailed Prairie Dog. *Cynomys ludovicianus*. SD; NC, NE, SE, SC; Habs. 1, 2.

Family Geomyidae (Pocket Gophers)
Wyoming Pocket Gopher. *Thomomys clusius*. HL (24, 25); AlSh, gravelly loose soils.
Idaho Pocket Gopher. *Thomomys idahoensis*. HL (22); AlSh, MoMe, shallow, stony soils.
Northern Pocket Gopher. *Thomomys talpoides*. Ub; Pan; Habs. 1, 2, 4–7.
Nebraska (Plains) Pocket Gopher. *Geomys (bursarius) lutescens*. HL (13, 21, 22, 28); Hab. 1.

Family Heteromyidae (Pocket Mice)
Olive-backed Pocket Mouse. *Perognathus fasciatus*. Wi (all); Habs. 1, 2.
Plains Pocket Mouse. *Perognathus flavescens*. HL (5, 19, 21, 27, 28); Habs. 1–3.
Silky Pocket Mouse. *Perognathus flavus*. HL (13, 14, 20, 21, 27, 28); NE, SE; Habs. 1–3.
Great Basin Pocket Mouse. *Perognathus (parvus) mollipilosa*. HL (22, 23, 24); Habs. 1–3.
Hispid Pocket Mouse. *Chaetodipus hispidus*. HL (7, 21, 27, 28); Habs. 1, 2.
Ord's Kangaroo Rat. *Dipodomys ordii*. Di; NC, NE, SW, SC, SE; Habs. 1–3.

Family Castoridae (Beavers)
Beaver. *Castor canadensis*. Ub; Pan; Hab. 8.

Family Cricetidae (New World Mice)
Northern Grasshopper Mouse. *Onychomys leucogaster*. Ub; Sca; Habs. 1, 2.
Western Harvest Mouse. *Reithrodontomys megalotis*. Di (all); Habs. 1, 2, 8.
Plains Harvest Mouse. *Reithrodontomys montanus*. Lo (*7*, 14, 21, 27, 28); Hab. 1.
Canyon Mouse. *Peromyscus crinitus*. Lo (23); Habs. 2, 3, juniper rock outcrops.
White-footed Mouse. *Peromyscus leucopus*. Lo (2, 4, *5*, *6*, 7, 8, 12, 13, 21, 26); Hab. 8.
Deer Mouse. *Peromyscus maniculatus*. Ub; Pan; Habs. 1–8.
Pinyon Mouse. *Peromyscus truei*. HL (23, 24); Habs. 2–4.
Muskrat. *Ondatra zibethicus*. Ub; Pan; Hab. 8.
Heather Vole. *Phenacomys intermedius*. SD (all); Habs. 5–7.
Southern Red-backed Vole. *Clethrionomys gapperi*. Wi (all); Habs. 5, 6.
Long-tailed Vole. *Microtus longicaudus*. Wi (all); Habs. 4–8.
Montane Vole. *Microtus montanus*. Wi; NW, NC, SW, SC, SE; Habs. 4–8.
Prairie Vole. *Microtus ochrogaster*. Di; NW, NC, NE, SC, SE; Habs. 1, 8.
Meadow Vole. *Microtus pennsylvanicus*. Wi (all); Habs. 1, 8.
Water Vole. *Microtus richardsoni*. Lo (1, 2, 4, 8, 9, *12*, 16, *17*); RiSS, RiHe, MoMe.
Sagebrush Vole. *Lemmiscus curtatus*. Di (all); Habs. 2, 4, sage obligate.
Bushy-tailed Woodrat. *Neotoma cinerea*. Ub; Pan; Habs. 3–7.

Family Muridae (Old World Mice and Rats)
House Mouse. *Mus musculus*. SD (all); HuCL, HuFR, HuTC.
Norway Rat. *Rattus norvegicus*. HL (6, 7, 19, 27, 28); HuCL, HuFR, HuTC.

Family Dipodidae (Jumping Mice)
Meadow Jumping Mouse. *Zapus hudsonicus*. HL (6, 7); Habs. 5, 8.
Western Jumping Mouse. *Zapus princeps*. Wi (all); Habs. 5, 8.

Family Erethizontidae (Porcupines)
Porcupine. *Erethizon dorsatum*. Ub; Pan; Habs. 3–8.

Order Carnivora (Carnivores)
Family Canidae (Dogs)
Coyote. *Canis latrans*. Ub; Pan; Habs. 1–8.
Gray Wolf. *Canis lupus*. HL (1, 2, 3, 8, 9, 10, 11, 16, 17; historically all); Habs. 1–8, federally listed as threatened in Wyoming until 2017, now legally hunted.
Swift Fox. *Vulpes velox*. Di; NC, NE, SW, SC, SE; GrSG; Hab. 1.
Red Fox. *Vulpes vulpes*. Ub; Pan; Habs. 4–8.
Gray Fox. *Urocyon cinereoargenteus*. SD (all); Habs. 2–5, 8.

Family Ursidae (Bears)
Black Bear. *Ursus americanus*. Di (all); Habs. 4–6, 8.
Grizzly Bear. *Ursus arctos horribilis*. Lo (1, 2, 3, 8, 9, 10, 15, 16, 17; historically all); Habs. 1–8, federally listed as threatened.

Family Procyonidae (Raccoons)
Northern Raccoon. *Procyon lotor*. Ub; Pan; Hab. 8.
Ringtail. *Bassariscus astutus*. HL (22, 23, *26*, 27, *28*); Habs. 2–5, 8.

Family Mustelidae (Weasels)
American Badger. *Taxidea taxus*. Ub; Habs. 1–4, 7.
Fisher. *Pekania (Martes) pennanti*. Lo (*1*, *2*, 5, 8, *9*, 11, 17, 26); FoMC, FoOG, RiDe, FoAs.
American (Pacific) Marten. *Martes (americana) caurina*. SD (all); Habs. 5–7.
American Mink. *Neovison vison*. Ub; Pan; Hab. 8.
Northern River Otter. *Lontra canadensis*. Di (all); Hab. 8.
Ermine (Short-tailed Weasel). *Mustela erminea*. Wi (all); Habs. 5–8.
Long-tailed Weasel. *Mustela frenata*. Ub; Pan; Habs. 1–8.
Black-footed Ferret. *Mustela nigripes*. Lo (19; many other historical and recent records from reintroductions); GrSG, prairie dog associate, federally endangered.
Least Weasel. *Mustela nivalis*. Lo (4, *5*, *6*, *7*, *14*); MoMe, RiDe SaSG.
Wolverine. *Gulo gulo*. Lo (*1*, *2*, 5, *8*, *9*, *15*, *16*, 17, *26*, 28); FOMC, FoOG, dense coniferous forest, candidate for federal listing as threatened/endangered.
Striped Skunk. *Mephitis mephitis*. Ub; Pan; Habs. 1–8.
Western Spotted Skunk. *Spilogale gracilis*. SD; NW, NC, SW, SC, SE; Habs. 2–5.
Eastern Spotted Skunk. *Spilogale putorius*. Lo (12, 19, 20, 21, 25, 26, 27, *28*); Hab. 8.

Family Felidae (Cats)
Bobcat. *Lynx rufus*. Ub; Habs. 2–6, 8.
Canada Lynx. *Lynx canadensis*. Lo (5, 8, 9, 15, 27; historically also 1, 2, 4); Habs. 6, 7.
Mountain Lion (Puma, Cougar). *Puma concolor*. Ub; Habs. 2–8.

Order Artiodactyla (Even-toed Ungulates)
Family Cervidae (Deer)
Elk. *Cervus canadensis*. Ub; Pan; Habs. 1–8.
Mule Deer. *Odocoileus hemionus*. Ub; Pan; Habs. 1–8.
White-tailed Deer. *Odocoileus virginianus*. Ub; Pan; Hab. 8.
Moose. *Alces alces shirasi*. Di; NW, NC, SW, SC, SE; Habs. 6, 8.

Family Antilocapridae (Pronghorns)
Pronghorn. *Antilocapra americana*. Ub; Pan; Habs. 1, 2.

Family Bovidae (Bison, Sheep, and Goats)
American Bison. *Bison bison*. Lo (1, 2, 8; historically more widespread in NC, NE, SW, SC, SE); Habs. 1, 2, 7.
Mountain Goat. *Oreamnos americanus*. Lo (1, 2, 8); Habs. 6, 7.
Bighorn (Mountain) Sheep. *Ovis canadensis*. Wi (all); Habs. 5–7.

Birds

This species list and abundance status categories are based mainly on Faulkner (2010), excluding hypothetical species, and with taxonomic sequence and nomenclature updated to 2018.

The sequence of symbols for nonvagrant species employs the first six sets of codes. Note that for breeding migrant species their breeding latilongs (for highly localized to slightly dispersed breeders), or their regional breeding distributions (for slightly dispersed or dispersed species), are parenthetically listed, followed by their broader nonbreeding distributions. Where breeding habitats differ from general habitats, the breeding habitats are listed first in parentheses. The latilong data are from Orabana et al. (2012), with some species additions or updates based on Faulkner and other more recent information, totaling 444 species.

Order Anseriformes (Waterfowl)

Family Anatidae (Ducks, Geese, and Swans)

Snow Goose. *Anser caerulescens*. Ea; Co; Mig; WeSW.

Ross's Goose. *Anser rossii*. Ea; Un; Mig; WeSW.

Greater White-fronted Goose. *Anser albifrons*. Ea; Ra; Mig; WeSW.

Brant. *Branta bernicla*. Vagrant.

Cackling Goose. *Branta hutchinsii*. Ea; Co; Mig; WeSW.

Canada Goose. *Branta canadensis*. Ub; Pan; Co; Res; WeSW.

Fulvous Whistling-Duck. *Dendrocygna bicolor*. Vagrant.

Mute Swan. *Cygnus olor*. Vagrant.

Trumpeter Swan. *Cygnus buccinator*. Lo (1, 2, 7, 8, 9); We; Un; Res; WeSW.

Tundra Swan. *Cygnus columbianus*. Pan; Un; Mig; WeSW.

Whooper Swan. *Cygnus cygnus*. Vagrant.

Ruddy Shelduck. *Tadorna ferruginea*. Vagrant.

Common Shelduck. *Tadorna tadorna*. Vagrant.

Wood Duck. *Aix sponsa*. SD; NW, NC, NE, SC, SE, Ea; Un; Su; Res; RiDe.

Garganey. *Spatula querquedula*. Vagrant.

Blue-winged Teal. *Spatula discors*. Ub; Pan; Co; Su; Res; WeSW.

Cinnamon Teal. *Spatula cyanoptera*. Ub; Pan; Un; Su; Res; WeSW.

Northern Shoveler. *Spatula clypeata*. Ub; Pan; Co; Su; Res; WeSW.

Gadwall. *Mareca strepera*. Pan; Co; Su; Res; WeSW.

Eurasian Wigeon. *Mareca penelope*. Vagrant.

American Wigeon. *Mareca americana*. Ub; Pan; Co; Su; Res; WeSW.

Mallard. *Anas platyrhynchos*. Ub; Pan; Co; Res; WeSW.

Mottled Duck. *Anas fulvigula*. Vagrant.

American Black Duck. *Anas rubripes*. Vagrant.

Northern Pintail. *Anas acuta*. Pan; Co; Su; Res; WeSW.

Green-winged Teal. *Anas crecca*. Ub; Pan; Co; Res; WeSW.

Canvasback. *Aythya valisineria*. Pan; Co; Mig; WeSW, WeDW.

Redhead. *Aythya americana*. Wi; Pan; Co; Res; WeSW, WeDW.

Ring-necked Duck. *Aythya collaris*. Lo (1, 2, 8, 16, 17, 18, 26); Pan; Co; Res; WeSW, WeDW.

Tufted Duck. *Aythya fuligula*. Vagrant.

Greater Scaup. *Aythya marila*. Ra; ?; Mig; WeDW.

Lesser Scaup. *Aythya affinis*. Di; Pan, NW, NC, SW, SC, SE; Un; Su; Res; WeSW, WeDW.

Harlequin Duck. *Histrionicus histrionicus*. Lo (1, 8, 9, 10, 15); NW; Ra; Su; Res; RiLe.

Surf Scoter. *Melanitta perspicillata*. Pan; Ra; Mig; WeDW.

White-winged Scoter. *Melanitta deglandi*. Pan; Ra; Mig; WeDW.

Black Scoter. *Melanitta americana*. Pan; Ra; Mig; WeDW.

Long-tailed Duck. *Clangula hyemalis*. Pan; Ra; Mig; WeDW.

Bufflehead. *Bucephala albeola*. Lo (1, 2, 8, 27); Pan; Un; Res; WeDW.

Common Goldeneye. *Bucephala clangula*. Wi; Pan; Co; Vis; WeDW.

Barrow's Goldeneye. *Bucephala islandica*. Pan; Co; Res; WeDW.

Hooded Merganser. *Lophodytes cucullatus*. Pan; Un; Mig; WeSW.

Common Merganser. *Mergus merganser*. Wi; Pan; Co; Res; RiDe, RiLe.

Red-breasted Merganser. *Mergus serrator*. Pan; Un; Mig; WeDW.

Ruddy Duck. *Oxyura jamaicensis*. Wi; Pan; Co; Su; Res; WeSW.

Order Galliformes (Gallinaceous Birds)

Family Odontophoridae (New World Quail)

Northern Bobwhite. *Colinus virginianus*. Lo (21, 28); SE; Ra; Res; RiSS, HuCL.

Family Phasianidae (Pheasants, Grouse, and Turkeys)

Chukar. *Alectoris chukar*. SD; NW, NC, SW, SC, SE, Ea; Un; Res; MoRo, AlSh.

Gray Partridge. *Perdix perdix*. Di (all); NE; Un; Res; HuCL.

Ring-necked Pheasant. *Phasianus colchicus*. Di; NW, NC, NE, SE; Un; Res; HuCL.

Ruffed Grouse. *Bonasa umbellus*. SD; NW, NC, NE, SW, SC; Co; Res; FoAs, FoMC.

Greater Sage-Grouse. *Centrocercus urophasianus*. Ub; Pan; Co; Res; SaSG, SaSh.

White-tailed Ptarmigan. *Lagopus leucura*. HL (26); NW, SE; Ra; Res; MoAl.

Dusky Grouse. *Dendragapus obscurus*. Wi; Co; Res; FoDC, FoMC, FoSF.

Greater Prairie-Chicken. *Tympanuchus cupido*. Vagrant.

Sharp-tailed Grouse. *Tympanuchus phasianellus*. SD; NW, NC, NE, SE; Un; Res; GrSG, GrMG.

Wild Turkey. *Meleagris gallopavo*. Di; NW, NC, NE, SC, SE; Co; Res; RiDe, FoPP, WoJu.

Order Podicipediformes (Grebes)

Family Podicipedidae (Grebes)

Pied-billed Grebe. *Podilymbus podiceps*. Ub; Pan; Un; Su; Res; WeSW.

Horned Grebe. *Podiceps auritus*. Un; Mig; WeSW.

Red-necked Grebe. *Podiceps grisegena*. Ra; Fa; Mig; WeSW.

Eared Grebe. *Podiceps nigricollis*. Wi (all); Pan; Co; Su; Res; WeSW.

Western Grebe. *Aechmophorus occidentalis.* Wi (all); Pan; Co; Su; Res; WeSW.

Clark's Grebe. *Aechmophorus clarkii.* HL (3, 10, 17, 19, 23); NC, SW, SC; Un; Su; Res; WeSW.

Order Columbiformes (Pigeons and Doves)

Family Columbidae (Pigeons and Doves)

Rock (Dove) Pigeon. *Columba livia.* Ub; Pan; Co; Res; HuTC, HuFR.

Band-tailed Pigeon. *Patiogioenas fasciata.* Vagrant.

African Collared-Dove. *Streptopelia roseogrisea.* Vagrant.

Eurasian Collared-Dove. *Streptopelia decaocto.* Wi; Pan; Un; Res; HuTC, HuFR.

White-winged Dove. *Zenaida asiatica.* Vagrant.

Mourning Dove. *Zenaida macroura.* Ub; Pan; Co; Su; Res; HuTC, HuFR.

Passenger Pigeon. *Ectopistes migratorius.* Extinct.

Order Cuculiformes (Cuckoos)

Family Cuculidae (Cuckoos)

Yellow-billed Cuckoo. *Coccyzus americanus.* HL (5, 7, 17, 21, 28); NE; Ra; Su; Res; RiDe, FoDe.

Black-billed Cuckoo. *Coccyzus erythropthalmus.* Lo (3, 4, 5, 6, 7, 10, 11, 17, 18, 19, 21); NE; Ra; Su; Res; RiDe.

Order Caprimulgiformes (Goatsuckers)

Family Caprimulgidae (Nightjars)

Common Nighthawk. *Chordeiles minor.* Ub; Pan; Co; Su; Res; GrSG; FoPP; HuTC.

Common Poorwill. *Phalaenoptilus nuttallii.* Di (all); Pan; Co; Su; Res; WoJu, FoPP.

Order Apodiformes (Swifts and Hummingbirds)

Family Apodidae (Swifts)

Chimney Swift. *Chaetura pelagica.* HL (20, 28); SE; Un; Su; Res; HuTC.

Vaux's Swift. *Chaetura vauxi.* Vagrant.

White-throated Swift. *Aeronautes saxatalis.* Di (all); Pan; Co; Su; Res; MoRo.

Family Trochilidae (Hummingbirds)

Rivoli's (Magnificent) Hummingbird. *Eugenes fulgens.* Vagrant.

Ruby-throated Hummingbird. *Archilochus colubris.* Vagrant.

Black-chinned Hummingbird. *Archilochus alexandri.* HL (23); NW, NC, SW, SC; Ra; Su; Res; FoPP.

Anna's Hummingbird. *Calypte anna.* Vagrant.

Broad-tailed Hummingbird. *Selasphorus platycercus.* Di; Pan; Co; Su; Res; RiDe, FoDe.

Rufous Hummingbird. *Selasphorus rufus.* W, SE; Co; Fa; Mig; FoAs, MoMe.

Calliope Hummingbird. *Selasphorus calliope.* Lo (1, 2, 4, 5, 8, 9, 11, 15, 17, 24); Co; Sca; Su; Res; FoAs, MoMe.

Order Gruiformes (Cranes, Rails, and Relatives)

Family Rallidae (Rails and Coots)

Yellow Rail. *Coturnicops noveboracensis.* Vagrant.

Black Rail. *Laterallus jamaicensis.* Vagrant.

Virginia Rail. *Rallus limicola.* SD (all); Un; Sca; Su; Res; WeSW.

Sora. *Porzana carolina.* Wi; Pan; Un; Su; Res; WeSW.

Purple Gallinule. *Porphyrio martinica.* Vagrant.

Common Gallinule. *Gallinula galeata.* Vagrant.

American Coot. *Fulica americana.* Ub; Pan; Co; Su; Res; WeSW.

Family Gruidae (Cranes)

Sandhill Crane. *Antigone canadensis.* Di; Pan; Un; Su; Res; WeSW, MoMe, HuFR, HuCL.

Whooping Crane. *Grus americana.* Vagrant.

Order Charadriiformes (Shorebirds, Gulls, and Auks)

Family Recurvirostridae (Stilts and Avocets)

Black-necked Stilt. *Himantopus mexicanus.* Lo (3, 15, 16, 22, 23, 24, 27); So, We; Un; Su; Res; WeSW.

American Avocet. *Recurvirostra americana.* Wi; Pan; Co; Su; Res; WeSW.

Family Charadriidae (Plovers)

Black-bellied Plover. *Pluvialis squatarola.* Ea; Un; Mig; HUCL.

American Golden-Plover. *Pluvialis dominica.* NE; Ra; Mig; HUCL.

Snowy Plover. *Charadrius nivosus.* Lo (23, 24); NC, NE, SE, SC, SW; VR; Su; Res; ShSM.

Semipalmated Plover. *Charadrius semipalmatus.* Pan; Un; Mig; ShSM.

Piping Plover. *Charadrius melodus.* Ea; VR; Mig; ShSM.

Killdeer. *Charadrius vociferus.* Ub; Pan; Co; Su; Res; ShGR, WeSW, HuFR.

Mountain Plover. *Charadrius montanus.* Wi; Pan; Un; Su; Res; SaSG.

Family Scolopacidae (Sandpipers, Snipes, and Phalaropes)

Upland Sandpiper. *Bartramia longicauda.* Lo (4, 5, 6, 7, 13, 14, 18, 19, 21, 28); NC, NE, EC, SC; Un; Su; Res; GrMG, GrSG.

Whimbrel. *Numenius phaeopus.* Ea; Ra; Mig; WeSW.

Long-billed Curlew. *Numenius americanus.* Wi; Pan; Un; Su; Res; GrMG, WeSW.

Eskimo Curlew. *Numenius borealis.* Probably extinct.

Hudsonian Godwit. *Limosa haemastica.* Ea; VR; Mig, WeSW.

Marbled Godwit. *Limosa fedoa.* Ea; Un; Mig; WeSW.

Ruddy Turnstone. *Arenaria interpres.* Ea; VR; Mig; WeSW.

Red Knot. *Calidris canutus.* Ea; VR; Mig; WeSW.

Stilt Sandpiper. *Calidris himantopus.* Ea; Un; Mig; WeSW.

Sanderling. *Calidris alba.* Ea; Un; Mig; WeSW.

Dunlin. *Calidris alpina.* Ea; VR; Mig; WeSW.

Baird's Sandpiper. *Calidris bairdii.* Ea; Co; Mig; WeSW.

Least Sandpiper. *Calidris minutilla.* Pan; Co; Mig; WeSW.

White-rumped Sandpiper. *Calidris fuscicollis.* Ea; Un; Mig; WeSW.

Buff-breasted Sandpiper. *Calidris subruficollis.* Ea; VR; Mig; HuCL.

Pectoral Sandpiper. *Calidris melanotos*. Ea; Un; Mig; WeSW.

Semipalmated Sandpiper. *Calidris pusilla*. SE, Ea; Un; Mig; WeSW.

Western Sandpiper. *Calidris mauri*. SW, We; Un; Mig; WeSW.

Short-billed Dowitcher. *Limnodromus griseus*. Ea; VR; Mig; WeSW.

Long-billed Dowitcher. *Limnodromus scolopaceus*. Ea; Co; Mig; WeSW.

Wilson's Snipe. *Gallinago delicata*. Wi; Pan; Co; Su; Res; WeSW.

American Woodcock. *Scolopax minor*. Vagrant.

Spotted Sandpiper. *Actitis macularius*. Ub; Pan; Co; Su; Res; RiHe, WeSW.

Solitary Sandpiper. *Tringa solitaria*. Pan; Un; Mig; WeSW.

Lesser Yellowlegs. *Tringa flavipes*. Pan; Co; Mig; WeSW.

Willet. *Tringa semipalmata*. Di (all); Pan; Un; Su; Res; WeSW.

Greater Yellowlegs. *Tringa melanoleuca*. Pan; Co; Mig; WeSW.

Wilson's Phalarope. *Phalaropus tricolor*. Ub; Pan; Co; Su; Res; WeSW.

Red-necked Phalarope. *Phalaropus lobatus*. Pan; Co; Mig; WeSW.

Red Phalarope. *Phalaropus fulicarius*. Vagrant.

Family Stercorariidae (Jaegers)
Pomarine Jaeger. *Stercorarius pomarinus*. Vagrant.
Parasitic Jaeger. *Stercorarius parasiticus*. Vagrant.

Family Alcidae (Auks)
Long-billed Murrelet. *Brachyramphus perdix*. Vagrant.
Ancient Murrelet. *Synthliboramphus antiquus*. Vagrant.

Family Laridae (Gulls and Terns)
Black-legged Kittiwake. *Rissa tridactyla*. Vagrant.

Sabine's Gull. *Xema sabini*. Ra; Fa; Mig; WeSW, WeDW.

Bonaparte's Gull. *Chroicocephalus philadelphia*. Un, Sca; Mig; WeSW, WeDW.

Black-headed Gull. *Chroicocephalus ridibundus*. SE; VR; Mig; WeSW, WeDW.

Little Gull. *Hydrocoloeus minutus*. Vagrant.

Ross's Gull. *Rhodostethia rosea*. Vagrant.

Laughing Gull. *Leucophaeus atricilla*. Vagrant.

Franklin's Gull. *Leucophaeus pipixcan*. HL (2, 15, 22); Pan; Ra; Su; Res; HuCL, WeSW.

Heermann's Gull. *Larus heermanni*. Vagrant.

Mew Gull. *Larus canus*. Vagrant.

Ring-billed Gull. *Larus delawarensis*. HL (1, 5, 10, 19); Pan; Ra; Su; Res; HuCL, WeDW.

California Gull. *Larus californicus*. HL (1, 10, 19, 27); Pan; Co; Su; Res; HuCL, WeDW.

Herring Gull. *Larus argentatus*. Pan; Un; Mig; HuCL, WeDW, WeSW.

Iceland Gull. Thayer's Gull subspecies. *Larus glaucoides thayeri*. Pan; VR; Mig; WeDW, WeSW.

Lesser Black-backed Gull. *Larus fuscus*. Vagrant.

Glaucous-winged Gull. *Larus glaucescens*. Vagrant.

Glaucous Gull. *Larus hyperboreus*. Pan; Ra; Fa; Mig; WeDW, WeSW.

Great Black-backed Gull. *Larus marinus*. Vagrant.

Least Tern. *Sternula antillarum*. Ea; VR; Mig; RiLe.

Caspian Tern. *Hydroprogne caspia*. HL (1, 10, 19, 27); Pan; Un; Su; Res; WeDW, WeSW.

Black Tern. *Chlidonias niger*. Lo (1, 13, 15, 17, 22, 27); Pan; Ra; Su; Res; WeSW.

Common Tern. *Sterna hirundo*. Pan; Ra; Mig; WeDW.

Arctic Tern. *Sterna paradisaea*. Vagrant.

Forster's Tern. *Sterna forsteri*. HL (10, 15, 22, 26, 27); Pan; Un; Su; Res; WeSW.

Order Gaviiformes (Loons)
Family Gaviidae (Loons)
Red-throated Loon. *Gavia stellata*. Pan; VR; Mig; WeDW.
Pacific Loon. *Gavia pacifica*. Pan; Ra; Mig; WeDW.
Common Loon. *Gavia immer*. Pa; Un; Mig; WeDW.
Yellow-billed Loon. *Gavia adamsii*. Vagrant.

Order Procellariiformes (Tube-nosed Birds)
Family Procellaridae (Petrels and Shearwaters)
Streaked Shearwater. *Calonectris leucomelas*. Vagrant.

Order Ciconiiformes (Storks)
Family Ciconiidae (Wood Storks)
Wood Stork. *Mycteria americana*. Vagrant.

Order Suliformes (Frigatebirds and Cormorants)
Family Frigatidae (Frigatebirds)
Lesser Frigatebird. *Fregata ariel*. Vagrant.

Family Phalacrocoracidae (Cormorants)
Double-crested Cormorant. *Phalacrocorax auritus*. All; Pan; Co; Su; Res; WeDW, WeSW.

Order Pelicaniformes (Pelicans, Herons, and Ibises)
Family Pelecanidae (Pelicans)
American White Pelican. *Pelecanus erythroryhnchos*. HL (1, 10, 19, 27); Pan; Co; Su; Res; WeSW, WeDW.

Brown Pelican. *Pelecanus occidentalis*. Vagrant.

Family Ardeidae (Herons and Egrets)
American Bittern. *Botaurus lentiginosus*. Lo (8, 10, 15, 18, 21, 22, 26, 27, 28); Pan; Un; Su; Res; WeSW.

Least Bittern. *Ixobrychus exilis*. Vagrant.

Great Blue Heron. *Ardea herodias*. Ub; Pan; Co; Su; Res; RiDe, WeSW, ShGR.

Great Egret. *Ardea alba*. Pan; Ra; Su; Vis; WeSW, SaSM, ShGR.

Snowy Egret. *Egretta thula*. Lo (10, 15, 16, 19, 22, 26, 27); Pan; Ra; Su; Res; WeSW.

Little Blue Heron. *Egretta caerulea*. Vagrant.

Tricolored Heron. *Egretta tricolor*. Vagrant.

Cattle Egret. *Bubulcus ibis*. HL (27); Pan; Ra; Su; Res; HuCL, HuFR.

Green Heron. *Butorides virescens*. Pan; Ra; Su; Vis; WeSW.

Black-crowned Night-Heron. *Nycticorax nycticorax*. Lo (10, 15, 16, 18, 19, 22, 23, 26, 27, 28); Pan; Un; Su; Res; WeSW.

Yellow-crowned Night-Heron. *Nyctanassa violacea*. Vagrant.

Family Threskiornithidae (Ibises and Spoonbills)

White Ibis. *Eudocimus albus*. Vagrant.

Glossy Ibis. *Plegadis falcinellus*. Vagrant.

White-faced Ibis. *Plegadis chihi*. Lo (10, 15, 16, 22, 26, 27); Pan; Un; Su; Res; WeSW.

Order Cathartiiformes (Vultures)
Family Cathartidae (New World Vultures)

Turkey Vulture. *Cathartes aura*. Wi; Pan; Co; Su; Res; AlSh, SaSG, HuCL, HuFR.

Order Accipitriiformes (Hawks, Eagles, and Ospreys)
Family Pandionidae (Ospreys)

Osprey. *Pandion haliaetus*. SD (all); Pan; Un; Su; Res; RiDe, WeDW.

Family Accipitridae (Hawks and Eagles)

White-tailed Kite. *Elanus leucurus*. Vagrant.

Golden Eagle. *Aquila chrysaetos*. Ub; Pan; Un; Res; AlSh, SaSG, SaSh, MoAl, MoMe, MoRo.

Bald Eagle. *Haliaeetus leucocephalus*. Di (all); Pan; Un; Res; RiDe, WeTr, WeDW.

Mississippi Kite. *Ictinia mississippiensis*. Vagrant.

Northern Harrier. *Circus hudsonius*. Pan; Un; Res; WeSW, HuCL.

Sharp-shinned Hawk. *Accipiter striatus*. Wi (all); Pan; Un; Res; FoAs, FoMC.

Cooper's Hawk. *Accipiter cooperii*. Di; Pan, NW, NC, NE, SW, SC; Un; Res; FoAs, FoMC, RiDe.

Northern Goshawk. *Accipiter gentilis*. Wi (all); Pan; Un; Res; FoOG, FoDC, FoDe.

Harris's Hawk. *Parabuteo unicinctus*. Vagrant.

Red-shouldered Hawk. *Buteo lineatus*. Vagrant.

Broad-winged Hawk. *Buteo platypterus*. Ea; Un; Mig; FoDe.

Swainson's Hawk. *Buteo swainsoni*. Ub; Pan; Co; Su; Res; GrSG, GrMG, HuFR.

Red-tailed Hawk. *Buteo jamaicensis*. Ub; Pan; Co; Res; FoDe, FoDC, GrVa.

Ferruginous Hawk. *Buteo regalis*. Ub; Pan; Un; Res; GrSG, SaSh, SaSG, AlSh.

Rough-legged Hawk. *Buteo lagopus*. Pan; Co; Win; Mig; SaSG, GrSG; HuCL.

Order Strigiformes (Owls)
Family Tytonidae (Barn Owls)

Barn Owl. *Tyto alba*. Lo (6, 16, 19, 21, 23, 28); Pan; Ra; Su; Res; HuFR, GrVa.

Family Strigidae (Typical Owls)

Flammulated Owl. *Psiloscops flammeolus*. HL (25); NW, SC; Ra; Su; Res; FoPP.

Eastern Screech-Owl. *Megascops asio*. HL (2, 19, 21); NW, NC, NE, SE, SC; Un; Res; HuPG, HuFR, FoAs, FoMC, RiDe.

Western Screech-Owl. *Megascops kennicottii*. HL (1, 2, 4, 5, 8, 9); NW, NC; Ra; Res; HuPG, HuFR, FoAs, FoMC, RiDe.

Great Horned Owl. *Bubo virginianus*. Ub; Pan; Co; Res; HuPG, HuFR, FoAs, FoDC, FoMC, FoPP, RiDe, WoJu.

Snowy Owl. *Bubo scandiacus*. No; Ra; Win; Mig; GrVa, HuCL, HuF.

Northern Hawk Owl. *Surnia ulula*. Vagrant.

Northern Pygmy-Owl. *Glaucidium gnoma*. HL (1, 8, 9, 15); We; Ra; Res; FoMC, FoOG, FoDC.

Burrowing Owl. *Athene cunicularia*. Ub; Pan; Co; Su; Res; SaSG, SaSh.

Barred Owl. *Strix varia*. Vagrant.

Great Gray Owl. *Strix nebulosa*. HL (1, 8, 15, 16) NW, NC, SC, SW; Un; Res; FoMC, MoMe.

Long-eared Owl. *Asio otus*. Di (all); Pan; Un; Res; RiDe, FoDC.

Short-eared Owl. *Asio flammeus*. Di (all); Pan; Un; Res; GrMG, WeSW.

Boreal Owl. *Aegolius funereus*. HL (1, 8, 26); NW, SC, SW; Un; Res; FoOG, FoSF.

Northern Saw-whet Owl. *Aegolius acadicus*. SD (all); Pan; Co; Res; FoDC, FoPP.

Order Coraciiformes (Kingfishers and Relatives)
Family Alcedinidae (Kingfishers)

Belted Kingfisher. *Megaceryle alcyon*. Ub; Pan; Co; Su; Res; RiDe, RiHe.

Order Piciformes (Woodpeckers)
Family Picidae (Woodpeckers)

Lewis's Woodpecker. *Melanerpes lewis*. Di (all); Pan; Un; Su; Res; FoPP; RiDe.

Red-headed Woodpecker. *Melanerpes erythrocephalus*. SD; Pan, NC, NE, SE, SC; Un; Su; Res; RiDe, FoPP.

Acorn Woodpecker. *Melanerpes formicivorous*. Vagrant.

Red-bellied Woodpecker. *Melanerpes carolinus*. Vagrant.

Williamson's Sapsucker. *Sphyrapicus thyroideus*. SD (all); Un, Sca; Su; Res; FoMC, FoAs.

Yellow-bellied Sapsucker. *Sphyrapicus varius*. Vagrant.

Red-naped Sapsucker. *Sphyrapicus nuchalis*. Wi; Pan; Co; Su; Res; FoAs.

Downy Woodpecker. *Dryobates pubescens*. Ub; Pan; Co; Res; FoAs, RiDe.

Hairy Woodpecker. *Dryobates villosus*. Ub; Pan; Co; Res; FoAs, RiDe.

White-headed Woodpecker. *Dryobates albolarvatus*. Vagrant.

American Three-toed Woodpecker. *Picoides dorsalis*. SD (all); Un, Sca; Res; FoSG.

Black-backed Woodpecker. *Picoides arcticus*. HL (1, 7, 8, 14); NW, NE, SC, SW; Un; Res; FoSG.

White-headed Woodpecker. *Dryobates albolarvatus*. Vagrant.

Northern Flicker. *Colaptes auratus*. Ub; Pan; Co; Res; FoAs, FoPP, RiDe

Pileated Woodpecker. *Dryocopus pileatus*. Vagrant.

Order Falconiformes (Falcons and Caracaras)

Family Falconidae (Falcons)

Crested Caracara. *Caracara cheriway*. Vagrant.

American Kestrel. *Falco sparverius*. Ub; Pan; Co; Su; Res; HuFR, RiDe.

Merlin. *Falco columbarius*. Di (all); Pan; Un; Res; RiDe, FoPP.

Gyrfalcon. *Falco rusticolus*. No; VR; WM; GrVa.

Peregrine Falcon. *Falco peregrinus*. Lo (1, 2, 3, 7, 8, 9, 10, 16, 17, 19); Pan; Un; Su; Res; GrVa, HuTC, MoRo.

Prairie Falcon. *Falco mexicanus*. Ub; Pan; Un; Res; MoRo, GrVa.

Order Passeriformes (Perching Birds)

Family Tyrannidae (New World Flycatchers)

Olive-sided Flycatcher. *Contopus cooperi*. Di (all); Pan; Un; Su; Res; FoSG, MoMe.

Western Wood-Pewee. *Contopus sordidulus*. Ub; Co; Su; Res; RiDe, FoAs, FoMC.

Eastern Wood-Pewee. *Contopus virens*. Vagrant.

Alder Flycatcher. *Empidonax alnorum*. Vagrant.

Willow Flycatcher. *Empidonax traillii*. Di (all); Pan; Un; Su; Res; RiSS.

Least Flycatcher. *Empidonax minimus*. SD (all); Co, Sca; Su; Res; RiDe.

Hammond's Flycatcher. *Empidonax hammondii*. SD; NW, NC, SC, SW; Un; Su; Res; FoSF.

Gray Flycatcher. *Empidonax wrightii*. SD; NC, SE, SC, SW; Un; Su; Res; WoJu.

Dusky Flycatcher. *Empidonax oberholseri*. Di (all); Pan; Co; Su; Res; FoAs, FoPP, WoJu.

Cordilleran Flycatcher. *Empidonax occidentalis*. Wi (all); Pan; Co; Su; Res; RiHe, HuFR, FoMC.

Eastern Phoebe. *Sayornis phoebe*. Ea; HL (7); VR; Su; Res; RiDe, RiSS.

Say's Phoebe. *Sayornis saya*. Ub; Pan; Co; Su; Res; HuFR.

Vermilion Flycatcher. *Pyrocephalus rubinus*. Vagrant.

Ash-throated Flycatcher. *Myiarchus cinerascens*. HL (23, 24, 25); SC, SW; Un; Su; Res; WoJu.

Great Crested Flycatcher. *Myiarchus crinitus*. Vagrant.

Cassin's Kingbird. *Tyrannus vociferans*. Lo (6, 7, 14, 17, 21, 27, 28); Ea; Un; Su; Res; FoPP.

Western Kingbird. *Tyrannus verticalis*. Wi; Pan; Co; Su; Res; HuFR, HuPG, WoJu.

Eastern Kingbird. *Tyrannus tyrannus*. Ub; Pan; Co; Su; Res; RiDe, HuFR, HuTC.

Scissor-tailed Flycatcher. *Tyrannus forficatus*. Vagrant.

Fork-tailed Flycatcher. *Tyrannus savana*. Vagrant.

Family Laniidae (Shrikes)

Loggerhead Shrike. *Lanius ludovicianus*. Ub; Pan; Un; Su; Res; GrVa, SaSh, HuFR.

Northern Shrike. *Lanius borealis*. Wi; No; Un; Mig; GrVa, HuFR.

Family Vireonidae (Vireos)

White-eyed Vireo. *Vireo griseus*. Vagrant.

Gray Vireo. *Vireo vicinior*. Vagrant.

Yellow-throated Vireo. *Vireo flavifrons*. Vagrant.

Cassin's Vireo. *Vireo cassinii*. We; VR; Mig; RiDe.

Blue-headed Vireo. *Vireo solitarius*. Vagrant.

Plumbeous Vireo. *Vireo plumbeus*. SD (all); Un, Sca; Su; Res; FoAs, FoMC.

Philadelphia Vireo. *Vireo philadelphicus*. Vagrant.

Warbling Vireo. *Vireo gilvus*. Ub; Pan; Co; Su; Res; RiDe, FoAs.

Red-eyed Vireo. *Vireo olivaceus*. Lo (3, 4, 5, 7, 8, 25); Ea; Un; Su; Res; RiDe, FoDe.

Family Corvidae (Crows, Jays, and Magpies)

Gray (Canada) Jay. *Perisoreus canadensis*. Di; Pan; Un; Res; FoSF, FoLP, FoMC.

Pinyon Jay. *Gymnorhinus cyanocephalus*. SD; Pan, We; Un; Res; WoJu.

Steller's Jay. *Cyanocitta stelleri*. SD; Co, Sca; Res; FoPP, FoMC, FoLP.

Blue Jay. *Cyanocitta cristata*. Lo (2, 4, 5, 7, 14, 19, 21, 27, 28); Un; Res; HuPG, HuFR, HuTC, RiDe.

Woodhouse's Scrub-Jay. *Aphelocoma woodhouseii*. HL (23, 24, 25); NE, SE, SC, SW; Ra; Res; WoJu.

Clark's Nutcracker. *Nucifraga columbiana*. Wi; Pan; Un; Res; FoPP, FoMC, FoLP, FoSF.

Black-billed Magpie. *Pica hudsonia*. Ub; Pan; Co; Res; AlSh, SaSh, WoJu, HuFR.

American Crow. *Corvus brachyrhynchos*. Wi; Pan; Co; Res; HuPG, HuFR, HuTC, RiDe.

Common Raven. *Corvus corax*. Di (all); Pan; Un; Res; MoRo, HuFR, HuTC, SaSh, FoMC, FoSG.

Family Alaudidae (Larks)

Horned Lark. *Eremophila alpestris*. Ub; Pan; Co; Res; GrVa, MoAl, MoMe.

Family Hirundinidae (Swallows)

Purple Martin. *Progne subis*. HL (25); Ea; Ra; Su; Res; HuTC, HuFR.

Tree Swallow. *Tachycineta bicolor*. Ub; Pan; Co; Su; Res; FoAs.

Violet-green Swallow. *Tachycineta thalassina*. Ub; Pan; Co; Su; Res; MoRo, FoAs, HuFR.

Northern Rough-winged Swallow. *Stelgidopteryx serripennis*. Ub; Pan; Co; Su; Res; RiHe, WeSW, HuFR.

Bank Swallow. *Riparia riparia*. Wi; Pan; Un; Su; Res; RiHe, WeSW.

Cliff Swallow. *Petrochelidon pyrrhonota*. Ub; Pan; Co; Su; Res; MoRo, HuFR.

Barn Swallow. *Hirundo rustica*. Pan; Co; Su; Res; HuFR, HuPG, HuTC.

Family Paridae (Chickadees and Titmice)

Black-capped Chickadee. *Poecile atricapillus*. Ub; Pan; Co; Res; HuPG, FoAs, FoDe, FoMC.

Mountain Chickadee. *Poecile gambeli*. Wi; Pan; Co; Res; FoMC, FoPP, FoSF.

Juniper Titmouse. *Baeolophus ridgwayi*. HL (22, 23, 24, 25); SE, SC, SW; Un; Res; WoJu.

Family Aegithalidae (Long-tailed Tits and Bushtits)
Bushtit. *Psaltriparus minimus*. HL (23, 24); SE, SC, SW; Un; Res; WoJu.

Family Sittidae (Nuthatches)
Red-breasted Nuthatch. *Sitta canadensis*. Ub; Pan; Un; Res; FoOG, FMC, FoAs.
White-breasted Nuthatch. *Sitta carolinensis*. Wi; Pan; Un; Res; RiDe, FoPP.
Pygmy Nuthatch. *Sitta pygmaea*. Lo (3, 7, 15, 18, 19, 20, 22, 23, 27); Un, Sca; Res; FoSF, FoOG, FoPP.

Family Certhiidae (Creepers)
Brown Creeper. *Certhia americana*. SD (all); Un, Sca; Res; FoSF, FoLP.

Family Troglodytidae (Wrens)
Rock Wren. *Salpinctes obsoletus*. Ub; Pan; Un; Su; Res; MoRo.
Canyon Wren. *Catherpes mexicanus*. SD (all); Un; Res; MoRo.
House Wren. *Troglodytes aedon*. Ub; Pan; Co; Su; Res; RiDe, HuTC, HuFR.
Bewick's Wren. *Thryomanes bewickii*. HL (22, 23, 24, 25, 26); SE, SC, SW; Un; Su; Res; WoJu.
Winter Wren. *Troglodytes hiemalis*. NE; VR; Win; Vis; FoMC.
Pacific Wren. *Troglodytes pacificus*. HL (8, 15); NW; Ra; Su; Res; FoOG; FoMC.
Sedge Wren. *Cistothorus platensis*. Vagrant.
Marsh Wren. *Cistothorus palustris*. Di (all); Un, Sca; Su; Res; WeSW.
Carolina Wren. *Thryothorus ludovicianus*. Vagrant.

Family Polioptilidae (Gnatcatchers)
Blue-gray Gnatcatcher. *Polioptila caerulea*. Lo (10, 19, 21, 23, 24, 25); Un, Sca; Su; Res; WoJu.

Family Cinclidae (Dippers)
American Dipper. *Cinclus mexicanus*. SD; NW, NC, SE, SC, SW; Un; Res; RiLe.

Family Regulidae (Kinglets)
Golden-crowned Kinglet. *Regulus satrapa*. Lo (1, 2, 8, 15, 16, 22, 23, 25, 26); Un, Sca; Su; Res; FoSF, FoOG, FoMC.
Ruby-crowned Kinglet. *Regulus calendula*. Wi; Pan; Co; Su; Res; FoSF, FoOG, FoMC.

Family Turdidae (Thrushes)
Eastern Bluebird. *Sialia sialis*. HL (7, 27, 28); Ea; Ra; Su; Res; HuPG, HuFR, FoDe.
Western Bluebird. *Sialia mexicana*. HL (13, 23, 26); Sca, VR; Su; Res; FoMC, FoAs.
Mountain Bluebird. *Sialia currucoides*. Ub; Pan; Co; Su; Res; HuCL, FoAs, MoMe, MoAl.

Townsend's Solitaire. *Myadestes townsendi*. Wi; Pan; Co; Res; FoPP, FoMC, FoLP, FoSF.
Veery. *Catharus fuscescens*. Di (all); Pan; Co; Su; Res; FoDe, FoDC, FoOG.
Gray-cheeked Thrush. *Catharus minimus*. Ea; VR; Mig; RiDe.
Swainson's Thrush. *Catharus ustulatus*. Di (all); Pan; Co; Su; Res; RiDe.
Hermit Thrush. *Catharus guttatus*. Pan; Co; Su; Res; FoMC, FoOG, FoSF, FoAs.
Wood Thrush. *Hylocichla mustelina*. Ea; Ra; Mig; FoDe.
American Robin. *Turdus migratorius*. Ub; Pan; Co; Res; HuPG, HuTC, HuTC, FoDe.
Varied Thrush. *Ixoreus naevius*. We; Ra; Su; Vis; FoMC, FoDe.

Family Mimidae (Thrashers and Catbirds)
Gray Catbird. *Dumetella carolinensis*. Ub; Pan; Un; Su; Res; RiSS.
Northern Mockingbird. *Mimus polyglottos*. HL (13, 24, 27, 28); Un; Su; Res; RiDe, RiSS, HuTC.
Sage Thrasher. *Oreoscoptes montanus*. Ub; Pan; Co; Su; Res; SaSG, SaSh.
Brown Thrasher. *Toxostoma rufum*. SD; Pan; Co; Su; Res; RiSS, RiDe, HuPG.

Family Bombycillidae (Waxwings)
Bohemian Waxwing. *Bombycilla garrulus*. Pan; Un; Win; Mig; WoJu, HuPG.
Cedar Waxwing. *Bombycilla cedrorum*. Pan; Un; Res; WoJu, HuPG.

Family Motacillidae (Pipits)
American Pipit. *Anthus rubescens*. Lo (1, 2, 4, 8, 9, 15, 17, 26); Pan; Un; Su; Res; MoAl, RiHe, MoMe.
Sprague's Pipit. *Anthus spragueii*. No; Ra; Mig; GrVa.

Family Fringillidae (Finches)
Brambling. *Fringilla montifringilla*. Vagrant.
Gray-crowned Rosy-Finch. *Leucosticte tephrocotis*. Wi; Pan; Co; Mig; HuCL, HuPG, HuFR.
Black Rosy-Finch. *Leucosticte atrata*. Lo (1, 2, 4, 8, 9, 15, 16); Un; Res; MoAl, MoMe.
Brown-capped Rosy-Finch. *Leucosticte australis*. HL (26); Sca, Ra; Res; MoAl, MoMe.
Pine Grosbeak. *Pinicola enucleator*. SD (all); Un, Sca; Res; FoMC, FoSF.
House Finch. *Haemorhous mexicanus*. Ub; Pan; Ab; Res; HuTC.
Purple Finch. *Haemorhous purpureus*. VR; WM; HuFR.
Cassin's Finch. *Haemorhous cassinii*. Di (all); Pan; Un; Res; FoMC.
Red Crossbill. *Loxia curvirostra*. Wi (all); Pan; Co; Res; FoPP, FoLP, FoSF.
White-winged Crossbill. *Loxia leucoptera*. HL (1, 8, 9); Sca, Ra; Res; FoSF.
Common Redpoll. *Acanthis flammea*. No; Ra; Win; Mig; HuFR, HuTC, FoDe.
Hoary Redpoll. *Acanthis hornemanni*. Vagrant.

Pine Siskin. *Spinus pinus*. Wi; Pan; Co; Res; FoSF, FoAs, RiDe, HuTC.

Lesser Goldfinch. *Spinus psaltria*. HL (21); SW, SE; Ra; Su; Res; HuTC, WoJu.

Lawrence's Goldfinch. *Spinus lawrencei*. Vagrant.

American Goldfinch. *Spinus tristis*. Pan; GrVa, HuCL, HuTC.

European Goldfinch. *Carduelis carduelis*. Vagrant.

Evening Grosbeak. *Coccothraustes vespertinus*. SD (all); Pan; Un; Res; FoMC, FoAs.

Family Calcariidae (Longspurs and Snow Buntings)

Lapland Longspur. *Calcarius lapponicus*. No; Un; Wi; Mig; GrVa.

Chestnut-collared Longspur. *Calcarius ornatus*. HL (6, 13, 21, 25, 28); NW, NC, NE, SE, SC; Un; Su; Res; GrMG, GrVa.

Smith's Longspur. *Calcarius pictus*. NE; VR; Mig; GrVa.

McCown's Longspur. *Rhynchophanes mccownii*. Di; Pan, NW, NC, NE, SE, SC; Co; Su; Res; GrSG.

Snow Bunting. *Plectrophenax nivalis*. No; Ra; Win; Mig; GrVa.

Family Emberizidae (Sparrows)

Green-tailed Towhee. *Pipilo chlorurus*. Ub; Pan; Co; Su; Res; SaSG, SaSh.

Spotted Towhee. *Pipilo maculatus*. Di (all); Pan; Co; Su; Res; SaSh, AlSh, WoJu, FoPP, FoAs.

Canyon Towhee. *Melozone fusca*. Vagrant.

Cassin's Sparrow. *Peucaea cassinii*. HL (21); NC, NE, SE; Ra; Su; Res; SaSG.

American Tree Sparrow. *Spizelloides arborea*. Pan; Co; Win; Mig; HuCL, HuFR.

Chipping Sparrow. *Spizella passerina*. Ub; Pan; Co; Su; Res; HuCL, HuFR, RiHe.

Clay-colored Sparrow. *Spizella pallida*. NE; Un; Mig; GrVa, HuFR.

Brewer's Sparrow. *Spizella breweri*. Ub; Pan; Co; Su; Res; SaSG, SaSh.

Field Sparrow. *Spizella pusilla*. HL (6); Ea; VR; Su; Res; GrVa, HuFR.

Vesper Sparrow. *Pooecetes gramineus*. Ub; Pan; Co; Su; Res; GrVa, GrSG, HuFR, HuCL.

Lark Sparrow. *Chondestes grammacus*. Wi; Pan; Un; Su; Res; GrVa, GrSG, HuFR.

Black-throated Sparrow. *Amphispiza bilineata*. NW, SE, SC, SW; VR; Su; Res, ?; AlSh, SaSh.

Sagebrush Sparrow. *Artemisiospiza nevadensis*. Di (All); Pan; Un; Su; Res; SaSh, SaSG.

Lark Bunting. *Calamospiza melanocorys*. Wi; Pan; Co; Su; Res; GrSG, GrMG, GrVa.

Savannah Sparrow. *Passerculus sandwichensis*. Wi; Pan; Un; Su; Res; MoMe, WeSW.

Grasshopper Sparrow. *Ammodramus savannarum*. SD (all); Un; Sca; Su; Res; GrTG, GrMG.

Baird's Sparrow. *Centronyx bairdii*. Ea; Ra; Mig; GrMG, GrSG.

LeConte's Sparrow. *Ammospiza leconteii*. Vagrant.

Nelson's Sparrow. *Ammospiza nelsoni*. Vagrant.

Fox Sparrow. *Passerella iliaca*. SD; NW, NC, SC, SW; Un; Su; Res; RiSS.

Song Sparrow. *Melospiza melodia*. Ub; Pan; Co; Res; RiSS, RiHe, WeSW.

Lincoln's Sparrow. *Melospiza lincolnii*. Wi; NW, NC, SE, SC, SW; Un; Su; Res; RiSS.

Swamp Sparrow. *Melospiza georgiana*. Ea; Ra; Mig; WeSW.

White-throated Sparrow. *Zonotrichia albicollis*. Ra; Mig.

Harris's Sparrow. *Zonotrichia querula*. VR; Win; Mig.

White-crowned Sparrow. *Zonotrichia leucophrys*. Wi; NW, NC, SE, SC, SW; Co; Su; Res.

Golden-crowned Sparrow. *Zonotrichia atricapilla*. Vagrant.

Dark-eyed Junco. *Junco hyemalis*. Wi; Pan; Co; Res; HuCL, HuFR, HuPG, FoPP, FoMC, FoSF, FoSG.

Family Icteriidae (Chats)

Yellow-breasted Chat. *Icteria virens*. SD; NC, NE, SE, SC, SW; Un, Sca; Su; Res; RiSS.

Family Icteridae (Blackbirds, Orioles, and Meadowlarks)

Bobolink. *Dolichonyx oryzivorus*. SD (all); Pan; Un; Su; Res; GrMG, GrTG, HuFR.

Red-winged Blackbird. *Agelaius phoeniceus*. Ub; Pan; Co; Su; Res; AlWe, WeSW, GrVa.

Eastern Meadowlark. *Sturnella magna*. Vagrant.

Western Meadowlark. *Sturnella neglecta*. Ub; Pan; Ab; Su; Res; GrMG, GrSG, GrVa.

Yellow-headed Blackbird. *Xanthocephalus xanthocephalus*. Ub; Pan; Co; Su; Res; WeSW.

Rusty Blackbird. *Euphagus carolinus*. No; VR; Mig; WeSW.

Brewer's Blackbird. *Euphagus cyanocephalus*. Ub; Pan; Co; Su; Res; HuFR, HuPG, MoMe, FoSG.

Common Grackle. *Quiscalus quiscula*. Wi (all); Pan; Co; Su; Res; RiDe, HuFR, HuPG, HuTC.

Great-tailed Grackle. *Quiscalus mexicanus*. HL (21, 27); NE, SE, SC, SW; Ra; Su; Res; HuTC, HuFR, WeSW.

Brown-headed Cowbird. *Molothrus ater*. Ub; Pan; Co; Su; Res; GrVa, GrSG, HuCL, HuFR.

Orchard Oriole. *Icterus spurius*. HL (5, 6, 19, 21, 27, 28); Ea; Un; Su; Res; RiDe, RiSS.

Bullock's Oriole. *Icterus bullockii*. Ub; Pan; Co; Su; Res; RiDe, HuFR.

Baltimore Oriole. *Icterus galbula*. Vagrant.

Scott's Oriole. *Icterus parisorum*. HL (23, 24, 25); SC, SW; Ra; Su; Res; WoJu.

Family Parulidae (New World Warblers)

Ovenbird. *Seiurus aurocapilla*. HL (4, 5, 7, 14, 19, 20); NW, NC, NE, SE. SC, SW; Un; Su; Res; FoDe, FoDC, FoAs, RiDe.

Worm-eating Warbler. *Helmitheros vermivorum*. Vagrant.

Northern Waterthrush. *Parkesia noveboracensis*. HL (8, 27); Sca, Ra; Su; Res; FoAs, RiDe.

Golden-winged Warbler. *Vermivora chrysoptera*. Ea; Ra; Mig; GrVa.

Blue-winged Warbler. *Vermivora cyanoptera*. Ea; VR; Mig; FoSG, RiSS.

Black-and-white Warbler. *Mniotilta varia*. Ea; Ra; Mig; FoDe.

Prothonotary Warbler. *Protonotaria citrea*. Vagrant.

Tennessee Warbler. *Oreothlypis peregrina*. NE; Ra; Mig; FoDe, HuPG.

Orange-crowned Warbler. *Oreothlypis celata*. Di; Pan, NW, NC, SE, SC, SW; Un; Su; Res; FoSG, FoAs, RiDe.

Nashville Warbler. *Oreothlypis ruficapilla*. Ea; Ra; Mig; FoDe, WoJu.

Virginia's Warbler. *Oreothlypis virginiae*. HL (4, 18, 19, 23, 24, 28); So; Ra; Su; Res; WoJu, RiSS, FoSG.

Connecticut Warbler. *Oporornis agilis*. Vagrant.

MacGillivray's Warbler. *Geothlypis tolmiei*. Wi (all); Pan; Co; Su; Res; RiSS, GrVa, HuPG.

Mourning Warbler. *Geothlypis philadelphia*. Vagrant.

Kentucky Warbler. *Geothlypis formosa*. Vagrant.

Common Yellowthroat. *Geothlypis trichas*. Ub; Pan; Un; Su; Res; WeSW.

Hooded Warbler. *Setophaga citrina*. Vagrant.

American Redstart. *Setophaga ruticilla*. Di (all); Pan; Un; Su; Res; FoDe, RiDe.

Cape May Warbler. *Setophaga tigrina*. Vagrant.

Northern Parula. *Setophaga americana*. Ea; Ra; Mig; RiDe.

Magnolia Warbler. *Setophaga magnolia*. Ea; Ra; Mig; FoMC.

Bay-breasted Warbler. *Setophaga castanea*. Ea; VR; Mig; RiDe, HuPG.

Blackburnian Warbler. *Setophaga fusca*. Vagrant.

Yellow Warbler. *Setophaga petechia*. Ub; Pan; Co; Su; Res; RiSS.

Chestnut-sided Warbler. *Setophaga pensylvanica*. Ra; Mig.

Blackpoll Warbler. *Setophaga striata*. Ea; Ra; Mig; RiDe, HuPG.

Palm Warbler. *Setophaga palmarum*. Ea; VR; Mig; RiDe, HuPG.

Pine Warbler. *Setophaga pinus*. Vagrant.

Yellow-rumped Warbler. *Setophaga coronata*. Ub; Pan; Co; Su; Res; FoMC, FoAs, RiDe.

Yellow-throated Warbler. *Setophaga dominica*. Vagrant.

Prairie Warbler. *Setophaga discolor*. Vagrant.

Black-throated Gray Warbler. *Setophaga nigrescens*. HL (4, 10, 18, 19); NW, NC, SE, SC, SW; Un; Su; Res; WoJu.

Hermit Warbler. *Setophaga occidentalis*. Vagrant.

Black-throated Green Warbler. *Setophaga virens*. Vagrant.

Canada Warbler. *Cardellina canadensis*. Vagrant.

Wilson's Warbler. *Cardellina pusilla*. SD; NW NC, SC, SW; Pan; Co; Su; Res; RiSS.

Red-faced Warbler. *Cardellina rubrifrons*. Vagrant.

Family Cardinalidae (Tanagers, Grosbeaks, and Buntings)
Hepatic Tanager. *Piranga flava*. Vagrant.

Summer Tanager. *Piranga rubra*. Vagrant.

Scarlet Tanager. *Piranga olivacea*. Vagrant.

Western Tanager. *Piranga ludoviciana*. Wi (all); Pan; Co; Su; Res; FoDe, FoDC, FoMC.

Northern Cardinal. *Cardinalis cardinalis*. Ea; Ra; Mig; HuTC, HuPG, FoDe.

Yellow Grosbeak. *Pheucticus chrysopeplus*. Vagrant.

Rose-breasted Grosbeak. *Pheucticus ludovicianus*. HL (19); Un, Sca; Su; Res; FoDe, HuTC, HuPG.

Black-headed Grosbeak. *Pheucticus melanocephalus*. Wi; Pan; Co; Su; Res; FoDe, FoAs, FoPP, HuPG.

Blue Grosbeak. *Passerina caerulea*. Lo (7, 8, 17, 19, 21, 23, 28); NW, NC, NE, SE, SC, SW; Un; Su; Res; RiDe, HuPG, HUFR.

Lazuli Bunting. *Passerina amoena*. Wi; Pan; Un; Su; Res; RiSS, FoSG, GrVa.

Indigo Bunting. *Passerina cyanea*. HL (4, 7); NW, NC, NE, SE, SC, SW; Ra; Su; Res; RiSS, FoSG, GrVa.

Painted Bunting. *Passerina ciris*. Vagrant.

Dickcissel. *Spiza americana*. HL (5, 7, 21); NW, NC, NE, SE, SC, SW; Ra; Su; Res; GrTG, GrVa.

REPTILES AND AMPHIBIANS

Wyoming has a species diversity of 12 species of amphibians (1 salamander and 11 frogs and toads) and 22 species of reptiles (4 turtles, 7 lizards, and 11 snakes). Koch and Peterson (1995) reported seven amphibians and nine reptiles from Yellowstone and Grand Teton National Parks, reflecting that region's markedly colder environments than are typical of most of Wyoming. By comparison, the more southerly latitudes of Nebraska support 14 amphibians (2 salamanders and 11 frogs and toads) and 48 reptiles (9 turtles, 10 lizards, and 29 snakes) (Ballinger, Lynch, and Smith, 2010).

Family Ambystomatidae (Salamanders and Newts)
Tiger Salamander. *Ambystoma tigrinum*. Ub; Pan; Co; WeSW, WeDW; Pla–Alp.

Family Pelobatidae (Spadefoots)
Plains Spadefoot Toad. *Spea bombifrons*. Lo (10, 11, 12, 13, 19, 20, 21, 26, 27); NW, NC, NE, SE, SC; Un; SaSG, SaSh; Pla.

Great Basin Spadefoot Toad. *Spea intermontana*. HL (16, 17, 18, 23, 24, 25); NC, NE, SE, SC; Co; SaSG, SaSh; Pla.

Family Bufonidae (Toads)
Boreal Toad. *Bufo boreas boreas*. Lo (1, 2, 8, 9, 16, 18, 22, 25, 26); NW, SW, SC; Co; WeSW, WeDW; Foo–Sub.

Great Plains Toad. *Bufo cognatus*. HL (7, 15); NE; Un; GrVa, WeSW; Pla.

Wyoming Toad. *Bufo baxteri*. HL (27); SE; Ra; GrSG, WeSW; Pla.

Woodhouse's (Rocky Mountain) Toad. *Bufo woodhousei*. Lo (2, 3, 4, 11, 12, 13, 14, 18, 20, 21, 25); NW, NC, NE, SE, SC; Co; GrSG, WeSW; Pla.

Family Ranidae (Frogs)
American Bullfrog. *Rana (Lithobates) catesbeiana*. Lo (1, 8, 13, 14, 17, 20, 21, 27, 28); NW, NE, SE, SC; Spr; WeSW, WeDW; Pla.

Northern Leopard Frog. *Rana (Lithobates) pipiens*. Wi (all); Spr; Co; WeSW; Foo–Mon.

(Columbia) Spotted Frog. *Rana (Lithobates) pretiosa*. Lo (1, 2, 4, 6, 9, 16, 22); NW, NC, SW; Co; WeSW, RiLe; Foo–Mon.

Wood Frog. *Rana (Lithobates) sylvatica*. HL (4, 26); NC, SC; Ra; WeSW, RiLe; Mon.

Family Hylidae (Chorus Frogs)
Boreal Chorus Frog. *Pseudacris triseriata maculata.* Wi (all); Spr; Co; WeSW; Pla–Alp.

Family Trionychidae (Softshell Turtles)
Western Spiny Softshell Turtle. *Apalone spinifera hartwegi.* SD; NC, NE, SE; Co; WeDW, RiLe; Pla.

Family Emydidae (Pond and Box Turtles)
Ornate Box Turtle. *Terrapene ornata ornata.* HL (21); SE; Un; GrSG; Pla.
Western Painted Turtle. *Chrysemys picta bellii.* HL (12, 15, 16, 17); NC, NE, SE; Co; WeSW, WeDW, RiLe; Pla.

Family Chelydridae (Snapping Turtles)
Snapping Turtle. *Chelydra serpentina.* Lo (4, 5, 6, 7, 13, 19, 21, 28); NC, Ea; Co; WeSW, WeDW; Pla.

Family Phrynosomatidae
(Family Iguanidae in Baxter and Stone, 1985)
Northern Sagebrush Lizard. *Sceloporus graciosus graciosus.* SD; NC, NE, SE, SC, SW; Co; SaSh, SaSG, MoRo; Pla, Foo.
Plateau (Prairie) Lizard. *Sceloporus (consobrinus) undulatus. Three subspecies:* Northern Plateau Lizard. *S. (c.) u. elongates.* HL (20, 22, 23, 27); SW; Co; SaSh, SaSG, MoRo; Pla, Foo. Red-lipped Prairie Lizard. *S. (c.) u. erythrocheilus.* HL (20, 27); SE; Co; MoRo; Pla, Foo. Northern Prairie Lizard. *S. (c.) u. garmani.* HL (21); SE; Co; MoRo, GrVa; Pla.
Northern Tree Lizard (Cliff Lizard). *Urosaurus ornatus wrighti.* HL (23); SW; Un; SaSh; Pla.
Greater Short-horned Lizard. *Phrynosoma hernandesi (douglassi).* Di (all); Co; SaSG, GrSG; Pla.
Great Plains (Northern) Earless Lizard. *Holbrookia maculata maculata.* HL (28); SE; Un; GrVa; Pla.

Family Teidae (Racerunners)
Six-lined (Prairie) Racerunner. *Cnemidophorus sexlineatus viridis.* HL (20, 28); SE; Un; GrTG, GrMG, GrSG; Pla.

Family Scincidae (Skinks)
Many-lined Skink. *Eumeces multivirgatus.* HL (21, 28); SW; Un; GrVa; Pla.

Family Boidae (Boas)
Rubber Boa. *Charina bottae.* Lo (1, 2, 4, 8, 10, 15, 16, 17); NW, NC, SC, SW; Ra; FoMC; Foo–Mon.

Family Colubridae (Harmless Egg-laying Snakes)
Plains Hognosed Snake. *Heterodon nasicus nasicus.* Lo (3, 6, 10, 13, 14, 17, 19, 21); NC, NE, SE, SC; Co; GrVa; Pla.
Eastern Yellow-bellied Racer. *Coluber constrictor flaviventris.* SD; NC, NE, SE, SC; Co; GrSG; Pla, Foo.
Striped Whipsnake. *Masticophis (Coluber) taeniatus.* HL (23); SW; Ra; AlSh; Pla.
Smooth Green Snake. *Opheodrys vernalis.* HL (7, 19, 21, 25, 26, 27); NE, EC, SC; Ra; GrVa, FoMC; Foo, Mon.
Black Hills Red-bellied Snake. *Storeria occipitomaculata pahasapae.* HL (7); NE; Un; FoPP; Pla.
Western (Pale) Milksnake. *Lampropeltis triangulum multistriata.* Lo (4, 17, 19, 20, 21, 27, 28); NC, NE, SE; Un; RiDe, GrVa, MoRo; Pla, Foo.
Bullsnake (Gophersnake). *Pituophis catenifer. Two subspecies:* Great Basin Bullsnake. *P. c. deserticola.* HL (17, 23, 24); SW; Co; AlSh, SaSh, SaSG; Pla. Say's Bullsnake. *P. c. sayi.* Di; NW, NC, NE, SE, SC; Co; All plains and foothills habitats; Pla, Foo.

Family Natricidae (Gartersnakes)
Western Terrestrial (Wandering) Gartersnake. *Thamnophis elegans vagrans.* Wi (all); Spr; Co; RiSS, RiHe, GrSG, GrVa; Pla–Sub.
Common Gartersnake. *Thamnophis sirtalis. Two subspecies:* Red-sided Gartersnake. *T. s. parietalis.* Lo (2, 3, 4, 7, 14, 21); NW, NC, NE, SE; Co; RiHe, RiSS, WeSW; Foo–Mon. Valley Gartersnake. *T. s. fitchi.* HL (15, 22); NW; Co; RiHe, RiSS, WeSW; Foo–Mon.
Western Plains Gartersnake. *Thamnophis radix haydeni.* HL (4, 6, 7); NE, SE; Co; RiHe, RiSS, WeSW; Pla.

Family Crotalidae (Rattlesnakes)
Prairie Rattlesnake. *Crotalus viridis. Two subspecies:* Prairie Rattlesnake. *C. v. viridis.* Wi; Co; GrSG, GrVa, MoRo; Pla, Foo. Midget Faded Rattlesnake. *C. v. concolor.* HL (23); SW; Ra; SaSh; Foo, Mon.

Information on wildlife-viewing and bird-finding locations in Wyoming can also be found in various publications that cover the entire state (Scott, 1993; Wyoming Game and Fish Department, 1996), the Rocky Mountain region (Johnsgard, 2011a), the Bighorn Mountains (Canterbury, Johnsgard, and Downing, 2013), Jackson Hole (Raynes and Wile, 1994), Yellowstone National Park (McEneaney, 1988), and the Yellowstone–Grand Teton region (Wilkinson, 2008). State park and general tourism information may be obtained from the Wyoming Office of Tourism: 5611 High Plains Road, Cheyenne, WY 82007. Phone: 307-777-7777. Fax: 307-777-2877. Website: https://www.travelwyoming.com

There are several local chapters of the National Audubon Society in Wyoming: at Casper (Murie Audubon Society, PO Box 2112, Casper, WY 82602), Cheyenne (Cheyenne–High Plains Audubon Society, PO Box 2502, Cheyenne, WY 82003), Laramie (Laramie Audubon Society, PO Box 878, Laramie, WY 82073), and Sheridan (Bighorn Audubon Society, PO Box 535, Sheridan, WY 82801). The state headquarters' address is Audubon Rockies, 116 N. College Ave., Suite 1, Ft. Collins, CO 80524 (http://rockies.audubon.org/). As of 2018 Wyoming's Audubon Society had identified 45 Important Bird Areas, which are listed on the society's website (https://www.audubon.org/important-bird-areas/state/wyoming).

Bighorn Canyon National Recreation Area. This steep-sided, 1,000-foot-deep canyon in northern Wyoming north of Lovell was partly impounded by the construction of Yellowtail Dam on the Bighorn River in Montana, which formed a 71-square-mile lake. The recreation area encompases 120,000 acres. From the top of the canyon, turkey vultures, golden eagles, and other raptors can often be seen in flight or perched along the canyon's steep walls. A visitor center is in Lovell. The Pryor Mountains Wild Horse Range is north of Lovell, a 40,000-acre area where horses that reportedly were descended from those of early Spanish conquistadors run free. There is also a good chance of seeing bighorn sheep in this region. The horse herd is managed by the Bureau of Land Management (contact 866-468-7826 or wildhorse@blm.gov). Address of recreation area headquarters: 5 Ave. B, PO Box 7458, Fort Smith, MT 59035 (307-548-5406).

Bighorn National Forest. This 1.1 million-acre national forest in the Bighorn Mountains includes a wilderness area, Cloud Peak, of 195,000 acres. The region's highest peak, Cloud Peak, rises to 13,165 feet of elevation, exceeded only by Garret Peak in the Wind River range. There are 300 lakes in the forest, as well as glacial fields and alpine tundra. Ranger stations are at Buffalo, Greybull, Lovell, and Worland. The 327 birds of the Bighorn region have been described by Canterbury, Johnsgard, and Downing (2013). Canterbury and Johnsgard (2017) also described the common birds of the Bighorn Mountains foothills and the vicinity of the Brinton Museum (600 acres), a historic ranch now converted into a major western art museum near Big Horn. The USDA Forest Service office for Bighorn is headquartered at 2013 Eastside 2nd St., Sheridan, WY 82801 (307-674-2600).

Black Hills National Forest. This 1.2 million-acre national forest is mostly located in South Dakota but extends into northeastern Wyoming. The forest's highest peak (in South Dakota) is at 7,242 feet of elevation, well below the subalpine zone. Much of the forest consists of ponderosa pines, which are now under severe attack by pine bark beetles. The majority of the Black Hills' bird life comprises eastern-oriented birds, such as the blue jay, red-headed woodpecker, eastern bluebird, brown thrasher, and indigo bunting. However, the forest also includes several endemic Rocky Mountain species, such as the red-naped sapsucker and the dusky and cordilleran flycatchers. Some species with general western orientations, such as the white-throated swift, Townsend's solitaire, and MacGillivray's warbler are also present. The Black Hills region is home to an endemic "white-winged" race of the dark-eyed junco. A comprehensive *Birds of the Black Hills* (Pettingill and Whitney, 1965) is long out of print, but an updated bird list of 195 species reported for the forest is available from the Black Hills National Forest, 1019 N. 5th St., Custer, SD 57730 (605-673-9200).

Bridger-Teton National Forest. This enormous 3.4-million acre national forest is an important part of the Greater Yellowstone ecosystem. The forest encompasses 1.2 million acres of wilderness, including the Teton (575,000 acres), Gros Ventre (287,000 acres), Winegar Hole (14,000 acres), and Bridger (428,000 acres) wilderness areas and ranges in altitude from about 6,000 to 13,785 feet (Gannet Peak, Wyoming's highest peak). The forest's vegetation varies from sagebrush and other lowland arid brush to alpine meadows and glaciers. Species such as the greater sage-grouse occupy sagebrush, while the harlequin duck, common merganser, and American dipper inhabit clear and swift mountain streams. Water and marsh birds such as the trumpeter swan, ring-necked duck, Barrow's goldeneye, bufflehead, and sandhill crane nest around beaver ponds. Several species designated by the US Forest Service as rare or sensitive also occur here, such as the northern goshawk, great gray owl, boreal owl, American three-toed woodpecker, and Hammond's flycatcher. The Forest Service's list of 355 birds represents more than 90 percent of the species that occur regularly in Wyoming and includes at least 157 summer residents, 80 permanent residents, 68 migrants, and 14 winter visitors. It and other information is available at ranger stations in Afton, Big Piney, Kemmerer, Moran, and Pinedale. (See also the Grand Teton National Park and Jackson Hole section.)

Grizzly bear, adult male

Besides the Bridger-Teton National Forest to the south of Yellowstone National Park, several other Wyoming national forests are also part of the Greater Yellowstone ecosystem. Part of the Caribou-Targhee National Forest borders the western boundary of Yellowstone and Grand Teton National Parks, and the Shoshone National Forest borders the eastern edge of Yellowstone. The Gallatin National Forest borders the northwestern and northern parts of Yellowstone, and the Custer National Forest barely reaches its northeastern corner. Collectively these seven national forests, three national wildlife refuges, two national parks, and other regional federal and state lands total about 18,000 square miles (a region equal in area to that of Vermont plus New Hampshire) and compose the Greater Yellowstone ecosystem, the largest nationally preserved and intact ecosystem south of Canada. Information on Wyoming's national forests can be obtained from the regional headquarters, USDA Forest Service, Rocky Mountain Region, 1617 Cole Blvd., Building 17, Lakewood, CO 80401 (303-275-5350).

Bureau of Land Management (BLM) Natural Areas. Wyoming has nearly 18 million acres of public-access land under management by the BLM. Several sites among the many BLM properties are of possible interest to birders. Muddy Mountain Environmental Education Area (12,000 acres, 10 miles south of Casper) includes peaks exceeding 8,000 feet with golden eagles and other raptors. Another BLM site of potential interest to birders is the Middlefork Recreation Area (48,239 acres, 11 miles west of Kaycee), with steep 1,000-foot canyon walls above the Powder River, with associated raptors and greater sage-grouse on the uplands. Other BLM sites of possible interest include Goldeneye Wildlife and Recreation Area (1,153 acres, including the 500-acre Goldeneye Reservoir, 20 miles west of Casper) and West Slope (448,300 acres, 15 miles east of Lovell on the western slope of the Bighorn Mountains).

Many of Wyoming's BLM properties are of special importance to the rapidly declining greater sage-grouse and other sage-adapted species. A couple are Shirley Basin (Carbon and Natrona Counties) and the Red Desert (Fremont and Sweetwater Counties), which have both been designated as state-level Important Bird Areas.

Two other BLM locations in the Red Desert region that are also state-level Important Bird Areas and that support many sage-grouse leks are Ninemile Draw and Little Sandy Landscape. The greater sage-grouse has been proposed for federal listing as threatened, and Wyoming offers the best opportunities for preserving it. Information on these BLM locations can be obtained from the Wyoming State BLM Office, PO Box 1828, Cheyenne, WY 82003 (307-775-6256).

Local information can also be obtained from regional field offices at Buffalo (307-684-1100), Casper (307-261-7600), Cody (307-578-5900), Kemmerer (307-828-4500), Lander (307-332-8400), Newcastle (307-746-6600), Pinedale (307-367-5300), Rawlins (307-328-4200), Rock Springs (307-352-0256), and Worland (307-347-5100). Online bird lists are available for the two following BLM districts: Northeastern Wyoming, Casper District (office at Casper).

Cokeville Meadows National Wildlife Refuge. This recently established (1993) refuge occupies 26,600 acres and is located about 20 miles south of Cokeville along the Bear River in extreme southwestern Wyoming. The refuge provides excellent breeding habit for waterfowl such as redheads and trumpeter swans, marsh birds such as white-faced ibis and snowy egrets, and upland birds such as sage-grouse, golden eagles, and migrant peregrine falcons. It is managed out of Seedskadee National Wildlife Refuge (see contact information in that section) and as of 2018 there were no visitor facilities at the refuge.

Devils Tower National Monument. This national monument covers 1,346 acres in northeastern Wyoming near the western limit of the Black Hills. Typical western-oriented species include the white-throated swift, violet-green swallow, black-headed grosbeak, spotted towhee and Bullock's oriole. The 1,200-foot column of volcanic basalt is surrounded by ponderosa pines, scrubby thickets, and grasslands. A bird species checklist (158 species, 75 nesters) is available. Burrowing owls have not been reported, but prairie dogs have long been common (19 colony acres in 1994). Address: Box 10, Devils Tower WY 82714 (307-467-5283), www.nps.gov/deto. For more information contact the headquarters at PO Box 10, Gillette, Devils Tower, WY 82714 (307-467-5283).

Flaming Gorge National Recreation Area. This large area (201,000 acres) in southwestern Wyoming and adjacent Utah is centered on the 90-mile-long Flaming Gorge Reservoir, and includes part of Ashley National Forest (which totals 1.38 million acres, mostly in Utah). The area's bird list includes 259 species, with notable numbers of raptors (15 species), waterfowl (24 species), and especially shorebirds and gulls (31 species). Its scrubby juniper habitats attract the juniper titmouse, gray flycatcher, Bewick's wren, and black-throated gray warbler, all of which are uncommon to rare species in Wyoming. A checklist is available from Red Canyon Lodge, 2450 Red Canyon Lodge, Dutch John, UT 84023 (435-889-3759) or the Ashley National Forest, Flaming Gorge Ranger District, PO Box 279, Manila, UT 84046 (435-784-3445). The Flaming Gorge National Recreation Area is partly in Utah and is part of the Ashley National Forest, headquartered at 355 N. Vernal Ave., Vernal, UT 84078 (801-789-1181).

Fort Laramie National Historic Site. This historical site has a trail along the Platte and Laramie Rivers that leads to their confluence, with good bird watching and wildlife viewing. For more information contact the site headquarters at 965 Gray Rocks Road, Fort Laramie WY 82212 (307-837-2221).

Fossil Butte National Monument. This 8,000-acre site is a 50-million-year-old Eocene lake bed in southwestern Wyoming that is world famous for its remarkable fish fossils but also has produced some notable bird fossils. For more information, contact the headquarters at PO Box 592, Kemmerer, WY 83101 (307-877-4450).

Grand Teton National Park and Jackson Hole. This national park includes the adjacent Teton Range and a section of the Snake River valley that extends from the vicinity of Jackson Lake south to about ten miles south of Jackson ("Jackson Hole"). The entire park region consists of more than 480 square miles of high mountains (with a maximum elevation of 13,766 feet), coniferous forests, and sage-covered plains and riparian valleys between 6,000 and 7,000 feet of elevation. All the roads are limited to lower elevations, but many hiking trails extend into the mountain forests and alpine zone. The park is recognized as one of Wyoming's Important Bird Areas; park specialties include breeding trumpeter swans, sandhill cranes, greater sage-grouse, and black rosy-finches. Follett (1986) provided a seasonally organized list of 226 Grand Teton National Park birds, 15 of which are very rare vagrants. Of the regularly occurring species, 170 were reported during summer, and most of these might be considered as prospective if not proven breeders. For information on Grand Teton National Park, contact the park headquarters at Moose, WY 83012 (307-733-2880). The Grand Teton Association sponsors field trips and lectures and offers book discounts to members. For more information, write Grand Teton Association, PO Box 70, Moose, WY 83012, call 307-739-3606, or see https://www.grandtetonassociation.org/18/home.htm.

The nearby Jackson Hole area, the National Elk Refuge, and the adjacent Bridger-Teton National Forest (see separate descriptions) are all part of the Greater Yellowstone ecosystem and support many additional species and habitats. Raynes and Wile (1994) provided individual accounts of 68 species of Jackson Hole/Grand Teton birds and also a seasonally organized checklist of 293 species. Raynes's "pocket guide" (2000) list provides relative seasonal abundance and habitat information on 301 species.

A "Birds of Jackson Hole" checklist published by the Wyoming Game and Fish Department contains 340 species and covers most of the Bridger-Teton National Forest. It is available from the Wyoming Game and Fish Department, 5400 Bishop Blvd., Cheyenne, WY 82006 and might also be available at the headquarters of Bridger-Teton National Forest, 340 N. Cache, PO Box 1888, Jackson, WY 83001 (307-739-5500).

Hutton Lake National Wildlife Refuge. This 1,900-acre refuge is located 12 miles south of Laramie at an elevation of 7,150 feet. The refuge consists of five small lakes, a marsh, and uplands and attracts at least 153 species of birds, including 61 breeders. Nesting marsh birds include the eared grebe, American avocet, Wilson's phalarope, Forster's tern, and black-crowned night-heron.

Notable land birds include the golden eagle, McCown's longspur, and sage thrasher. Several small lakes that vary in size and depth are present, with the one nearest the refuge entrance the largest and deepest. It often has a variety of migrant ducks, plus grebes (eared and western). The most distant wetlands include a marshy area that attracts many migrant shorebirds and dabbling ducks. Managed from the Arapahoe National Wildlife Refuge in Colorado, the refuge roads are barely passable rutted trails, and there are no available water or toilet facilities. In the same general region (off Highway 230) is Hattie Lake, a flood-control reservoir. Depending on the amount of water present, it may attract large numbers of diving ducks, and many shorebirds, gulls, and American white pelicans. This NWR is also managed through Colorado's Arapaho NWR. Address: PO Box 457, Walden, CO 80480 (303-482-5155). The nearby Bamforth National Wildlife Refuge was established to protect the endangered Wyoming toad and is off limits to the public.

Medicine Bow National Forest. This 1.6-million acre national forest in south-central Wyoming consists of three rather small but beautiful mountain ranges, the Medicine Bow, Snowy, and Sierra Madre. The forest includes Encampment River (10,400 acres), Houston Park (31,000 acres), Platte River (23,000 acres), and Savage Run (15,000 acres) wilderness areas and reaches a maximum elevation of 12,013 feet (Medicine Bow Peak). Two high-elevation passes are the Snowy Range Pass at 10,846 feet and Battle Pass at 9,955 feet. No bird list for this forest is yet available, but Brooklyn Lake and Lewis Lake in the Snowy Range are good places for finding alpine species such as American pipits and brown-capped rosy finches. The white-tailed ptarmigan has not been seen in these mountains for several decades. Battle Creek Campground in the Sierra Madre Range is at a lower altitude in mixed coniferous-deciduous forest. This campground has some scrub oaks that rarely have attracted band-tailed pigeons and support an assortment of western flycatchers, towhees, and jays (Scott, 1993). Ranger stations are at Douglas, Encampment, Laramie, and Saratoga. For information, contact the US Forest Service office, 2468 Jackson St., Laramie, WY 82070 (307-745-2300).

National Elk Refuge. This 25,000-acre refuge is located just north of Jackson and is part of the Greater Yellowstone ecosystem. Although primarily dedicated to providing winter habitat for up to nearly 10,000 elk, the largely grassland- and sage-covered refuge has attracted at least 219 bird species, including 125 breeders. The greater sage-grouse, golden eagle, sandhill crane, trumpeter swan, and long-billed curlew are among the refuge's notable nesting species. For more information, contact the refuge manager at PO Box C, Jackson, WY 83001 (307-733-9212).

Ocean Lake Wildlife Habitat Management Area. Located 17 miles northwest of Riverton, off Highway 134, about half of the 12,750-acre area consists of a large, shallow lake with marshy edges that attract many migrating ducks and grebes. The site also attracts sandhill cranes, about 400 of which stage here during spring and fall, and some remain to nest. Up to 3,000 geese and 10,000 ducks stop here during migration. Ocean Lake is owned by the Wyoming Game and Fish Department, 5400 Bishop Blvd., Cheyenne, WY 82006 (307-777-4600).

Seedskadee National Wildlife Refuge. This refuge of 14,455 acres in southwestern Wyoming lies at about 6,000 feet elevation, in the Green River valley. The refuge is dominated by sagebrush and other arid-adapted plants and has an associated bird fauna, together with bluff-associated and riparian woodland birds. A graveled auto tour route parallels the Green River and provides many overlook views. The refuge attracts at least 227 species, including 120 breeders. Breeding raptors include the osprey, golden eagle, prairie falcon, merlin, and short-eared owl. Breeding marsh birds include the white-faced ibis, sora, Virginia rail, Forster's tern and great blue heron. Land birds include the Woodhouse's scrub-jay, pinyon jay, blue-gray gnatcatcher, western bluebird, and green-tailed towhee. Further information can be obtained from the refuge manager at PO Box 67, Green River, WY 82935 (307-875-2187).

Shoshone National Forest. This 2.4 million-acre forest extends over much of central and southern Wyoming and includes five wilderness areas: Absaroka (24,000 acres), Fitzpatrick (199,000 acres), North Absaroka (360,000 acres, with peaks to 11,700 feet), Popo Agie (101,000 acres, with 20 peaks exceeding 12,000 feet), and Washakie (714,000 acres, with peaks exceeding 13,000 feet). Ranger stations are at Cody, Dubois, Lander, Meeteetse, and Powell. No bird list for this forest is yet available, but the Jackson Hole list may be applicable. For information, contact the US Forest Service office, 808 Meadow Lane Ave., Cody, WY 82414 (307-527-6241).

Table Mountain Wildlife Habitat Management Area. Located about 15 miles southeast of Torrington, this 1,716-acre, state-owned area is a major migration stopover point for waterfowl in spring. During fall it is open to controlled waterfowl hunting. The marsh attracts thousands of snow geese and hundreds of American white pelicans as well as Canada, Ross's, and greater white-fronted geese, plus dozens of duck species. Migrant shorebirds include many "peep" (*Calidris*) sandpipers, especially stilt sandpipers. About ten miles to the west (and about five miles south of Yoder) is Bump Sullivan Reservoir and nearby Springer Lake, the latter owned by the Wyoming Game and Fish Department. Bump Sullivan Reservoir's migrant birds are much like those of Table Mountain, but the reservoir lacks marshy habitats. Springer Lake is an alkaline wetland that is notable for its migrant shorebirds, American white pelicans, double-crested cormorants, Canada and snow geese, sandhill cranes, and grebes. For more information, contact the Wyoming Game and Fish

Department headquarters at 5400 Bishop Blvd., Cheyenne, WY 82006 (307-777-4600).

Thunder Basin National Grassland. This federally owned site of 572,211 acres consists of shortgrass and shrub-steppe plains at 3,600–5,200 feet elevation. This is the largest area of federally protected grasslands in Wyoming, which support the state's biggest herd of pronghorns as well as mule deer and other sage-steppe species. Prairie dog colonies in this grassland once had the largest collective colony acreage of any of the national grasslands (about 18,200 acres in 1998). However, sylvatic plague in 2001 caused great losses here. The Douglas office of the US Forest Service has maps of the prairie dog colonies at Thunder Basin. Among the birds of the grassland, reputed breeders include 13 raptors, 6 owls, and 5 woodpeckers, of which many are forest-dependent. The burrowing owl, greater sage-grouse, mountain plover, long-billed curlew, and both longspurs are among the regular breeders. In addition, there are 72 species of mammals, including elk, pronghorn, mule deer, mountain lion, bobcat, coyote, and swift fox. For more information, contact the US Forest Service, 2550 E. Richards St., Douglas, WY 82633 (307-358-4690).

Uinta-Wasatch-Cache National Forest. This 1.2 million-acre national forest is mostly located in Utah, but it extends into southwestern Wyoming along the northern edge of the Uinta Range. The forest's maximum elevation is 13,528 feet and includes the 456,000-acre High Uintas Wilderness. Some of the birds of special interest are the northern goshawk, golden eagle, dusky grouse, American three-toed woodpecker, and northern saw-whet owl. Notable finches include the black rosy-finch, pine grosbeak, and red crossbill. The bird list contains more than 200 species and is available from the National Forest office, 8230 Federal Building, 125 South St., Salt Lake City, UT 84138 (801-524-5030).

Wyoming State Parks. Wyoming has 12 state parks that total more than 100,000 acres in area: Bear River (Evanston, 290 acres, 307-789-6540); Boysen (Shoshoni, 34,705 acres, 307-876-2796), Buffalo Bill, (west of Cody, 12,000 acres, 307-587-9227), Curt Gowdy (Cheyenne, 1,960 acres, 307-632-7946), Edness K. Wilkins (Casper, 319 acres, 307-577-5150), Glendo (Glendo, 19,000 acres), Guernsey (Guernsey, 8,638 acres, 307-836-1942), Hawk Springs (Hawk Springs, 2,000 acres, 307-836-2334), Hot Springs (Thermopolis, 1,029 acres, 307-864-2176), Keyhole (Moorcroft, 15,674 acres, 307-322-2220), Seminoe (north of Sinclair, 21,741 acres, 307-320-3013), and Sinks Canyon (southwest of Lander, 600 acres, 307-332-6333). Bird lists are not yet available for these parks.

Yellowstone National Park. This, the oldest US national park, is also the largest south of Canada, covering more than 3,400 square miles. Most of the roads are at elevations in excess of 7,000 feet, with passes as high as 8,850 feet. The park's highest point (Eagle Peak) is 11,358 feet. Mount Washburn is lower (at 10,243 feet) but is fairly accessible by trail and supports alpine birds such as American pipits and black rosy-finches. The majority of the land area of the park is covered with lodgepole pine forests that are now regenerating from massive 1988 forest fires that blackened 36 percent of the park's land area. Douglas-fir, Engelmann spruce, and subalpine fir occur at the higher elevations. Some fairly extensive areas of grassland are present in the northern parts of the park, which is an important year-round habitat for large grazing mammals and their predators. Yellowstone Lake has the only nesting colony of the American white pelicans in any national park, and the park is also notable for its nesting populations of the bald eagle, osprey, sandhill crane, and trumpeter swan. National park breeding rarities include the Caspian tern, harlequin duck, and pinyon jay.

In their survey of Montana wildlife-watching sites, Carol and Hank Fischer (1995) described 12 outstanding birding areas in the park, which has been recognized as one of Wyoming's Important Bird Areas. Terry McEneaney's 1988 book *Birds of Yellowstone* is a comprehensive birding reference and includes a great deal of information on local birds and bird-finding. A 2014 park checklist of birds is available from the park at PO Box 168, Yellowstone National Park, WY 82190 (307-344-7381) or can be downloaded at https://www.nps.gov/yell/learn/nature/upload/BirdChecklist2014.pdf. The Yellowstone Forever Institute (https://www.yellowstone.org/experience/yellowstone-forever-institute/) offers short courses on regional birds and other nature-oriented subjects as well as field seminars and nature tours. Contact Yellowstone Forever at PO Box 117, Yellowstone National Park, WY 82190 (406-848-2400).

Burrowing owl

Chapter 7 • In Praise of Special Places and Irreplaceable Species

About fifty thousand years ago, as the northern hemisphere was locked in a global deep-freeze and the continental glaciers of the Pleistocene were at a maximum, a large land bridge that connected Asia and North America existed in the general region now occupied by the Bering Sea and Alaska, the so-called Beringia region. Across that corridor many mammals migrated from Asia over the millennia, including North America's ancestral brown bears and, much more recently, the first humans. One early influx of bears arrived in North America from Asia less than fifty thousand years ago. Some of these ancestral Alaskan brown bears apparently became isolated in island and coastal habitats by the last of the great glaciers, and the polar bear evolved from them. A later influx of bears from Asia produced the modern brown and grizzly bears. A much smaller bear that had evolved in the Old World about 1.5 million years ago and arrived at least eight thousand years ago in North America became the modern black bear.

From their original area of North American occupation in Beringia, the ancestral brown bears moved south into central North America toward the end of the Pleistocene epoch, or about twenty thousand years ago. At their peak, their range extended south into northern Mexico, and east to the edge of the prairies in Canada's Prairie Provinces and the Great Plains states. In Alaska, these huge bears are called Alaska brown bears, or sometimes Kodiak bears. Alaskan bears weigh on average up to a third more than the more southern populations and can sometimes exceed one thousand pounds. The generally accepted name for the populations south of Canada is grizzly bear, in reference to the adults' gray-tipped ("grizzled") pelage. Transitional populations link these two extreme genetic types and even a few recent hybrids between Alaska brown bears and polar bears.

During presettlement times, the grizzly was widespread in western North America, from the Cascade and Sierra Nevada mountains east across the Rocky Mountains to the high plains grasslands. During 1804–5, grizzlies were encountered by the Lewis and Clark expedition in what is now North Dakota and Montana, and were seen again by Clark on his return trip down the Yellowstone River in 1806. The group's narratives of meeting grizzly bears, which they variously called white bears or gray bears, still provide for exciting reading material. Clark's account of chasing a grizzly for two miles while on horseback provides the first evidence of grizzlies in the Greater Yellowstone ecosystem.

More recently, as firearms and ammunition have improved, killing a grizzly bear as a unique hunter's trophy has increasingly become one of the ultimate icons of manhood for the most thoroughly gun-addicted Americans. As a result, nearly all the grizzly populations of western North America have been extirpated, except in remote areas such as Alaska, and within a few well-monitored sanctuaries, such as our western national parks.

With the 1973 passage of the Endangered Species Act, the grizzly bear was classified as a threatened species throughout the lower forty-eight states. In a corollary action, and while I was doing field research in the Tetons, the Greater Yellowstone region was proposed as critical habitat for grizzlies in 1976. This recommendation initiated nearly as much anger among ranchers, landowners, and developers as do current federal attempts to impose national jurisdiction over so-called states' rights. As a result of these pressures, the critical habitat designation for Yellowstone was never officially adopted.

Since Yellowstone National Park's formation in 1872, its bears theoretically have been secure because, according to its official 1883 management principles, the only animals that can be legally killed within park boundaries are fish. However, the park's principals have rarely followed these principles. For example, to satisfy fishermen, park personnel regularly destroyed the eggs in a nesting colony of American white pelicans on a small island in Yellowstone Lake, although white pelicans consume almost no fish of sporting value. Until the early decades of the twentieth century, thousands of coyotes, nearly all of the park's mountain lions, and all of the park's wolves were shot or poisoned. The loss of these predators resulted in large population increases in prey species such as elk, and their overgrazing produced widespread habitat deterioration.

I first saw wild bears on a trip to Yellowstone with my parents during the post-war recovery years of the late 1940s; we observed more than fifty black bears during a memorable two-day trip through the park. My teenage introduction to bears had occurred during the period when roadside feeding of animals by tourists in national parks was the norm. Two decades later, Yellowstone National Park began a campaign to separate bears from all human encounters and dealt harshly with any bears that failed to cooperate.

This draconian policy had its origins in 1967, after two grizzlies in Glacier National Park killed two young women campers. The women had both been wearing perfume, leading park officials to claim that the bears had been attracted to them by odor, rather than the attack being a result of the park's inadequate bear management. At that time, bear feeding by tourists was a well-established practice at both Glacier and Yellowstone parks, and in both locations the animals had lost all fear of humans.

The two deaths in Glacier represented only the fourth and fifth lethal attacks by grizzlies on humans in the entire history of the national parks, but it caused the administrators of national parks to reevaluate their bear policy. Yellowstone Park modified its garbage dumps by eliminating anything that might be attractive to bears. This change forced the bears to search elsewhere for food, such as around campgrounds. During 1966, before the garbage dump policy took effect, a total of nine bears that visited a campground near Yellowstone Lake were trapped and removed or killed. In 1968, after a nearby dump had eliminated all access to garbage by bears, the number of them removed or killed there had risen almost fourfold to thirty-three.

Accurate estimates of bear mortality associated with Yellowstone's control actions are impossible to obtain. For example, in the 13 years between 1970 and 1982, Yellowstone officials reported an average annual loss of 18 grizzly bears that died accidentally, were trapped and euthanized, or were transported to remote locations. However, Frank Craighead reported in his 1979 book *Track of the Grizzly* that over the four-year period 1969–72 an average of 32 Yellowstone grizzlies were killed annually.

In 1971 alone, well over 40 grizzlies were killed near the wild-west town of West Yellowstone, located just outside the western boundary of Yellowstone Park, where snowmobiles have priority over cars and owning lots of firearms is a status symbol. The grizzly deaths included 18 radio-tagged bears that had been part of the Craighead brothers' long-term and monumental study on Yellowstone's grizzly populations and ecology. Park officials did not receive the Craigheads' research results well and tried hard to restrict or terminate their studies.

A 1975 National Academy of Sciences report estimated a Greater Yellowstone population of about 300 grizzlies, a total that was lowered by a team of independent scientists to possibly fewer than 200 by 1982. By then the Park Service had reassessed and reduced its control activities. It is also becoming increasingly apparent that, because of the very large home ranges of grizzlies, illegal killing of the animals outside the park strongly influences regional bear numbers. Grizzly pelts and other body parts, such as their claws and bile, have high commercial value, making the bears attractive targets for poachers.

In recent decades the regional prospects for grizzlies have improved through better-informed park management and slightly improved control of illegal killings. However, in 2012 a record number of 56 bears were known to have been killed by humans in the Greater Yellowstone region, representing about 10 percent of their estimated total population. By comparison, a total of seven human deaths have been caused by Yellowstone's grizzlies during the park's entire 146-year history. Glacier National Park has likewise had seven lethal grizzly attacks over its 109-year history. Yellowstone Park averages well over 3 million visitors per year, and Glacier slightly under 2 million, so the chances of being killed by a bear at either park are much less likely than of becoming an astronaut.

By comparison, Yellowstone typically has up to ten bison attacks on humans per year, and during the fifteen years from 1979 to 1994 there were two fatalities and 56 injuries caused by bison in Yellowstone Park. Thus, the park's seemingly tame and lethargic bison are hundreds of times more likely to attack visitors than are its grizzly bears. Closer to home, domestic dogs and cattle each kill an average of roughly 20 Americans annually, while bees, hornets, and wasps average more than 60.

Yellowstone's grizzly populations have markedly improved lately, in spite of high cub mortality rates and an undisclosed number of bears being euthanized by the park. Of seven females with cubs that a friend monitored in 2013, only two still had any yearlings present in 2014. The bears' regional annual growth rate from 1983 to 2001 has been estimated at 4 percent to 7 percent, and James Halfpenny estimated in his 2007 book *Yellowstone Bears in the Wild* that the Greater Yellowstone ecosystem then held 500 to 600 grizzly bears. Grizzlies have also recently expanded their ranges south out of Yellowstone into Grand Teton National Park, where visitors are now increasingly likely to see them. There may now be about 700 grizzlies in the region.

In September 2013 I visited Grand Teton National Park and with Tom Mangelsen saw many of the places I had come to love during the 1970s. There were many obvious changes. For example, the Teton bison herd, which had consisted of a few dozen animals when I first saw it in the 1940s, had multiplied to nearly 1,000 head, and Yellowstone Park had about 3,500.

Besides seeing all the common Teton birds and mammals, Tom and I also extensively watched three subadult grizzly bears peacefully foraging on plant roots in grassy subalpine meadows near Togwotee Lodge. The bears also scavenged the carcass of a moose that a trophy hunter had killed and, except for the head and antlers, left behind to rot. Fall grizzly foods in the Yellowstone area often consist mostly of the seeds of whitebark pines dug out of squirrel caches, army cutworm moths, and a wide variety of plant leaves and roots. For three days Tom and I watched the bears, and at times more than 20 carloads of tourists and local wildlife photographers lined the roadsides. None bothered the bears, and the bears paid little attention to the onlookers. On a few occasions a bear would cross the highway, patiently waiting for the traffic to thin out and provide a safe crossing. One even wandered to within a stone's throw of our parked car, providing me with a heart-stopping sense of awe at seeing such a beautiful animal in its element and imprinting on my mind an incredible lifetime memory.

Tom recently told me that there is now an all-out effort to trap most of Yellowstone's regional grizzlies. One male (#760) was trapped twice in less than nine months and fitted with a radio collar. He is one of the few bears that was often seen by park visitors during 2014 and probably has been observed by hundreds of thousands over the past four years. With no history of being aggressive, he nevertheless now conspicuously wears two large yellow ear tags and a big radio collar, reminding a wildlife watcher more of a decorated Christmas tree than a wild bear. As Tom said of Yellowstone's grizzlies, "The American public does not want to come to their national parks to see Christmas-tree bears!"

More ominously, the grizzly became potentially legally hunted as a trophy species in Wyoming, Idaho, and Montana after it was delisted from its threatened status in 2017, a decision made by the Trump administration's US Fish and Wildlife Service (USFWS). The Wyoming Game and Fish Department immediately made plans for a grizzly bear hunt in 2018, with a lottery system set for 22 kill permits to be made available at a bargain price of $600 each for Wyoming residents and $6,000 for nonresidents. More than 7,000 would-be hunters (and some nonhunting activists) applied for the lottery opportunity at $35 per lottery ticket. The hunt was canceled at the last moment, thanks to the decision of a Montana District Court judge, who concluded that the USFWS was "arbitrary and capricious" in removing protection for the bear. So the bears have at least a temporary reprieve.

Beyond the ethical question of killing some of North America's rarest bears for "sport," there are the dangers posed by big-game hunters to other humans, other nontarget wildlife, and themselves. Not all hunters are expert marksmen, and a wounded grizzly bear is an extremely dangerous animal. With that thought in mind, the fact that the Greater Yellowstone region may yet become a much more hazardous place for both bears and humans, and a far sadder one, in which the sight of free-roaming, relatively tame grizzlies will become nothing but a memory. Grizzly bears might not be so appealing if they were not so powerful and so little understood. Baba Dioum, a great African ecologist, once wrote, "In the end, we will conserve only what we love; we will love only what we understand; and we will understand only what we are taught." It is my fervent hope that this book might serve as a teaching tool to all who love Wyoming and its wildlife.

White-tailed ptarmigan

Glossary

accipiter   A member of the hawk genus *Accipiter*.

adaptation   A genetic trait that increases the ability of an individual organism to better survive and reproduce within its environment and is thus favored (selected for) by natural selection.

allopatric   Descriptive of two populations having nonoverlapping geographic ranges. *See also* sympatric.

biomass   The total living weight (mass) of a species, or a collection of species that occupy a particular location.

biome   A geographically large-scale ecosystem of plants, animals, and their environment that share important common characteristics (e.g., climate, soil, dominant species), such as the tallgrass prairie biome.

biota   All the living organisms of a particular place, region, or some larger geographic entity.

brood parasitism   The laying of eggs by an individual female into the nest of another bird of the same or a different species. Also called egg parasitism and nest parasitism.

buteo   A member of the hawk genus *Buteo*.

carapace   The dorsal bony "shell" of a turtle. *See also* plastron.

Cenozoic era   The geologic period encompassing the past 65 million years, the so-called Age of Mammals.

cirque   A steep U-shaped hollow at the upper end of a glacially formed valley.

class   In taxonomy, the category above the order and below the phylum. *See also* family, order.

community   In ecology, interacting populations of plants and animals occupying a specific site. Plant communities are often named after one or two of their most dominant species. *See also* ecosystem, dominant.

coevolution   The reciprocal evolutionary influences on two interactive species on one another over time, typically to their mutual (symbiotic) benefit.

conspecific   Two or more populations belonging to the same species. *See also* interspecific.

cranial crest   Bony enlargements centered along the dorsal midline of the skull in toads.

crepuscular   Active during the periods of dusk and dawn. *See also* diurnal, nocturnal.

dimorphism   Occurring in two forms or appearances, such as sexual dimorphism. Such forms are sometimes called morphs. *See also* monomorphism.

diurnal   Active during daylight hours. *See also* crepuscular, nocturnal.

dominant   In ecology, those species having the strongest environmental effects in a biological community and, if removed, would have the greatest disruption on the remaining community structure.

double-brooding   The undertaking of two reproductive cycles during a single season; sometimes also called multiple brooding. *See also* renesting.

ecosystem   An interacting group of plants, animals, and their physical and chemical environment, as limited by energy flows and nutrient cycles. Ecosystems might be as small and transient as a mud puddle or as large, complex, and permanent as the entire earth. *See also* community, habitat.

ecotone   An ecological transition zone, such as a prairie-woodland transition.

endangered   Descriptive of taxa existing in such small numbers as to be in direct danger of extinction.

endemic   Refers to a population native to and confined to a particular area or region. *See also* pandemic.

exotics   Taxa that have been accidentally or purposefully introduced into an area where they are not native.

extirpated   Descriptive of taxa (usually a species or subspecies) that have been locally eliminated from an area or region but still persist elsewhere within their overall range.

facultative   Refers to unconstrained activities or behavior, such as being able to parasitize or exploit a variety of other organisms rather than a single kind. *See also* obligatory.

family   In taxonomy, the category above that of the genus (or the subfamily, if present) and below that of the order. In animal nomenclature, family names consistently end in "idae," subfamily names in "inae." Neither category is italicized. *See also* genus, order.

fauna   (adj., faunal) The collective animal life of a specific area or region.

fledging period   The amount of time required for an individual bird to pass from hatching to its first flight. *See also* nestling period.

flora   The collective plant life of a specific area or region. The term also applies to a listing or description of all the plants comprising a particular taxonomic group, or of those occurring in a specific region.

forb   A broad-leaved herbaceous plant (thus, herbaceous plants other than grasses and sedges). *See also* herb, sedge.

forest   A plant community having closely spaced, relatively tall trees, the distance between the trees typically being less than their height. *See also* savanna, woodland.

genotype   The genetic makeup of an individual plant or animal. *See also* phenotype.

genus   (pl., genera; adj., generic) One or more species having genetic characteristics indicating, if two or more are members, that they are a closely related evolutionary assemblage. Single-species genera have important unique genetic traits and are called monotypic genera. The genus is also the taxonomic category immediately above the species and below the family levels and when printed is italicized and capitalized. *See also* family, species, systematics, taxonomy.

guild   A group of species occupying the same habitat and sharing certain important niche characteristics. *See also* niche.

habitat   The physical and biological environment of a specific place or general area.

halophyte   A plant that is adapted to live in soils having high salt content. *See also* xerophyte.

herb   (adj., herbaceous) A nonwoody vascular plant.

herpetiles   (informally, herps) A collective term for reptiles and amphibians. Herpetology is the study of herpetiles.

hertz   (abbreviated Hz) A unit of sound frequency equal to one cycle per second; 1,000 Hz equals one kilohertz (kHz). *See also* ultrasonic.

hibernaculum   An overwintering (hibernating) site.

hibernation   A period of prolonged winter dormancy, marked by greatly reduced metabolic activity. Some arid-adapted rodents undergo similar periods of summer dormancy, termed estivation.

home range   The area occupied, but not defended, by an animal over a defined time period (e.g., daily, monthly, annually). *See also* territory.

hybrid   The offspring of a mating between two genetically very different organisms, usually species. *See also* species, subspecies.

interspecific   Refers to interactions between two or more species, such as interspecific hybridization or competition. *See also* conspecific.

krummholz   Low, wind-shaped, and often twisted woody vegetation at timberline elevations.

larva   (pl., larvae; adj., larval): A free-living embryonic stage during the life cycle of an animal that is structurally different from that of the adult, and which occurs by a gradual process of anatomical transformation. *See also* metamorphosis.

larynx   The vocal structure of mammals. *See also* syrinx, song, vocalization.

lek   A location at which sexually active males of a local mating population meet to compete with one another for access to mating opportunities, either by establishing social dominance over other males or being able to attract females effectively. Male behavior at leks is called lekking; leks are sometimes also described as arenas.

lentic   Describes an environment of flowing water. *See also* lotic.

leucistic   (n., leucism) A variant ("morph") plumage in birds in which the pigment pattern is very pale because unusually little melanin is present. *See also* melanistic.

lotic   Describes an environment of standing water. *See also* lentic.

melanistic   (n., melanism) A variant ("morph") plumage or pelage, which is very dark, with unusually large amounts of melanin present. *See also* leucistic.

mesic   Environmental moisture conditions intermediate between dry (xeric) and wet (hydric). *See also* xeric.

metamorphosis   A process of structural or anatomical change in an animal's life cycle, typically from a larva to an adult. *See also* larva.

mixed-grass prairie   (or mid-grass) Perennial grasslands that are dominated by intermediate-stature grasses, often 0.5 to 1.5 meters high. These typically occur in environments that are more mesic (relatively moist) than those that support only short grasses but too xeric (moisture deficient) to support tall grasses. *See also* short-grass prairie, tall-grass prairie.

monogamous   A mating system in which females and males pair-bond and mate with a single member of the opposite sex, either over a single breeding cycle or season, or indefinitely longer. *See also* polyandrous, polygynous, promiscuous.

monomorphism   Occurring as a single form or appearance, such as sexual monomorphism, in which the sexes are externally identical (monomorphic). *See also* dimorphism.

moraine   A glacially modified landscape, the result of local movement, removal, or deposition of substrate materials by glacial action.

morph   A specific gene-based phenotype of a population in which two or more different and stable phenotypes ("morphs") occur. Synonymous with the commonly used term "phase," which inaccurately implies a temporary or seasonal appearance. *See also* genotype, leucistic, melanistic, phenotype.

mutualism   An ecological situation in which two species mutually benefit from each other's presence, which might involve facultative or obligatory interactions. Often also called symbiosis. *See also* coevolution.

mycophagy   The consumption of fungi, such as mushrooms; common in many forest-dwelling rodents.

nesting period   As used here, the duration of a species' or individual's breeding time from the start of the first nest until the last nest is abandoned.

nestling period   The amount of time an individual young bird is a nestling, from hatching until it leaves the nest. In most species this is the same as the fledging period, the age of first flight.

niche   An organism's specific role or biological "profession" within its ecological habitat, as defined by its behavioral, morphological, and physiological adaptations to that environment. *See also* adaptation, habitat.

nocturnal   Active during nighttime. *See also* crepuscular, diurnal.

nomenclature   The process of naming things systematically. Binomial nomenclature consists of the naming and describing of organisms by the combination of two words. *See also* systematics, taxonomy.

nonpasserine   A collective term for all the birds that do not belong to the order Passeriformes (the so-called perching birds). *See also* passerine.

obligatory   Refers to constrained activities or behavior, such as an obligatory host species or symbiotic partner that requires the presence of another species. *See also* facultative.

order   (adj., ordinal) In taxonomy, the category above that of the family and below that of the class. In animal nomenclature, at least in bird taxonomy, the names of orders consistently end in "iformes," as in Passeriformes; in other vertebrate groups ordinal names often end in "ia," as in Rodentia and Lacertilia. *See also* family, class.

orographic precipitation   A precipitation pattern that is influenced (reduced or increased) by the presence of relatively nearby mountains.

oviparous   Descriptive of animals that give birth to shelled embryos (eggs) that complete their development outside the mother's body. When a shelled egg is hatched before leaving the female's oviduct the reproductive process is called ovovivipary. *See also* viviparous.

oviposition   The act of laying (ovopositing) a shelled egg externally. The ovarian release of an unfertilized ovum is ovulation.

pandemic   Refers to a population that occurs widely over a region, continent, or the world. *See also* endemic.

parotoid gland   A large secretory gland located behind each eye in toads.

passerine   A collective term for all the birds that belong to the order Passeriformes, the so-called perching birds. "Perching" birds collectively have long hind toes that permit clinging to small branches and so forth. *See also* nonpasserine, songbird.

phenotype   The outward characteristics or appearance of an organism as opposed to its genetic traits, or genotype. *See also* genotype.

pheromones   Chemicals that are produced by an animal, released into the air, and received by other conspecific individuals upon which they have physiological or behavioral influences. Corresponding substances in plants are called allomones.

plastron   The ventral bony "shell" of a turtle. *See also* carapace.

Pleistocene epoch   The geologic time interval ("epoch") extending from about 1.6 million years ago (the end of the Pliocene epoch) to 12,000 years ago (the start of the Holocene epoch). Popularly known as the "Ice Age," this epoch included several major glacial periods and associated interglacial intervals. Several Wyoming mountain ranges were strongly influenced by glaciers, but most lower landscapes were only indirectly affected.

polyandrous   A mating system in which females often pair-bond and mate with two or more males simultaneously or successively. *See also* monogamous, polygynous, promiscuous.

polygynous   A mating system in which males often pair-bond and mate with two or more females simultaneously or successively. *See also* monogamous, polyandrous, promiscuous.

prairie   A native plant community that is dominated by perennial grasses. Prairies may be further defined by the relative stature of the dominant species (tallgrass, mixed-grass, shortgrass), by their dominant plant taxa (e.g., big bluestem), or by the life form of these grasses (bunchgrasses vs. sod-forming grasses).

Precambrian era   Refers to the geologic period in Earth's history prior to the Cambrian era, which occurred more than 544 million years ago.

promiscuous   A mating system in which both sexes may mate with multiple members of the opposite sex, often without any other prolonged social interactions. *See also* monogamous, polyandrous.

rain shadow   An area of reduced rainfall where nearby mountains intercept the normal regional precipitation pattern.

refugium   (pl., refugia) A location or region where certain taxa have been able to survive after the general environment or climate changed and eliminated those taxa from surrounding areas. *See also* relicts.

relicts   Plant or animal taxa that persist locally in an otherwise markedly altered habitat or climate. *See also* refugium.

renesting   A second breeding effort undertaken following the failure of a first attempt. *See also* double-brooding.

riparian   Descriptive of a land habitat that borders water, such as a riverbank.

savanna   A grass-dominated plant community that is interspersed with widely scattered trees. *See also* woodland, forest.

sedge   A general term for members of the plant family Cyperaceae, which are grasslike plants that have solid, triangular stems.

shortgrass prairie   Perennial grasslands that occur in regions too dry (arid) to support mixed-grass prairie. *See also* mixed-grass prairie, steppe.

song   In birds, complex vocalizations that are produced by vibratory membranes in the syrinx, and which are variably modulated in frequency, loudness, and harmonic content by the trachea, oral cavity, and related structures. *See also* passerine, songbird, syrinx.

songbird   A general term for those passerine birds that utter complex vocalizations considered to be esthetically appealing, or "songs." Some passerine birds that are more anatomically "primitive" are often highly vocal but lack the complex vocal structures that occur in species of the more advanced groups and that enable the utterance of acoustically highly complex vocalizations called songs. *See also* passerine, song, syrinx.

species   (adj., specific) A population of plants or animals that has definable and stable morphological traits. More specifically, a population of individuals capable of interbreeding with others of that population but reproductively isolated from all other populations. The term "species" is also a taxonomic category, below that of the genus and above the subspecies. Latin names of species and subspecies are italicized but not capitalized; generic names (e.g., *Passer*) are always italicized and capitalized. When written, the organism's specific Latin name follows the generic name (e.g., *Passer domesticus*). The higher taxonomic categories, such as families (e.g., Passeridae) and orders (e.g., Passeriformes) are capitalized but not italicized. *See also* genus, family, order, subspecies, systematics, taxonomy.

steppe A Russian term for shortgrass prairie. *See also* shortgrass prairie.

stromatolite A cone-shaped or rounded fossil rock formation composed of ancient (Precambrian) fossils of blue-green bacteria, the oldest known life forms on earth.

subspecies Geographically defined populations of a species that are geographically distinct but not genetically isolated from other such populations. Subspecies are often called races. *See also* allopatric, species.

succession In ecology, refers to changes in the botanical and zoological composition of the plant and animal life of a biological community over time, from early ("pioneer") through transitional ("seral") stages until a stabilized ("climax") biotic composition is attained, as influenced by climate, substrate, and other environmental factors.

syrinx (adj., syringeal): The unique vocal structure of birds, located at the junction of the two bronchi and the trachea and usually consisting of two pairs of vibratory membranes (tympanic membranes). *See also* songbird.

symbiosis The association of two organisms of different species to the benefit of one or both species. *See also* coevolution, mutualism.

sympatric Descriptive of two species having significant geographic overlap, at least during their breeding season. Also called syntopic. *See also* allopatric.

systematics Biological study that consists of the erection of nomenclatural categories that best reflect the apparent evolutionary relationships of two or more groups of organisms. *See also* nomenclature, taxonomy.

tallgrass prairie Perennial grasslands that are dominated by tall-stature grasses, often at least two meters high. These typically occur in climates that are more mesic (relatively moist) than those that support intermediate-height grasses but too xeric (moisture-deficient)—or too frequently burned—to support forests. Also referred to as "true" prairie. *See also* mixed-grass prairie, shortgrass prairie.

talus Rock rubble at the base of a mountain slope.

taxon (pl., taxa) A biological unit in the taxonomic hierarchy, from subspecies to kingdom. Taxonomy is the study and systematic organization of taxa according to their evolutionary relationships (phylogenies), including their technical names (nomenclature).

taxonomy (adj., taxonomic) The process of systematically organizing organisms in an orderly series of standardized categories that best reflect their evidence-based evolutionary relationships. Also called systematics. *See also* nomenclature.

territory An area advertised and defended by individuals of one sex (usually the males) against others of the same sex and species (territoriality), thus controlling its resources and attracting the opposite sex of that species. *See also* home range.

tragus A structure at the external opening of the ear canal that, at least in some bats, controls the passage of sound into the inner ear. The external ear of mammals is called the pinna.

tympanum The eardrum of vertebrates. In amphibians, the eardrums are located on the head surface behind the eyes, and in reptiles other than snakes (which lack them), birds, and mammals, they are located internally at the end of auditory canals. Tympanic membranes may also be sound producers, as in the syrinx of birds. *See also* syrinx.

ultrasonic Sound frequencies that exceed the range of human hearing, generally those above approximately 15,000 to 20,000 hertz. *See also* hertz.

vascular plants A collective term for plants that have more complex internal structures (such as a nutrient transportation system) than do "lower" plants, which include the mosses and algae.

viviparous Descriptive of animals that give birth to living offspring (larvae, etc.) rather than producing eggs that develop outside the body. *See also* oviparous.

vocalization Sound made by animals through specialized structures for moving air past vibratory membranes (mainly the syrinx in birds and the larynx in mammals and amphibians). Vocalizations are usually acoustically complex sounds with varied aspects of pitch, loudness, duration, cadence, and harmonic structure that allow for specific information transfer. *See also* larynx, song, songbird, syrinx.

woodland A term variously used to describe any wooded landscape or, more precisely, a plant community of low scattered trees in a nontree vegetational matrix; the distance between the trees is usually greater than the height of the trees. *See also* forest, savanna.

xeric Dry environmental conditions. *See also* mesic.

xerophyte A plant that is adapted to live in soils with low water availability. *See also* halophyte.

References

Wyoming

Cary, M. 1917. *Life Zone Investigations in Wyoming*. Washington, DC: USDA Bureau of Biological Survey, North American Fauna No. 40.

Cerovski, A. O., M. Grenier, B. Oakleaf, L. Van Fleet, and S. Patla. 2004. *Atlas of Birds, Mammals, Amphibians, and Reptiles in Wyoming*. Cheyenne: Wildlife Division, Wyoming Game and Fish Department. 155 pp. (*See also* Orabona et al., 2012)

Keinath, D. A., B. Heidel, and G. P. Beauvais. 2003. *Wyoming Plant and Animal Species of Concern*. Laramie: Wyoming Natural Diversity Database, University of Wyoming.

Keinath, D. A., M. D. Andersen, and G. P. Beauvais. 2010. *Range and Modeled Distribution of Wyoming's Species of Greatest Conservation Need*. Laramie: Wyoming Natural Diversity Database, University of Wyoming.

Knight, D. H., G. P. Jones, W. A. Reiners, and W. H. Romme. 2014. *Mountains and Plains: The Ecology of Wyoming Landscapes*. New Haven, CT: Yale University Press, 352 pp.

Luce, B., B. Oakleaf, A. Cervoski, L. Hunter, and J. Friday. 1997. *Atlas of Birds, Mammals, Amphibians and Reptiles in Wyoming*. Lander: Wyoming Game and Fish Department. 163 pp.

Martner, R. E. 1986. *Wyoming Climate Atlas*. Lincoln: University of Nebraska Press.

Orabona, A., C. Rudd, M. Grenier, Z. Walker, S. Patla, and B. Oakleaf. 2012. *Atlas of Birds, Mammals, Amphibians, and Reptiles in Wyoming*. Lander: Wyoming Game and Fish Department. 227 pp. (Since 2012, updates are added to the department's website: https://wgfd.wyo.gov/WGFD/media/content/PDF/Wildlife/Nongame/WILDLIFE_ANIMALATLAS.pdf)

Stephenson, S. 2011. Under the bark: Pine beetles are changing the face of Wyoming mountains. *Wyoming Wildlife* 73(1): 17–21.

Wyoming Game and Fish Department. 2010. *Wyoming State Wildlife Action Plan*. Cheyenne: Wyoming Game and Fish Department.

Wyoming Geography and Regional Botany

Alexander, R. R. 1985. *Major Habitat Types, Community Types, and Plant Communities in the Rocky Mountains*. Ft. Collins, CO: US Forest Service Rocky Mountain Forest and Range Experiment Station, General Technical Report RM-123. 105 pp. https://www.biodiversitylibrary.org/bibliography/99752#/summary

———. 1986. *Classification of the Forest Vegetation of Wyoming*. Ft. Collins, CO: US Forest Service Rocky Mountain Forest and Range Experiment Station, Research Note RM-466. 10 pp. https://www.biodiversitylibrary.org/bibliography/99117#/summary

Alexander, R. R., G. R. Hoffman, and J. M. Wirsing. 1986. *Forest Vegetation of the Medicine Bow National Forest in Southeastern Wyoming: A Habitat Type Classification*. Fort Collins, CO: US Forest Service Rocky Mountain Forest and Range Experiment Station, Research Paper RM-271.

Beetle, A. A. 1960. *A Study of Sagebrush: The Section Tridentatae of Artemisia*. Laramie: University of Wyoming Agricultural Experiment Station Bulletin 368. 83 pp.

Beetle, A. A., and K. L. Johnson. 1982. *Sagebrush in Wyoming*. Laramie: University of Wyoming Agricultural Experiment Station Bulletin 779. 68 pp.

Bowman, W. D., and T. R. Seastedt, eds. 2001. *Structure and Function in an Alpine Ecosystem*. New York: Oxford University Press.

Choate, G. A. 1963. *The Forests of Wyoming*. Ogden, UT: US Forest Service Intermountain Forest and Range Experiment Station, Resource Bulletin INT-2. 45 pp.

Clark, T. W., and R. D. Dorn. 1979. *Rare and Endangered Vascular Plants and Vertebrates of Wyoming*. Published by the authors. 78 pp.

Craighead, J. J., F. C. Craighead, Jr., and R. J. Davis. 1963. *A Field Guide to Rocky Mountain Wildflowers*. Boston: Houghton Mifflin. (Includes more than 200 color photographs, 118 line drawings, and descriptions of 590 species occurring from Arizona and New Mexico to southern Canada)

Cushman, R. C., and S. R. Jones. 1988. *The Shortgrass Prairie*. Boulder, CO; Pruett.

Daubenmire, R. F. 1943. Vegetational zonation in the Rocky Mountains. *The Botanial Review* 9(6): 325–393. https://www.jstor.org/stable/4353289 (Timberline is 10,000–11,000 feet in Wyoming latitudes 40–45°.)

Despain, D. G. 1990. *Yellowstone Vegetation: Consequences of Environment and History in a Natural Setting*. Boulder, CO: Roberts Rinehart. 239 pp.

Dorn, R. D. 1977a. *Manual of the Vascular Plants of Wyoming*. 2 vols. New York: Garland. 1,498 pp.

———. 1977b. *Flora of the Black Hills*. Cheyenne, WY: Published by the author. 377 pp.

———. 2001. *Vascular Plants of Wyoming*. 3rd ed. Missoula, MT: Mountain West. 412 pp. (A successor to Dorn's 1977a publication with 200 additional species)

Fertig, W. 1992. A floristic survey of the west slope of the Wind River Range, Wyoming. MS thesis, University of Wyoming, Laramie.

Green, A. W., and R. C. Conner. 1989. *The Forests of Wyoming*. Ogden, UT: US Forest Service Intermountain Research Station, Resource Bulletin INT-12. 91 pp.

Guennel, G. K. 1995. *Guide to Colorado Wildflowers*. 2 vols. Englewood, CO: Westcliffe. (Volume 1 includes color paintings and photographs of more than 300 species of trees, shrubs, forbs, and grasses of the plains and foothills. Volume 2 includes more than 300 montane species. Both are organized by flower color.)

Habec, J. R. 1987. Present-day vegetation in the northern Rocky Mountains. *Annals of the Missouri Botanical Garden* 74(4): 804–840. 37 pp. https://www.jstor.org/stable/2399451

Hallsten, G. P., Q. D. Skinner, and A. A. Beetle. 1987. *Grasses of Wyoming*. 3rd ed. Laramie: University of Wyoming Agricultural Experiment Station Research Journal 202. 432 pp.

Hoffman, G. R., and R. R. Alexander. 1976. *Forest Vegetation of the Bighorn Mountains, Wyoming: A Habitat Type Classification*. Ft. Collins, CO: USDA Forest Service Rocky Mountain Forest and Range Experiment Station, Research Paper RM-170. 38 pp. https://archive.org/details/CAT92273542

———. 1987. *Forest Vegetation of the Black Hills National Forest of South Dakota and Wyoming: A Habitat Type Classification*. Ft. Collins, CO: USDA Forest Service Rocky Mountain Forest and Range Experiment Station, Research Paper RM-276. 48 pp. https://www.fs.fed.us/rmrs/publications/forest-vegetation-black-hills-national-forest-south-dakota-and-wyoming-habitat-type

Hurd, R. M. 1961. Grassland vegetation in the Big Horn Mountains, Wyoming. *Ecology* 42(3): 459–467.

Kershaw, L., A. MacKinnon, and J. Pojar. 1998. *Plants of the Rocky Mountains*. Edmonton, AB: Lone Pine. 384 pp. (Identification guide to more than 1,300 plant species from Mexico to southern Canada)

Larson, G. E., and J. R. Johnson. 1999. *Plants of the Black Hills and Bear Lodge Mountains*. Brookings: South Dakota State University Agricultural Experiment Station Research Bulletin B732. (Describes 600 species of ferns, forbs, grasses, sedges, shrubs, and trees, illustrated with color photographs)

Leopold, E. B., and M. F. Denton. 1987. Comparative age of grassland and steppe east and west of the northern Rocky Mountains. *Annals of the Missouri Botanical Garden* 74(1): 841–867.

Lukas, L. E., B. E. Nelson, and R. L. Hartman. 2012. A floristic inventory of vascular plants of the Medicine Bow National Forest and vicinity, southeastern Wyoming, USA. *Journal of the Botanical Research Institute of Texas* 6(2): 759–787.

McDougall, W. B., and H. A. Baggley. 1956. *The Plants of Yellowstone National Park*. Rev. ed. National Park Service: Yellowstone Library and Museum Association. 186 pp. (Keys, line drawings, and color and black-and-white photographs of Yellowstone's flora)

Mutel, C. F., ad J. C. Emerick. 1992. *From Grassland to Glacier: The Natural History of Colorado and the Surrounding Region*. 2nd ed. Boulder, CO: Johnson. 280 pp.

Nelson, B. E. 1984. *Vascular Plants of the Medicine Bow Range*. Bozeman, MT: Jelm Mountain. 357 pp.

Parsons, W. H. 1979. *Field Guide: The Middle Rockies and Yellowstone*. Dubuque, IA: Kendall-Hunt.

Pfister, R. D., B. L. Kovalchik, S. F. Arno, and R. C. Presby. 1977. *Forest Habitat Types of Montana*. Ogden, UT: USDA Forest Service Intermountain Forest and Range Experiment Station, General Technical Report INT–34. 174 pp. https://www.fs.usda.gov/treesearch/pubs/41077

Phillips, H. W. 1999. *Central Rocky Mountain Wildflowers: Including Yellowstone and Rocky Mountain National Parks*. Helena, MT: Falcon Press. 272 pp. (Identification guide to 260 species of Greater Yellowstone plants, with color photographs organized by flower color)

Porter, C. L. 1962. *A Flora of Wyoming, Part I*. University of Wyoming Agricultural Experiment Station Bulletin 402. 42 pp. http://repository.uwyo.edu/ag_exp_sta_bulletins/415/ (Introduction and floral list through gymnosperms)

Reed, R. M. 1969. A study of vegetation in the Wind River Mountains, Wyoming. PhD diss., Washington State University, Pullman. 77 pp.

———. 1976. Coniferous forest habitat types of the Wind River Mountains, Wyoming. *American Midland Naturalist* 95(1): 159–173.

Shaw, R. J. 2008. *Plants of Yellowstone and Grand Teton National Parks*. 2nd ed. Camano Island, WA: Wheelwright. 160 pp. (Photographic guide to 213 species of regional trees, shrubs, and wildflowers)

Skinner, Q. D. 2014. *A Field Guide to Wyoming Grasses*. Laramie, WY: Education Resources Publishing, University of Wyoming. 596 pp.

Steele, R., S. V. Cooper, D. M. Ondov, D. W. Roberts, and R. D. Pfister. 1983. *Forest Habitat Types of Eastern Idaho–Western Wyoming*. Ogden, UT: USDA Forest Service, Intermountain Forest and Range Experiment Station, General Technical Report INT-144. 122 pp. https://www.fs.fed.us/rm/pubs_int/int_gtr144.pdf

Tilt, T. W. 2015. *Flora of the Yellowstone: A Guide to the Wildflowers, Shrubs, Trees, Ferns, and Grass-Like Plants of the Greater Yellowstone Region of Idaho, Montana, and Wyoming*. Bozeman, MT: Gallatin Valley Land Trust and Sunbelt Publications. 408 pp. (Includes approximately 400 species, illustrated with color photographs)

Vizgirdas, R. S. 2007. *A Guide to Plants of Grand Teton and Yellowstone National Parks*. Salt Lake City: University of Utah Press. 408 pp. (A comprehensive identification guide to Greater Yellowstone's vascular plants)

Weber, W. A. 1991. *Rocky Mountain Flora*. Boulder: University Press of Colorado. 320 pp. (Drawings and color photographs of more than 700 wildflower species)

Whitlock, C. 1993. Postglacial vegetation and climate in Grand Teton and southern Yellowstone national parks. *Ecological Monographs* 63:173–198.

Winward, A. H. 2004. *Sagebrush of Colorado: Taxonomy, Distribution, Ecology, and Management*. Denver: Colorado Division of Wildlife.

Zwinger, A. H., and B. E. Willard. 1972. *Land Above the Trees: A Guide to American Alpine Tundra*. New York: Harper and Row. 467 pp.

Regional Geology, Paleontology, and History

Anderson, E. 1974. A survey of late Pleistocene and Holocene mammal fauna of Wyoming. Pp. 79–90, in *Applied Geology and Archeology: The Holocene History of Wyoming* (M. Wilson, ed.). Laramie: Geological Survey of Wyoming Report of Investigations No. 10. https://www.wsgs.wyo.gov/products/wsgs-1974-ri-10.pdf

Blackstone, D. L., Jr. 1988. *Traveler's Guide to the Geology of Wyoming*. 2nd ed. Laramie: Geological Survey of Wyoming Bulletin 67. 130 pp.

Blackwelder, E. 1909. Cenozoic history of the Laramie region, Wyoming. *Journal of Geology* 17: 429–444.

Chronic, H. 1984. *Pages of Stone: Geology of Western Parks and Monuments*. Seattle, WA: Mountaineer Press.

Dalton, N. H. 1906. *Geology of the Bighorn Mountains*. Washington, DC: US Geological Survey Professional Paper 51.

Dorn, R. D. 1986. *The Wyoming Landscape, 1805–1878*. Cheyenne, WY: Mountain West.

Howard, A. D. 1937. *History of the Grand Canyon of the Yellowstone*. Geological Society of America Special Paper 6. 169 pp.

Knight, D. H. 1994. *Mountains and Plains: The Ecology of Wyoming Landscapes*. New Haven, CT: Yale University Press.

Knight, S. H. 1990. *Illustrated Geologic History of the Medicine Bow Mountains and Adjacent Areas, Wyoming*. Laramie: Geological Survey of Wyoming Memoir 4. 48 pp. http://sales.wsgs.wyo.gov/illustrated-geologic-history-of-the-medicine-bow-mountains-and-adjacent-areas-wyoming-1979/

Kurten, B., and E. Anderson. 1980. *Pleistocene Mammals of North America*. New York: Columbia University Press. 442 pp.

Lageson, D. R., and D. Spearing. 1988. *Roadside Geology of Wyoming*. Missoula, MT: Mountain Press. 271 pp.

Love, J. D., and J. C. Reed, Jr. 1971. *Creation of the Teton Landscape*. Moose, WY: Grand Teton Natural History Association.

Snoke, A. W., J. R. Steidtmann, and S. M. Roberts, eds. 1993. *Geology of Wyoming*. Laramie: Geological Survey of Wyoming Memoir 5. http://sales.wsgs.wyo.gov/geology-of-wyoming-1993/

Voorhies, M. R. 1981. Ancient ashfall entombed prehistoric animals. *National Geographic* 159(1): 66–75.

———. 1990. Nebraska wildlife: Ten million years ago. *NEBRASKAland* 68(5): 8–17.

Regional Recent History and Ecology

Black, G. 2012. *Empire of Shadows: The Epic Story of Yellowstone*. New York: St. Martin's.

Craighead, F. C., Jr. 1994. *For Everything There Is a Season: The Sequence of Natural Events in the Grand Teton–Yellowstone Area*. Helena, MT: Falcon Press. 206 pp.

Finch, D. M. 1992. *Threatened, Endangered, and Vulnerable Species of Terrestrial Vertebrates in the Rocky Mountain Region*. Fort Collins, CO: USDA Forest Service Rocky Mountain Forest and Range Experiment Station, General Technical Report RM-215. 38 pp. https://www.fs.fed.us/rm/pubs_rm/rm_gtr215.pdf

Fishbein, S. 1989. *Yellowstone Country: The Enduring Wonder*. Washington, DC: National Geographic Society.

Flores, D. 2003. *The Natural West: Environmental History in the Great Plains and Rocky Mountains*. Norman: University of Oklahoma Press.

Froiland, S. G. 1978. *Natural History of the Black Hills*. Sioux Falls, SD: Center for Western Studies.

Hejl, S. J., R. L. Hutto, C. R. Preston, and D. M. Finch. 1995. Effects of silvicultural treatments in the Rocky Mountains. Pp. 220–244, in *Ecology and Management of Neotropical Migratory Birds: A Synthesis and Review of Critical Issues* (T. E. Martin and D. M. Finch, eds.). New York: Oxford University Press.

Houston, D. B. 1973. Wildfires in northern Yellowstone National Park. *Ecology* 54: 1111–1117.

Johnsgard, P. A. 1982. *Teton Wildlife: Observations by a Naturalist*. Boulder: Colorado Associated University Press. 128 pp. (Electronic edition published 2009 by the University of Nebraska–Lincoln Libraries Digital Commons: http://digitalcommons.unl.edu/biosciornithology/52/)

———. 2013. *Yellowstone Wildlife: Ecology and Natural History of the Greater Yellowstone Ecosystem*. Boulder: University Press of Colorado. 228 pp.

Keiter, R. B., and M. S. Boyce, eds. 1991. *The Greater Yellowstone Ecosystem: Redefining America's Wilderness Heritage*. New York: Princeton University Press.

Koel, T. M., P. E. Bigelow, P. D. Doepke, B. D. Ertel, and D. L. Mahony. 2005. Nonnative lake trout result in Yellowstone cutthroat trout decline and impacts to bears and anglers. *Fisheries* 30: 10–19.

Kotliar, N. B., S. J. Hejl, R. L. Hutto, V. A. Saab, C. P. Melcher, and M. E. McFadzen. 2002. Effects of fire and post-fire salvage logging on avian communities in conifer-dominated forests of the western United States. Pp. 49–64, in *Effects of Habitat Fragmentation on Birds in Western Landscapes: Contrasts with Paradigms for the Eastern United States* (T. George and D. S. Dobkin, eds.). Cooper Ornithological Society: Studies in Avian Biology No. 25.

Loope, L. L., and G. E. Gruell. 1973. The ecological role of fire in the Jackson Hole area, northwestern Wyoming. *Quaternary Research* 3: 425–443.

Mathews, D. 2003. Rocky Mountain Natural History: Grand Teton to Jasper. Portland, OR: Raven Editions. (Detailed descriptions of the region's landscapes, animals, and plants)

Morrison, M. 1993. *Fire in Paradise: The Yellowstone Fires and the Politics of Environmentalism*. New York: Harper Collins.

Pritchard, J. A. 1999. *Preserving Yellowstone's Natural Conditions*. Lincoln: University of Nebraska Press.

Raventon, E. 1994. *Island in the Plains: A Black Hills Natural History*. Boulder, CO: Johnson.

Ripple, W. J., and E. J. Larson. 2000. Historic aspen recruitment, elk, and wolves in northern Yellowstone National Park. *Biological Conservation* 95(3): 361–370.

Romme, W. H., and M. G. Turner. 1991. Implications of global climatic changes for biogeographic patterns in the Greater Yellowstone Ecosystem. *Conservation Biology* 5: 373–386.

Romme, W. H., M. G. Turner, L. L. Wallace, and J. Walker. 1995. Aspen, elk, and fire in northern Yellowstone National Park. *Ecology* 76: 2097–2106.

Smucker, K. M., R. L. Hutto, and R. M. Steele. 2005. Changes in bird abundance after wildfire: Importance of fire severity and time since fire. *Ecological Applications* 15(5): 1535–1549.

Mammals

Adams, R. A. 2003. *Bats of the Rocky Mountain West: Natural History, Ecology, and Conservation*, Boulder: University Press of Colorado. 328 pp.

Agnew, W. D. 1983. Flora and fauna associated with prairie dog ecosystems. MS thesis, Colorado State University, Ft. Collins. 47 pp.

Aldus C. M. 1937. Notes on the life history of the snowshoe hare. *Journal of Mammology* 28: 46–57.

Allen, G. M. 1939. *Bats*. Cambridge, MA: Harvard University Press. 368 pp.

Allen, J., M. Bekoff, and R. Crabtree. 1999. An observational study of coyote (*Canis latrans*) scent-marking and territoriality in Yellowstone National Park. *Ethology* 105: 289–302.

Allred, S. 2010. *The Natural History of Tassel-Eared Squirrels*. Albuquerque: University of New Mexico Press.

Altmann, M. 1959. Group dynamics in Wyoming moose during the rutting season, *Journal of Mammology* 40: 420–424.

Anderson, A. E. 1983. *A Critical Review of Literature on Puma (Felis concolor)*. Ft. Collins: Colorado Division of Wildlife Special Report 54. 91 pp.

Anderson, C. C. 1958. *The Elk of Jackson Hole*. Cheyenne: Wyoming Game and Fish Commission Bulletin 10. 184 pp.

Anderson, C. R., Jr., and F. G. Lindzey. 2005. Experimental evaluation of population trend and harvest composition in a Wyoming cougar population. *Wildlife Society Bulletin* 33(1): 179–188.

Anderson, D. E., T. R. Laurion, J. R. Cary, R. S. Sikes, M A. McLeod, and E. M. Gese. 2003. Aspects of swift fox ecology in southeastern Colorado. Pp. 139–148, in *Swift Fox Symposium: Ecology and Conservation of Swift Foxes in a Changing World* (M. A. Sovada and L. Carbyn, eds.). Canadian Plains Proceedings Book 34. Regina: University of Regina Press. (Proceedings of Canadian Plains Research Center held in Regina, Saskatchewan, February 18–9, 1998)

Anderson, E., S. C. Forrest, T. W. Clark, and L. Richardson. 1986. Paleobiology, biogeography, and systematics of the black-footed ferret, *Mustela nigripes* (Audubon and Bachman), *1851. Great Basin Naturalist Memoirs* 8: 11–62.

Anderson, S., and D. Inkley, eds. 1985. *Black-footed Ferret Workshop Proceedings*. Cheyenne: Wyoming Game and Fish Publications.

Armitage, K. B. 1962. Social behaviour of a colony of the yellow-bellied marmot (*Marmota flaviventris*). *Animal Behaviour* 10(3–4): 319–320. (Research conducted at Jackson Hole, WY)

———. 1998. Reproductive strategies of yellow-bellied marmots. *Journal of Mammology* 79: 385–393.

Armstrong, D. M. 1972. *Distribution of Mammals in Colorado*. Lawrence: University of Kansas Museum of Natural History Monograph No 3. 415 pp.

———. 1975. *Rocky Mountain Mammals*. Denver, CO: Rocky Mountain Nature Association. 174 pp.

———. 2006. *Lions, Ferrets, and Bears: A Guide to the Mammals of Colorado*. Denver: Colorado Division of Wildlife.

———. 2008. *Rocky Mountain Mammals: A Handbook of Mammals of Rocky Mountain National Park and Vicinity*. 3rd ed. Boulder: Colorado Associated University Press. 265 pp.

Armstrong, D. M., J. P. Fitzgerald, and C. A. Meaney. 2011. *Mammals of Colorado*. 2nd ed. Boulder: University Press of Colorado. 704 pp.

Armstrong, D. M., R. A. Adams, and J. Freeman. 1994. *Distribution and Ecology of Bats of Colorado*. Natural History Inventory of Colorado, University of Colorado Museum 15: 1–82.

Atkeson, T. D., R. L. Marchinton, and K. V. Miller. 1988. Vocalizations of white-tailed deer. *American Midland Naturalist* 120(1): 194–200.

Aubry, K. B., W. J. Zielinski, M. G. Raphael, G. Proulx, and S. W. Buskirk, eds. 2012. *Biology and Conservation of Martens, Sables, and Fishers: A New Synthesis*. Ithaca, NY: Cornell University Press.

Avery, S. R. 1990. Vocalizations and behavior of the swift fox (*Vulpes velox*). MA thesis, University of Northern Colorado, Greeley. 104 pp.

Baker, R. J., L. C. Bradley, R. D. Bradley, J. W. Dragoo, M. D. Engstrom, R. S. Hoffman, C. A. Jones, F. Reid, D. W. Rice, and C. Jones. 2003. *Revised Checklist of North American Mammals North of Mexico, 2003*. Museum of Texas Tech University Occasional Papers No. 229. 24 pp.

Bakko, E. B., and L. N. Brown. 1967. Breeding biology of the white-tailed prairie dog (*Cynomys leucurus*) in Wyoming. *Journal of Mammology* 57: 576–578.

Banko, V. A., J. H. Shaw, and D. M. Leslie, Jr. 1999. Birds associated with black-tailed prairie dog colonies in southern shortgrass prairie. *Southwestern Naturalist* 44: 484–489.

Barber-Meyer, S. M., L. D. Mech, and P. J. White. 2008. Elk calf survival and mortality following wolf restoration to Yellowstone National Park. *Wildlife Monographs* 169: 1–30.

Barbour, R. W., and W. H. Davis. 1969. *Bats of America*. Lexington: University of Kentucky Press. 286 pp.

Barclay, R. M. R., M. B. Fenton, and D. W. Thomas. 1979. Social behavior in the little brown bat, *Myotis lucifugus*: II. Vocal communication. *Behavioral Ecology and Sociobiology* 6(2): 137–146.

Barnes, V. G., and O. E. Bray. 1967. *Population Characteristics of Black Bears in Yellowstone National Park*. Ft. Collins: Colorado Cooperative Wildlife Research Unit and Colorado State University. 199 pp.

Barrash, D. P. 1973. Territorial and foraging behavior of pika (*Ochotona princeps*) in Montana. *American Midland Naturalist* 89: 202–207.

Beckoff, M. 1974. *A General Bibliography of the Coyote* (Canis latrans). Boulder, CO: Coymar Press. 26 pp.

———, ed. 1978. *Coyotes: Biology, Behavior, and Management*. New York: Academic Press.

Beckoff, M., and C. B. Lowe, eds. 2007. *Listening to Cougar*. Boulder: University Press of Colorado.

Beckoff, M. C., and M. C. Wells. 1980. Social ecology and behavior of coyotes. *Scientific American* 242: 130–148.

Bee, J. W., G. E. Glass, R. S. Hoffmann, and R. R. Patterson. 1981. *Mammals in Kansas*. Lawrence: Museum of Natural History, University of Kansas. 300 pp.

Beier, P. 1991. Cougar attacks on humans in the United States and Canada. *Wildlife Society Bulletin* 19: 405–412.

Belan, I., P. N. Lehner, and T. W. Clark. 1978. Vocalizations of the American pine marten, *Martes americana. Journal of Mammology* 59: 871–874.

Belitsky, D. 1981. *Small Mammals of the Salt Wells*. Bureau of Land Management, Pilot Butte Planning Unit. 104 pp.

Biggins, D. E., B. J. Miller, L. R. Hanebury, R. Oakleaf, A. H. Farmer, R. Crete, and A. Dood. 1993. A technique for evaluating black-footed ferret habitat. Pp. 73–88, in *Proceedings of the Symposium on the Management of Prairie Dog Complexes for the Reintroduction of the Black-footed Ferret* (J. L. Oldemeyer, D. E. Biggins, B. J. Miller, and R. Crete, eds.). Washington, DC: US Fish and Wildlife Service Biological Report 13.

Bissell, S. I978. *Colorado Mammal Distribution, Latilong Study*. Denver: Colorado Division of Wildlife. 18 pp.

Bogan, M., and P. Cryan. 2000. The bats of Wyoming. Pp. 71–94, in *Reflections of a Naturalist: Papers Honoring Professor Eugene D. Fleharty* (J. R. Choate, ed.). Fort Hays Studies, Special Issue No. 1. Hays, KS: Fort Hays State University, Sternberg Museum of Natural History. 241 pp.

Bouvais, G. P., and R. Smith. 2004. *Predictive Distribution Maps of 54 Species of Management Concern in the Rocky Mountain Region of the US Forest Service*. Laramie: Wyoming Natural Diversity Database. 21 pp.

Boyce, M. S., and L. D. Hayden-Wing, eds. 1979. *North American Elk: Ecology, Behavior, and Management*. Laramie: University of Wyoming.

Broadbooks, H. E. 1970. Populations of the yellow-pine chipmunk (*Eutamias amoenus*). *American Midland Naturalist* 83: 472–488.

Broderick, H. J. 1954. *Wild Animals of Yellowstone National Park*. Mammoth, WY: National Park Service, Yellowstone Library and Museum Association.

Brown, D. J., W. A. Hubert, and S. H. Anderson. 1996. Beaver ponds create wetland habitat for birds in mountains of southeastern Wyoming. *Wetlands* 16(2): 127–133.

Brown, L. N. 1967a. Ecological distribution of mice in the Medicine Bow Mountains of Wyoming. *Ecology* 48: 677–680.

———. 1967b. Ecological sampling of six species of shrews and comparison of sampling methods in the central Rocky Mountains. *Journal of Mammology* 48: 617–623. (*Sorex cinereus, S. hoyi, S. merriami, S. nanus, S. palustris,* and *S. vagrans*)

———. 1967c. Seasonal activity patterns and breeding of the western jumping mouse (*Zapus princeps*) in Wyoming. *American Midland Naturalist* 78: 460–470.

———. 1970. Population dynamics of the western jumping mouse (*Zapus princeps*) during a four-year study. *Journal of Mammology* 51: 651–658.

Burnett, G. W. 1981. Movements and habitat use of American marten in Glacier National Park Montana. PhD diss., University of Montana, Missoula.

Buskirk, S. W. 2016. *Wild Mammals of Wyoming and Yellowstone Park*. Berkeley: University of California Press.

Buskirk, S. W., A. S. Harestad, M. G. Rapheal, and R. A. Powell, eds. 1994. *Martens, Sables, and Fishers: Biology and Conservation*. Ithaca, NY: Cornell University Press. 496 pp.

Butts, K. O., and J. C. Lewis. 1982. The importance of prairie dog towns to burrowing owls in Oklahoma. *Proceedings of the Oklahoma Academy of Science* 62: 46–52.

Byers, J. A. 1997. *American Pronghorn: Social Adaptations and the Ghosts of Predators Past*. Chicago: University of Chicago Press. 300 pp.

Caire, W., J. D. Tyler, B. P. Glass, and M. A. Mares. 1989. *Mammals of Oklahoma*. Norman: Oklahoma University Press. 567 pp.

Call, M. W. 1970. Beaver ecology and beaver-trout relationships in southeastern Wyoming. PhD diss., University of Wyoming, Laramie. 204 pp.

Camenzind, F. J. 1978. Behavioral ecology of coyotes (*Canis latrans*) on the National Elk Refuge, Jackson, Wyoming. PhD diss., University of Wyoming, Laramie.

Cameron, M. W. 1984. The swift fox (*Vulpes velox*) on the Pawnee National Grassland: Its food habits, population dynamics, and ecology. MA thesis, University of Northern Colorado, Greeley. 117 pp.

Campbell, T. M., III, and T. W. Clark. 1981. Colony characteristics and vertebrate associates of white-tailed and black-tailed prairie dogs in Wyoming. *American Midland Naturalist* 105: 269–276.

Campbell, T. M., III, T. W. Clark, L. Richardson, S. C. Forrest, and B. R. Houston. 1987. Food habits of Wyoming black-footed ferrets. *American Midland Naturalist* 117: 208–210.

Carbyn, L. N., S. H. Fritts, and D. R. Seip, eds. 1995. *Ecology and Conservation of Wolves in a Changing World*. Edmonton, AB: Canadian Circumpolar Institute, Occasional Publication 35.

Casey, D. E., J. DuWaldt, and T. W. Clark. 1986. Annotated bibliography of the black-footed ferret. *Great Basin Naturalist Memoirs* 8: 185–208.

Catania, K. C. 2013. The neurobiology and behavior of the American water shrew (*Sorex palustris*). *Journal of Comparative Physiology A* 119: 545–554.

Chapman, J. A., and G. A. Feldhamer, eds. 1982. *Wild Mammals of North America: Biology, Management and Economics*. Baltimore, MD: Johns Hopkins University Press. 1,147 pp.

Cid, M. S. 1987. Prairie dog and bison grazing effects on maintenance of attributes of prairie dog colony. PhD diss., Colorado State University, Fort Collins. 112 pp.

Clark, T. W. 1968. Ecological roles of prairie dogs. *Wyoming Range Management* 261: 102–104.

———. 1970a. Richardson's ground squirrel (*Spermophilus richardsonii*) in the Laramie Basin, Wyoming. *Great Basin Naturalist* 30: 55–70. (Now classified as a race of the Wyoming ground squirrel)

———. 1970b. Some prairie dog–range relationships in the Laramie Plains of Wyoming. *Wyoming Range Management* 282: 40–51.

———. 1971a. Towards a literature review of prairie dogs. *Wyoming Journal of Range Management* 286: 29–44.

———. 1971b. Ecology of the western jumping mouse in Grand Teton National Park. *Northwest Science* 45: 229–238.

———. 1973a. Distribution and reproduction of shrews in Grand Teton National Park, Wyoming. *Northwest Science* 47:128–131.

———. 1973b. A field study of the ecology and ethology of the white-tailed prairie dog (*Cynomys leucurus*), with a model for *Cynomys* evolution. PhD diss., University of Wisconsin, Madison. 215 pp.

———. 1973c. Local distribution and interspecies interactions in microtines, Grand Teton National Park, Wyoming. *Great Basin Naturalist* 33: 205–217.

———. 1977. Ecology and ethology of the white-tailed prairie dog (*Cynomys leucurus*). Milwaukee Public Museum, Publication in Biology and Geology No. 3. 97 pp.

———. 1979. The hard life of the prairie dog. *National Geographic* 156: 270–281.

———. 1986a. Annotated prairie dog bibliography, 1973 to 1985. Montana BLM Wildlife Technical Bulletin No. 1. Billings, MT: Montana State Office, Bureau of Land Management. 32 pp. (201 references)

———, ed. 1986b. The black-footed ferret. *Great Basin Naturalist Memoirs* 8: 1–308.

———. 1989. *Conservation Biology of the Black-footed Ferret*. Special Scientific Report No. 3. Philadelphia: Wildlife Preservation Trust International. 175 pp.

Clark, T. W., and M. R. Stromberg. 1987. *Mammals of Wyoming*. Lawrence: University of Kansas Museum of Natural History Public Education Series No. 10.

Clark, T. W., D. K. Hinckley, and T. Rich, eds. 1989. *The Prairie Dog Ecosystem: Managing for Biological Diversity*. Billings, MT: Bureau of Land Management Technical Bulletin 2. 55 pp.

Clark, T. W., S. C. Forrest, L. Richardson, D. Casey, and T. M. Campbell III. 1986. Descriptive ethology and activity patterns of black-footed ferrets. *Great Basin Naturalist Memoirs* 8: 72–84.

Clark, T. W., T. M. Campbell III, D. G. Socha, and D. E. Casey. 1982. Prairie dog colony attributes and associated vertebrate species. *Great Basin Naturalist* 42: 572–582.

Clark, T. W., V. Saab, and D. Casey. 1980. A partial bibliography of Wyoming mammals. *Northwest Science* 54: 55–67.

Clothier, R. J. 1955. Contribution to the life history of *Sorex vagrans* in Montana. *Journal of Mammology* 36: 214–221.

Cole, G. F. 1969. *The Elk of Grand Teton and Southern Yellowstone National Parks*. US Department of the Interior, National Park Service Research Report GRTE-N-1. 192 pp.

Conaway, C. H. 1952. Life history of the water shrew (*Sorex palustris navigator*). *American Midland Naturalist* 48: 219–248. (A Montana study)

Conner, D. A. 1985. Analysis of the vocal repertoire of adult pikas: Ecological and evolutionary perspectives. *Animal Behaviour* 33(1): 124–134.

Coppock, D. L., J. K. Detling, J. L. Dodd, and M. I. Dyer. 1980. Bison–prairie dog–plant interactions in Wind Cave National Park, South Dakota. *Proceedings of the Second Conference on Scientific Research in the National Parks*, vol. 12: 184.

Corcoran, A. J., and T. J. Weller. 2018. Inconspicuous echolocation in hoary bats (*Lasiurus cinereus*). *Proceedings of the Royal Society B: Biological Sciences* 285(1878): 20180441. 7 pp.

Corcoran, A. J., J. R. Barber, and W. E. Conner. 2009. Tiger moth jams bat radar. *Science* 325: 325–327.

Covell, D. F. 1992. Ecology of the swift fox (*Vulpes velox*) in southeastern Colorado. MS thesis, University of Wisconsin, Madison. 111 pp.

Craighead, F. C., Jr. 1979. *Track of the Grizzly*. San Francisco: Sierra Club Books.

Craighead, J. J., and J. A. Mitchell. 1982. Grizzly bear. Pp. 515–556, in *Wild Mammals of North America* (J. A. Chapman and G. E. Feldhamer, eds.). Baltimore, MD: Johns Hopkins University Press.

Craighead, J. J., J. S. Sumner, and J. A. Mitchell. 1995. *The Grizzly Bears of Yellowstone: Their Ecology in the Yellowstone Ecosystem, 1959–1992*. Washington, DC: Island Press.

Creel, S., and D. Christianson. 2009. Wolf presence and willow consumption by Yellowstone elk: Implications for a trophic cascade. *Ecology* 90: 2454–2466.

Crowe, D. M. 1986. *Furbearers of Wyoming*. Cheyenne: Wyoming Game and Fish Department.

Danz, P. 1997. *Of Bison and Man*. Boulder: University Press of Colorado. 231 pp.

Darden, S. K., T. Dabelsteen, and S. B. Pedersen. 2003. A potential tool for swift fox (*Vulpes velox*) conservation: Individuality of long-range barking sequences. *Journal of Mammalogy* 84: 1417–1427.

Dark-Smiley, D. N., and D. A. Keinath. 2003. *Species Assessment for Swift Fox* (Vulpes velox) *in Wyoming*. Cheyenne: US Department of the Interior, Bureau of Land Management. 51 pp. https://www.uwyo.edu/wyndd/_files/docs/reports/speciesassessments/swiftfox-dec2003.pdf

Dary, D. A. 1989. *The Buffalo Book: The Full Saga of the American Animal*. Athens: Swallow Press/Ohio University Press.

Davis, W. B., and D. J. Schmidly. 1994. *The Mammals of Texas*. Austin: Texas Parks and Wildlife. 338 pp.

Dearing, M. D. 1997. The function of haypiles of pikas (*Ochotona princeps*). *Journal of Mammalogy* 78: 1156–1163.

Demaris, S., and P. R. Krausman, eds. 2000. *Ecology and Management of Large Mammals in North America*. New York: Pearson.

Desmond, M. J., J. A. Savidge, and K. M. Eskridge. 2000. Correlations between burrowing owl and black-tailed prairie dog declines: A 7-year analysis. *Journal of Wildlife Management* 64: 1067–1075.

Dobie, J. F. 2006. *The Voice of the Coyote*. 2nd ed. Edison, NJ: Castle Books. 386 pp.

Dubay, S. A. 2000. Mycophagy as a nutritional strategy for small mammals in the Rocky Mountains. PhD dissertation, University of Wyoming, Laramie.

Duerbrouck, J., and D. Miller. 2001. *Cat Attacks: True Stories and Hard Lessons from Cougar Country*. Seattle, WA: Sasquatch Books. 256 pp.

Duffield, J. W., C. J. Neher, and D. A. Patterson. 2008. Wolf recovery in Yellowstone: Park visitor attitudes, expenditures, and economic impacts. *Yellowstone Science* 16(1): 20–25.

Durrant, S. D. 1952. *Mammals of Utah: Taxonomy and Distribution*. Lawrence: University of Kansas Publications, Museum of Natural History, vol. 6. 549 pp. https://www.biodiversitylibrary.org/item/21911#page/5/mode/1up

Eberhardt, L.L., P. J. White, R.A. Garrott, and D. B. Houston. 2007. A seventy-year history of trends in Yellowstone's northern elk herd. *Journal of Wildlife Management* 71: 594–602.

Egoscue, H. J. 1960. Laboratory and field studies of the northern grasshopper mouse. *Journal of Mammalogy* 41: 99–110.

Eisenberg, J. F. 1963. *The Behavior of Heteromyid Rodents*. University of California Publications in Zoology, vol. 69. Oakland: University of California Press. 111 pp.

Eller, C., and S. A. Banack. 2004. Variability in the alarm call of golden-mantled ground squirrels (*Spermophilus lateralis* and *S. saturatus*). *Journal of Mammalogy* 85: 43–50.

Escherich, P. 1981a. *Social Behavior of the Bushy-tailed Woodrat*, Neotoma cinerea. University of California Publications in Zoology, vol. 110. Oakland: University of California Press. 131 pp.

———. 1981b. Studies on the behavior of *Sorex vagrans*. *American Midland Naturalist* 72: 417–425.

Evans S. B., L. D. Mech, P. J. White, and G. A Sargeant. 2006. Survival of adult female elk in Yellowstone following wolf restoration. *Journal of Wildlife Management* 70: 1372–1378.

Fahnestock, J.T., and J. K. Detling. 2002. Bison–prairie dog–plant interactions in a North American mixed-grass prairie. *Oecologia* 132: 86–95.

Farentino, R. C. 1972. Social dominance and mating activity in the tassel-eared squirrel (*Sciurus aberti fereus*). *Animal Behaviour* 20: 316–326.

Feldhamer, G. A., B. C. Thompson, and J. A. Chapman. 2003. *Wild Mammals of North America: Biology, Management, and Conservation*. Baltimore, MD: John Hopkins University Press. 1,216 pp.

Findley, J. S. 1951. Habitat preference of four species of *Microtus* in Jackson Hole, Wyoming. *Journal of Mammalogy* 32: 118–120.

———. 1987. *The Natural History of New Mexican Mammals*. Albuquerque: University of New Mexico Press. 164 pp.

Finley, R. B., Jr. 1958. *The Wood Rats of Colorado: Distribution and Ecology*. University of Kansas Publications, Museum of Natural History, vol. 10, no. 6: 213–552. https://www.biodiversitylibrary.org/item/21921#page/249/mode/1up

Fitzgerald, J. P., C. A. Meanley, and D. M. Armstrong. 1994. *Mammals of Colorado*. Niwot: University of Colorado Press.

Foresman, K. R. 2001. *The Wild Mammals of Montana*. Special Publication of the American Society of Mammalogists No. 12. 278 pp.

Forrest, S. C., D. E. Biggans, L. Richardson, T. W. Clark, T. M. Campbell III, K. A. Fagerstone, and E. T. Thorne. 1988. Black-footed ferret (*Mustela nigripes*) population attributes at Meeteetse, Wyoming, 1981–1985. *Journal of Mammalogy* 59: 261–275.

Forsyth, A. 1999. *Mammals of North America: Temperate and Arctic Regions*. Buffalo, NY: Firefly Books. 350 pp.

Franzman, A. W., and C. C. Schwartz, eds. 2007. *Ecology and Management of the North American Moose*. 2nd ed. Boulder: University Press of Colorado.

Frederiksen, J. K., and C. N. Slobodchikoff. 2007. Referential specificity in the alarm calls of the black-tailed prairie dog. *Ethology, Ecology, and Evolution* 19: 87–99.

Fullard, J. H., and M. B. Fenton. 1979. Jamming bat echolocation: The clicks of arctiid moths. *Canadian Journal of Zoology* 57: 647–649.

Fuller, J. A., R. A. Garrott, P. J. White, K. E. Aune, T. J. Roffe, and J. C. Rhyan. 2007. Reproduction and survival of Yellowstone bison. *Journal of Wildlife Management* 71: 2365–2372.

Garner, H. W. 1974. Population dynamics, reproduction, and activities of the kangaroo rat, *Dipodomys ordii*, in western Texas. Graduate Studies, Texas Tech University 7: 1–28.

Gasson, W., R. G. Grogan, and L. Kruckenberg. 2003. *Black Bear Management in Wyoming*. Cheyenne: Wyoming Game and Fish Department. 95 pp.

Geist, V. 1971. *Mountain Sheep: A Study in Behavior and Evolution*. Chicago: University of Chicago Press. 383 pp.

Genoways, H. H., and J. H. Brown, eds. 1993. *Biology of the Heteromyidae*. Special Publication of the American Society of Mammalogists No. 10. 719 pp.

Gese, E. M., T. E. Stotts, and S. Grothe. 1996. Interactions between coyotes and red foxes in Yellowstone National Park, Wyoming. *Journal of Mammalogy* 77: 377–382.

Getz, L. L., C. S. Carter, and L. Gavish. 1981. The mating system of the prairie vole, *Microtus ochrogaster*: Field and laboratory evidence for pair-bonding. *Behavioral Ecology and Sociobiology* 8: 189–194.

Gilbert, F. F. 1969. Analysis of basic vocalizations of the ranch mink. *Journal of Mammalogy* 50: 625–627.

Goldingay, R. L., and J. S. Scheibe, eds. 2000. *Biology of Gliding Squirrels*. Fürth, Germany: Filander Verlag.

Goodrich, J. M., and S. W. Buskirk. 1998. Spacing and ecology of North American badgers (*Taxidea taxus*) in a prairie-dog (*Cynomys leucurus*) complex. *Journal of Mammalogy* 79: 171–179.

Gould, E., N. C. Negus, and A. Novicki. 1964. Evidence for echolocation in shrews. *Journal of Experimental Biology* 156: 19–38.

Grady, W. 1994. *The Nature of Coyotes*. Vancouver, BC: Douglas and McIntyre.

Graves, R. A. 2001. *The Prairie Dog: Sentinel of the Plains*. Lubbock: Texas Tech University Press. 133 pp.

Grenier, M. B., ed. 2011. "Threatened, Endangered, and Nongame Bird and Mammal Investigations." Lander: Wyoming Game and Fish Department, Nongame Program, Annual Completion Report (unpublished).

Grogan, R. G. 1997. Black bear ecology in southeast Wyoming: The Snowy range. MS thesis, University of Wyoming, Laramie.

Guenzel, R. J. 1986. Pronghorn ecology in south-central Wyoming. MS thesis, University of Wyoming, Laramie.

Gunderson, H. L., and B. R. Mahan. 1980. Analysis of sonagrams of American bison (*Bison bison*). *Journal of Mammalogy* 61: 379–381.

Haines, F. 1995. *The Buffalo*. Norman: University of Oklahoma Press. 244 pp.

Halfpenny, J. C. 2003. *Yellowstone Wolves in the Wild*. Helena, MT: Riverbend Press. 103 pp.

———. 2007. *Yellowstone Bears in the Wild*. Helena, MT: Riverbend Press. 123 pp.

Hall, E. R. 1981. *The Mammals of North America*. 2 vols. New York: Wiley. 1,181 pp.

Harlow, H. J., and G. E. Menkens. 1986. A comparison of hibernation in the black-tailed prairie dog, white-tailed prairie dog, and Wyoming ground squirrel. *Canadian Journal of Zoology* 64: 793–796.

Haroldson, M. A., C. G. Schwartz, and G. C. White. 2006. Survival of independent grizzly bears in the Greater Yellowstone Ecosystem, 1983–2001. *Wildlife Monographs* 161: 33–43.

Harrison, R. L. 2003. Swift fox demography, movements, denning, and diet in New Mexico. *Southwestern Naturalist* 48(2): 261–273.

Hart, F. M., and J. A. King. 1966. Distress vocalizations of young in two subspecies of *Peromyscus maniculatus*. *Journal of Mammalogy* 47(2): 287–293.

Hassien, F. D. 1973. Prairie dogs: A partial bibliography. Pp. 178–205, in *Proceedings of the Black-footed Ferret and Prairie Dog Workshop, September 4–6, 1973, Rapid City, South Dakota* (R. L. Linder and C. N. Hillman, eds.). Brookings: South Dakota State University. (Contains 437 references; see also Clark, 1986a)

Hatt, R. T. 1943. The pine squirrel in Colorado. *Journal of Mammalogy* 24: 311–345. (Research near Boulder)

Hayward, G. D., and P. H. Hayward. 1995. Relative abundance and habitat associations of small mammals in Chamberlain Basin, central Idaho. *Northwest Science* 69: 114–125.

Heffner, H. 2005. Hearing and sound localization in the kangaroo rat (*Dipodomys merriami*). *Journal of the Acoustical Society of America* 61: S59.

Henderson, F. R., P. Springer, and R. Adrian. 1969. *The Black-footed Ferret in South Dakota*. Technical Bulletin 4. Pierre: South Dakota Department of Game, Fish, and Parks.

Hill, J. E., and J. D. Smith. 1984. *Bats: A Natural History*. Austin: University of Texas Press.

Hodges, K. E., L. S. Mills, and K. M. Murphy. 2009. Distribution and abundance of snowshoe hares in Yellowstone National Park. *Journal of Mammalogy* 90: 870–878.

Hoffman, R. S., and D. Lattie. 1968. *A Guide to Montana Mammals: Identification, Habitat, Distribution, and Abundance*. Missoula: University of Montana Printing Services. 113 pp.

Hoffmeister, D. F. 1986. *Mammals of Arizona*. Tucson: University of Arizona Press. 602 pp.

Holm, G. W., F. G. Lindzey, and D. S. Moody. 1999. Interactions of sympatric black and grizzly bears in northwest Wyoming. *Ursus* 11: 99–108.

Hoogland, J. L. 1995. *The Black-tailed Prairie Dog: Social Life of a Burrowing Animal*. Chicago: University of Chicago Press. 562 pp.

———, ed. 2006. *Conservation of the Black-tailed Prairie Dog: Saving North America's Western Grasslands*. Washington, DC: Island Press. 342 pp.

Houston, D. B. 1968. *The Shiras Moose in Jackson Hole, Wyoming*. Moose, WY: Grand Teton Natural History Association Technical Bulletin 1. 110 pp.

———. 1982. *The Northern Yellowstone Elk: Ecology and Management*. New York: Macmillan.

Irby, L. R., and J. E. Knight, eds. 1998. *Bison Ecology and Management in North America*. Bozeman: Montana State University.

Jackson, V. L., and J. R. Choate. 2000. Dens and den sites for the swift fox, *Vulpes velox*. *Southwestern Naturalist* 45: 212–220.

Johnsgard, P. A. 2005. *Prairie Dog Empire: A Saga of the Shortgrass Prairie*. Lincoln: University of Nebraska Press. 244 pp.

———. 2006. The howdy owl and the prairie dog. *Birding*, January/February 2006, 40–44. http://publications.aba.org/birding_archive_files/v38n1p40.pdf

———. 2014. The lives and deaths of Yellowstone's grizzlies. *Prairie Fire*, August 2014, 1–3.

Johnson, D. R. 1967. Diet and reproduction of Colorado pikas. *Journal of Mammalogy* 48: 311–315.

Kaufman, D. W., and E. D. Fleharty. 1974. Habitat selection by nine species of rodents in north-central Kansas. *Southwestern Naturalist* 18: 443–451.

Kawamichi, T. 1976. Hay territoriality and dominance rank of pikas (*Ochotona princeps*). *Journal of Mammalogy* 57: 133–148.

Kays, R. W., and D. E. Wilson. 2002. *Mammals of North America*. Princeton, NJ: Princeton University Press. 248 pp.

Kazial, K. A., S. Pacheco, and K. N. Zielinski. 2008. Information content of sonar calls of little brown bats (*Myotis lucifugus*): Potential for communication. *Journal of Mammalogy* 89: 25–33.

Keinath, D. A. 2003. *Species Assessment for Fringed Myotis* (Myotis thysanodes*) in Wyoming*. Cheyenne: US Department of the Interior, Bureau of Land Management. 71 pp. https://www.uwyo.edu/wyndd/_files/docs/reports/speciesassessments/fringedmyotis-dec2003.pdf

———. 2004. *Species Assessment for White-tailed Prairie Dog* (Cynomys leucurus) *in Wyoming*. Cheyenne: US Department of the Interior, Bureau of Land Management. 47 pp. http://www.blm.gov/style/medialib/blm/wy/

Keinath, D. A., H. R. Griscom, and M. D. Andersen. 2014. Habitat and distribution of the Wyoming pocket gopher (*Thomomys clusius*). *Journal of Mammalogy* 95: 803–813.

Keith, J. O. 1965. The Abert squirrel and its dependence on ponderosa pine. *Ecological Monographs* 34: 383–401.

Kendall, K. C. 1983. Use of pine nuts by grizzly and black bears in the Yellowstone area. Pp. 166–173, in *Bears: Their Biology and Management*, vol. 5, A Selection of Papers from the Fifth International Conference on Bear Research and Management, Madison, Wisconsin, February, 1980.

Kilgore, D. L., Jr. 1969. An ecological study of the swift fox (*Vulpes velox*) in the Oklahoma panhandle. *American Midland Naturalist* 81: 512–534.

Kirkland, G. L., Jr., R. R. Parmenter, and R. E. Skoog. 1997. A five-species assemblage of shrews from the sagebrush-steppe of Wyoming. *Journal of Mammalogy* 28: 83–89. (*Sorex cinereus, S. merriami, S. monticolus, S. nanus*, and *S. preblei*)

Kitchen, A. M., E. M. Gese, and E. R. Schauster. 1999. Resource partitioning between coyotes and swift foxes: Space, time, and diet. *Canadian Journal of Zoology* 77: 1645–1656.

Kitchen, D. W. 1974. Social behavior and ecology of the pronghorn. *Wildlife Monographs* 38: 1–96.

Klinghammer, E. 1978. *The Behavior and Ecology of Wolves*. New York: Garland.

Knight, R. R., B. M. Blanchard, and L. L. Eberhardt. 1988. Mortality patterns and population sinks for Yellowstone grizzly bears, 1973–1985. *Wildlife Society Bulletin* 16: 121–125.

Knowles, C. J. 1982. Habitat affinity, populations, and population dynamics in a prairie dog town in the Black Hills of South Dakota. PhD diss., University of Montana, Missoula. 171 pp.

———. 1986. Some relationships of black-tailed prairie dogs to livestock grazing. *Great Basin Naturalist* 46: 198–203.

———. 1987. Reproductive ecology of black-tailed prairie dogs in Montana. *Great Basin Naturalist* 47: 202–206.

Krumm, R. E. 1969. An ecological study of the Audubon's cottontail on the Laramie Plains. MS thesis, University of Wyoming, Laramie.

Kruuk, H. 2006. *Otters: Ecology and Conservation*. Oxford, UK: Oxford University Press. 280 pp.

Larrison, E. J., and D. R. Johnson. 1981. *Mammals of Idaho*. Moscow, ID: Northwest Naturalist Books. 166 pp.

Lawrence, I. E. 1955. An ecological study of the snowshoe hare, *Lepus*

americanus bairdii Hayden, in the Medicine Bow National Forest of Wyoming. MS thesis, University of Wyoming, Laramie.

Laydet, F. 1988. *The Coyote*. Norman: University of Oklahoma Press.

Lechleitner, R. 1969. *Wild Mammals of Colorado: Their Appearance, Habits, Distribution, and Abundance*. Boulder, CO: Pruett. 254 pp.

Lemke, T. O. 2004. Origin, expansion, and status of mountain goats in Yellowstone National Park. *Wildlife Society Bulletin* 32: 532–541.

Lepri J. J., M. Theodorides, and C. J. Wysocki. 1988. Ultrasonic vocalizations by adult prairie voles, *Microtus ochrogaster*. *Cellular and Molecular Life Sciences* 44: 271–273.

Linder, R. L., and C. N. Hillman, eds. 1973. *Proceedings of the Black-footed Ferret and Prairie Dog Workshop, September 4–6, 1973*. Brookings: South Dakota State University.

Logan, K. A. 1983. Mountain lion population and habitat characteristics in the Big Horn Mountains of Wyoming. MS thesis, University of Wyoming, Laramie.

Long, C. A. 1965. *The Mammals of Wyoming*. Lawrence: University of Kansas Museum of Natural History Publication 14(18): 493–754.

Lott, D. F. 2002. *American Buffalo: A Natural History*. Berkeley: University of California Press.

Luce, R. J., M. A. Bogan, M. J. O'Farrell, and D. A. Keinath. 2004. *Species Assessment for Spotted Bat* (Euderma maculatum) *in Wyoming*. Cheyenne: US Department of the Interior, Bureau of Land Management. 60 pp.

Lynch, W. 1993. *Bears: Monarchs of the Northern Wilderness*. Vancouver: Douglas and McIntyre.

MacClintock, S. 1964. *Squirrels of North America*. New York: Van Nostrand Reinhold.

Macdonald, D. W. 2009. *The Princeton Encyclopedia of Mammals*. Princeton, NJ: Princeton University Press. 976 pp.

Mattson, D. J., B. M. Blanchard, and R. R. Knight. 1991. Food habits of Yellowstone grizzly bears, 1977–1987. *Canadian Journal of Zoology* 69: 1619–1629.

Maxell, M. H., and L. N. Brown. 1968. Ecological distribution of rodents on the high plains of eastern Wyoming. *Southwestern Naturalist* 13: 143–158.

McCarley, H. 1966. Annual cycle, population dynamics, and adaptive behavior of *Citellus tridecemlineatus*. *Journal of Mammalogy* 47: 294–316.

McHugh, T. 1958. Social behavior of the American buffalo (*Bison bison bison*). *Zoologica* 43: 1–40.

———. 1972. *The Time of the Buffalo*. New York: Knopf. 383 pp.

McKeever, S. 1964. The biology of the golden-mantled ground squirrel. Ecological Monographs 34: 383–401. (Studies in California)

Mech, D. L. 1970. *The Wolf: Ecology and Behavior of an Endangered Species*. Minneapolis: University of Minnesota Press.

———. 1999. Alpha status and the division of labor in wolf packs. *Canadian Journal of Zoology* 77: 1196–1203.

Mech, D. L., and L. Boitani, eds. 2003. *Wolves: Behavior, Ecology, and Conservation*. Chicago: Chicago University Press. 448 pp.

Miller, B., S. Forrest, and R. P. Reading. 1996. *Prairie Night: Black-Footed Ferrets and the Recovery of Endangered Species*. Washington, DC: Smithsonian Institution Press. 320 pp.

Miller, J. R. 2010. Stereotypic vocalizations in harvest mice (*Reithrodontomys*): Harmonic structure contains prominent and distinctive audible, ultrasonic, and non-linear elements. *Journal of Acoustical Society of America* 128: 1501.

Moulton, M. P., J. R. Choate, S. J. Bissell, and R. A. Nicholson. 1981. Associations of small mammals on the central High Plains of eastern Colorado. *Southwestern Naturalist* 26: 53–57.

Müller-Schwarze, D. 2011. *The Beaver*. 2nd ed. Ithaca, NY: Cornell University Press.

Murie, A. 1940. *Ecology of the Coyote in the Yellowstone*. Washington, DC: US National Park Fauna Series 4. 206 pp.

Murie, O. J. 1935. *Food Habits of the Coyote in Jackson Hole, Wyoming*. Washington, DC: USDA Circular 362. 24 pp.

———. 1951. *The Elk of North America*. New York: Knopf.

———. 1954. *A Field Guide to Animal Tracks*. Boston: Houghton Mifflin. 400 pp.

Murrant, M. N., J. Bowman, C. J. Garroway, B. Prinzen, H. Mayberry, and P. A. Faure. 2013. Ultrasonic vocalizations emitted by flying squirrels. *PLoS One* 8(8): e73045.

Musiani, M., L. Boitani, and P. Paquet, eds. 2010. *The World of Wolves: New Perspectives on Ecology, Behaviour, and Management*. Calgary, SK: University of Calgary Press.

Neal, E. G. 1996. *The Natural History of Badgers*. New York: Facts on File.

Negus, N. C., and J. S. Findley. 1959. The mammals of Jackson Hole, Wyoming. *Journal of Mammalogy* 31: 371–381.

Novakowski, N. S. 1969. The influence of vocalization on the behavior of beaver, *Castor canadensis* Kuhl. *American Midland Naturalist* 81: 198–204.

Oakleaf, B., A. O. Cerovski, and B. Luce. 1996. *Nongame Bird and Mammal Plan*. Cheyenne: Wyoming Game and Fish Department. 183 pp.

O'Gara, B. W., and J. D. Yoakum, eds. 2004. *Pronghorn: Ecology and Management*. Boulder: University Press of Colorado.

O'Gara, B. W., J. D. Yoakum, and R. E. McCabe, eds. 2004. *Pronghorn Ecology and Management*. Boulder: University Press of Colorado. 894 pp.

Oldemeyer, J. L. 1966. Winter ecology of bighorn sheep in Yellowstone National Park. MS thesis, Colorado State University, Fort Collins. 107 pp.

Olson, R., and W. A. Hubert. 1994. *Beaver: Water Resources and Riparian Habitat Manager*. Laramie: University of Wyoming. 48 pp.

Olson, T. L., and F. G. Lindzey. 2002a. Swift fox survival and production in southeastern Wyoming. *Journal of Mammalogy* 83: 199–206.

———. 2002b. Swift fox (*Vulpes velox*) home-range dispersal patterns in southeastern Wyoming. Canadian Journal of Zoology 80: 2024–2039.

Orr, R. T. 1977. *The Little Known Pika*. New York: Macmillan.

Pattie, D. L., and N. A. M. Verbeek. 1967. Alpine mammals of the Beartooth Mountains. *Northwest Science* 41: 110–117.

Pease, C. M., and D. J. Mattson. 1999. Demography of the Yellowstone grizzly bears. *Ecology* 80: 957–975.

Pechacek, P., F. G. Lindzey, and S. H. Anderson. 2000. Home range size and spatial organization of swift fox *Vulpes velox* (Say, 1823) in southeastern Wyoming. *Zeitschrift für Säugetierkunde* 65: 209–215.

Peterson, R. I. 1955. *North American Moose*. Toronto, ON: Greystone Books.

Pheifer, S. 1980. Aerial predation on Wyoming ground squirrels. *Journal of Mammalogy* 61: 368–371.

Pomerantz, S. M., and L. G. Clemens. 1981. Ultrasonic vocalizations in male deer mice (*Peromyscus maniculatus bairdi*): Their role in male sexual behavior. *Physiology and Behavior* 27: 869–872.

Powell, R. A. 1982. *The Fisher: Life History Ecology and Behavior*. Minneapolis: University of Minnesota Press.

Pringle, L. 1977. *The Controversial Coyote: Predation, Politics, and Ecology*. New York: Harcourt Brace Jovanovich.

Proulx, G., H. N. Bryant, and P. M. Woodard, eds. 1997. *Martes: Taxonomy, Ecology, Techniques, and Management*. Edmonton: Provincial Museum of Alberta. 496 pp.

Purcell, M. J. 2006. Pygmy rabbit (*Brachylagus idahoensis*) distribution and habitat selection in Wyoming. MS thesis, University of Wyoming, Laramie.

Putnam, R. 1988. *The Natural History of Deer*. Ithaca, NY: Comstock Press. 191 pp.

Quimby, D. C. 1951. The life history and ecology of the jumping mouse, *Zapus hudsonicus*. *Ecological Monographs* 21: 61–95.

Rafeal, M. R. 1988. Habitat associations of small mammals in a subalpine forest southeastern Wyoming. Pp. 359–367, in *Management of Amphibians, Reptiles, and Small Mammals in North America* (R. C. Szaro, K. E. Stevenson, and D. R. Patton, tech. coordinators). Denver, CO: USDA Forest Service Rocky Mountain Forest and Range Management Experiment Station, General Technical Report RM-166. 458 pp.

Reed-Eckert, M., C. Meaney, and G. P. Beauvais. 2004. *Species Assessment for Grizzly Bear* (Ursus arctos) *in Wyoming*. Cheyenne: US Department of the Interior, Bureau of Land Management. 54 pp. https://www.uwyo.edu/wyndd/_files/docs/reports/speciesassessments/grizzlybear-sep2004.pdf

Reid, F. A. 2006. *A Field Guide to Mammals of North America*. 4th ed. Boston, MA: Houghton Mifflin. 579 pp.

Rennicke, J. 1990. *Colorado Wildlife*. Helena, MT: Falcon Press. 138 pp.

Rinella, S. 2009. *America Buffalo: In Search of a Lost Icon*. New York: Spiegel and Grau.

Ripple, W. J., and R. L. Beschta. 2007. Restoring Yellowstone's aspen with wolves. *Biological Conservation* 138: 514–519.

Robinson, L. D. 1986. The vegetation and small mammals of the juniper zone in north central Wyoming. MS thesis, University of Wyoming, Laramie.

Roe, F. G. 1970. *The North American Buffalo: A Critical Study of the Species in its Wild State*. 2nd ed. Toronto, ON: University of Toronto Press. 957 pp.

Rompola, K. M. 2000. Small mammals of a juniper woodland and sagebrush-grassland mosaic in southwestern Wyoming. MS thesis, University of Wyoming, Laramie.

Rompola, K. M., and S. H. Anderson. 2004. Habitat of three rare species of small mammals in juniper woodlands of southwestern Wyoming. *Western North American Naturalist* 64(1): 86–92. (Cliff chipmunk, canyon mouse, and pinyon mouse)

Rothwell, R. G., G. Skutches, J. Straw, C. Sax, and H. Harju. 1978. *A Partial Bibliography of the Mammals of Wyoming and Adjacent States with Special Reference to Density and Habitat Affinity*. Cheyenne, WY: Bureau of Land Management Contract Number YA-512-CT8-126. 172 pp.

Royce, M. S. 1989. *The Jackson Elk Herd: Intensive Wildlife Management in North America*. New York: Cambridge University Press.

Roze, U. 1989. *The North American Porcupine*. Washington, DC: Smithsonian Institution Press.

Ruffer, D. G. 1965. Sexual behavior of the northern grasshopper mouse (*Onychomys leucogaster*). *Animal Behaviour* 13: 447–472.

Ruggiero, L. F., ed. 1994. *American Marten, Fisher, Lynx and Wolverine in the Western United States*. Denver, CO: USDA Forest Service, General Technical Report RM-254. 184 pp.

Ruth, T. K. 2004. Ghost of the Rockies: The Yellowstone Cougar Project. *Yellowstone Science* 12: 13–24.

Ryden, H. 1977. *God's Dog: A Celebration of the North American Coyote*. New York: Coward, McCann, and Geoghagen.

Sawyer, H., F. Lindzey, and D. McWhirter. 2005. Mule deer and pronghorn migration in western Wyoming. *Wildlife Society Bulletin* 33: 1266–1273.

Schauster, E. R., E. M. Gese, and A. M. Kitchen. 2002. Population ecology of swift foxes (*Vulpes velox*) in southeastern Colorado. *Canadian Journal of Zoology* 80: 307–319.

Schmidt, C. A. 2003. *Conservation Assessment for the Silver-Haired Bat in the Black Hills National Forest South Dakota and Wyoming*. Custer, SD: USDA Forest Service, Black Hills National Forest. 22 pp.

Schmidt-French, B., E. Gillam, and M. B. Fenton. 2006. Vocalizations emitted during mother-young interactions by captive eastern red bats *Lasiurus borealis* (Chiroptera: Vespertilionidae). *Acta Chiropterologica* 8(2): 477–484.

Schullery, P. 1992. *The Bears of Yellowstone*. Worland, WY: High Plains.

———. 1996. *The Yellowstone Wolf: A Guide and Sourcebook*. Worland, WY: High Plains.

Schwartz, C. C., M. A. Haroldson, and G. C. White. 2006. Survival of cub and yearling grizzly bears in the Greater Yellowstone ecosystem, 1983–2001. *Wildlife Monographs* 161: 23–31.

Scott, V. E., and G. L. Crouch. 1988. Summer birds and mammals of aspen-conifer forests in west-central Colorado. Fort Collins, CO: USDA Forest Service Rocky Mountain Forest and Range Experiment Station, Research Paper RM-280. 6 pp.

Seal, U., E. T. Thorne, M. Bogan, and S. Anderson, eds. 1989. *Conservation Biology and the Black-footed Ferret*. New Haven, CT: Yale University Press.

Skryja, D. D. 1970. Some aspects of the ecology of the least chipmunk (*Eutamias minimus operarius*) in the Laramie Mountains of southeast Wyoming. MS thesis, University of Wyoming, Laramie.

———. 1974. Reproductive biology of the least chipmunk (*Eutamias minimus*) in southwestern Wyoming. Journal of Mammalogy 55: 221–224.

Slobodchikoff, C. N., B. S. Perla, and J. L. Verdolin. 2009. *Prairie Dogs: Communication and Community in an Animal Society*. Cambridge, MA: Harvard University Press.

Smith, C. C. 1970. The coevolution of pine squirrels (*Tamiasciurus*) and conifers. *Ecological Monographs* 40: 349–371.

Smith, D. W. 2005. Ten years of Yellowstone wolves, 1995–2005. *Yellowstone Science* 15:17–19.

Smith, D. W., and G. Ferguson. 2005. *Decade of the Wolf: Returning the Wild to Yellowstone*. Guiford, CT: Lyons Press. 246 pp.

Smith, W. J., S. L. Smith, J. G. deVilla, and E. C. Oppenheimer. 1976. The jump-yip display of the black-tailed prairie dog *Cynomys ludovicianus*. *Animal Behavior* 24: 609–621.

Somers, P. 1973. Dialects in southern Rocky Mountain pikas, *Ochotona princeps* (Lagomorpha). *Animal Behaviour* 21: 124–137.

Spencer, A. W., and D. Pettus. 1966. Habitat preferences of five sympatric species of long-tailed shrews. *Ecology* 47(4): 677–683. (*Sorex cinereus*, *S. hoyi*, *S. nanus*, *S. palustris*, and *S. vagrans*)

Stahler, D. R., D. W. Smith, and D. S. Guernsey. 2006. Foraging and feeding ecology of the gray wolf (*Canis lupus*): Lessons from Yellowstone National Park, Wyoming, USA. *Journal of Nutrition* 136: 1923S–1926S.

Steelquist, R. U. 1998. *Field Guide to the North American Bison*. Seattle, WA: Sasquatch Books.

Streble, A. M. 1939. An ecological study of the mammals of the Badlands and Black Hills of South Dakota and Wyoming. *Ecology* 20: 382–393.

Streubel, D. P. 1995. *Small Mammals of the Yellowstone Ecosystem*. Boulder, CO: Roberts-Rinehart. 152 pp.

Stromberg, M. R., and M. S. Boyce. 1986. Systematics and conservation of the swift fox, *Vulpes velox*, in North America. *Biological Conservation* 35: 97–110.

Sundstrom, C., W. G. Hepworth, and K. L. Diem. 1973. Abundance, distribution, and food habits of the pronghorn. Wyoming Game and Fish Department Bulletin 12: 1–61.

Svenden, G. F. 1976. Vocalizations of the long-tailed weasel (*Mustela frenata*). *Journal of Mammalogy* 57: 398–399.

Szaro, R. C., K. E. Stevenson, and D. R. Patton, tech. coordinators. 1988. *Management of Amphibians, Reptiles, and Small Mammals in North America*. Fort Collins, CO: USDA Rocky Mountain Forest and Range Experiment Station, General Technical Report RM-166. 458 pp. https://www.fs.fed.us/rm/pubs_rm/rm_gtr166.pdf

Turbak, G. 1995. *Pronghorn: Portrait of the American Antelope*. New York: Cooper Square. 138 pp.

Turner, R. W. 1974. *Mammals of the Black Hills of South Dakota and Wyoming*. Lawrence: University of Kansas, Museum of Natural History Miscellaneous Publication No. 60. 178 pp.

Ulrich, T. J. 1990. *Mammals of the Northern Rockies*. Missoula, MT: Mountain Press. 155 pp. (Includes species that occur in Grand Teton, Yellowstone, and Watertown-Glacier National Parks)

Van Wormer, J. 1969. *World of the American Elk*. Philadelphia, PA: J. B. Lippincott. 159 pp.

Vaughan, T. A. 1969. Reproduction and population densities in a montane small mammal fauna. Pp. 51–74, in *Contributions in Mammalogy: A Volume Honoring Professor E. Raymond Hall* (J. Knox Jones, Jr., ed.). Lawrence: University of Kansas, Museum of Natural History, Miscellaneous Publication No. 51. (Research in Colorado)

Wambolt, C. 1998. Sagebrush and ungulate relationships in Yellowstone's northern range. *Wildlife Society Bulletin* 26: 429–437.

Waring, G. H. 1970. Sound communications of black-tailed, white-tailed, and Gunnison's prairie dogs. *American Midland Naturalist* 83:167–185.

———. 1996. Sound and communication in the yellow-bellied marmot (*Marmota flaviventris*). *Animal Behaviour* 14: 177–183.

Wassink, J. L. 1993. *Mammals of the Central Rockies*. Missoula, MT: Mountain Press. 157 pp. (Includes Colorado, Wyoming, Montana, and Idaho)

Wengeler, W. R., D. A. Kelt, and M. L. Johnson. 2010. Ecological consequences of invasive lake trout on river otters in Yellowstone National Park. *Biological Conservation* 143: 1144–1153.

Whitaker, J. O. 1996. *National Audubon Society Field Guide to North American Mammals*. Rev. ed. New York: Knopf. 935 pp.

White, P. J., and R. A. Garrot. 2005. Northern Yellowstone elk after wolf restoration. *Wildlife Society Bulletin* 33: 942–955.

Wilkinson, T. 2013. *Grizzlies of Pilgrim Creek*. New York: Rizzoli.

Willey, R. B., and R. E. Richards. 1981. Vocalizations of the ringtail (*Bassariscus astutus*). *Southwestern Naturalist* 26: 23–30.

Wilson, D. E., and S. Ruff. 1999. *The Smithsonian Book of North American Mammals*. Washington, DC: Smithsonian Institution Press. 816 pp.

Woods, S. E., Jr. 1980. *The Squirrels of Canada*. Ottawa, ON: National Museums of Canada.

Wyoming Game and Fish Department. 1978. *The Mule Deer of Wyoming*. Wyoming Game and Fish Department Bulletin 15. Cheyenne: Wyoming Game and Fish Department. 149 pp.

———. 2016. *Wyoming Grizzly Bear Management Plan*. Cheyenne: Wyoming Game and Fish Department. 61 pp.

Zevelloff, S. I., and F. R. Collett. 1988. *Mammals of the Intermountain West*. Salt Lake City: University of Utah Press. 363 pp.

Birds

Alderfer, J., ed. 2008. *Complete Birds of North America*. Washington, DC: Smithsonian Institution Press. 664 pp.

American Ornithologists' Union. Various dates. *The Birds of North America* (A. Poole and F. Gill, eds.). Philadelphia: The Birds of North America. (Now available online from the Cornell Lab of Ornithology and the American Ornithological Society at https://birdsna.org/Species-Account/bna/home.)

American Ornithologists' Union (AOU). 1998. *The AOU Checklist of North American Birds*. 7th ed. Washington, DC: American Ornithologists' Union. (See Chesser et al., 2017 for 2017 supplement.)

Anderson, S. H. 1980. Habitat selection, succession, and bird community organization. Pp. 13–225, in *Workshop Proceedings: Management of Western Forests and Grasslands for Nongame Birds* (R. M. DeGraaf and N. G. Tilghman, eds.). Ogden, UT: USDA Forest Service, General Technical Report INT–86.

Andrews, R., and R. Righter. 1992. *Colorado Birds: A Reference to their Distribution and Habitat*. Denver, CO: Denver Museum of Natural History.

Augustine, D. J., and S. K. Skagen. 2014. Mountain plover nest survival in relation to prairie dog and fire dynamics in shortgrass steppe. *Journal of Wildlife Management* 78: 595–602.

Baicich, P. J., and C. J. O. Harrison. 1997. *A Guide to the Nests, Eggs, and Nestlings of North American Birds*. 2nd. ed. New York: Academic Press. 346 pp.

Bailey, A. M., R. J. Niedrach, and A. L. Bailey. 1953. *The Red Crossbills of Colorado*. Museum Pictorial No. 9. Denver, CO: Denver Museum of Natural History. 64 pp.

Baker, M. C., and J. T. Boylan. 1999. Singing behavior, mating associations, and reproductive success in a population of hybridizing lazuli and indigo buntings. *Condor* 181: 493–503.

Balda, R. P., and N. Masters. 1980. Avian communities in the pinyon-juniper woodlands: A descriptive analysis. Pp. 146–167, in *Workshop Proceedings: Management of Western Forests and Grasslands for Nongame Birds* (R. M. DeGraaf and N. G. Tilghman, eds.). Ogden, UT: USDA Forest Service, General Technical Report INT–86.

Banko, W. 1960. *The Trumpeter Swan*. North American Fauna No. 63. Washington, DC: US Fish and Wildlife Service. 214 pp.

Beason, R. C., and E. C. Franks. 1974. Breeding behavior of the horned lark. *Auk* 91: 65–74.

Beck, J. L., J. W. Connelly, and K. P. Reese. 2009. Recovery of greater sage-grouse habitat features in Wyoming big sagebrush following prescribed fire. *Restoration Ecology* 17: 393–403.

Beecham, J. J., and M. N. Kochert. 1975. Breeding biology of the golden eagle in southwestern Idaho. *Wilson Bulletin* 87: 506–513.

Benkman, C. W. 1993. Adaptation to single resources and the evolution of crossbill (*Loxia*) diversity. *Ecological Monographs* 63: 305–325.

Benkman, C. W., T. L. Parchman, and E. Mezquida. 2010. Patterns of coevolution in the adaptive radiation of crossbills. *Annals of the New York Academy of Sciences* 1206: 1–16.

Bennett, J., and D. Keinath. 2001. *Distribution and Status of the Yellow-billed Cuckoo* (Coccyzus americanus) *in Wyoming*. Sheridan, WY: Wolf Creek Charitable Foundation. 54 pp.

Bicak, T. K. 1977. Some eco-ethological aspects of a breeding population of long-billed curlews (*Numenius americanus*) in Nebraska. MA thesis, University of Nebraska–Omaha, Omaha.

Bock, C. E. 1970. *The Ecology and Behavior of the Lewis's Woodpecker*. University of California Publications in Zoology 92. 100 pp.

Bock, C. E., and J. H. Bock. 1987. Avian occupancy following fire in a Montana shrubsteppe. *Prairie Naturalist* 19: 153–158. (Western meadowlark, chestnut-collared longspur, and grasshopper, lark, and Brewer's sparrows)

Bohne, J., T. Rinkes, and S. Kilpatrick. 2007. *Sage-Grouse Habitat Management Guidelines for Wyoming*. Alpine: Wyoming Game and Fish Department.

Braun, C. E. 1980. Alpine bird communities of western North America: Implications for management and research. Pp. 280–291, in *Workshop Proceedings: Management of Western Forests and Grasslands for Nongame Birds* (R. M. DeGraaf and N. G. Tilghman, eds.). Ogden, UT: USDA Forest Service, General Technical Report INT–86.

Braun, C. E., O. O. Oedekoven, and C. L. Aldridge. 2002. Oil and gas development in western North America: Effects on sagebrush steppe avifauna with particular emphasis on sage-grouse. Pp. 337–349, in *Transactions of the 67th North American Wildlife and Natural Resources Conference*. Wildlife Management Institute.

Brodrick, H. J. 1952. *Birds of Yellowstone National Park*. Yellowstone Interpretive Series No. 2. US National Park Service: Yellowstone Library and Museum Association. 58 pp.

Brown, J. L. 1964. The integration of agonistic behavior in the Steller's jay, *Cyanocitta stelleri* (Gmelin). *University of California Publications in Zoology* 60: 223–328.

Buseck, R. S., D. A. Keinath, and M. H. McGee. 2004. *Species Assessment for Sage Thrasher* (Oreoscoptes montanus) *in Wyoming*. Cheyenne, WY: US Department of the Interior, Bureau of Land Management. 72 pp. http://www.uwyo.edu/wyndd/_files/docs/reports/speciesassessments/sagethrasher-dec2004.pdf

Butterfield, J. D. 1969. Nest-site requirements of the lark bunting in Colorado. MS thesis, Colorado State University, Fort Collins.

Cable, T. T., S. Seltman, and K. J. Cook. 1996. *Birds of Cimarron National Grassland*. Ft. Collins, CO: US Forest Service, General Technical Report RM–GTR–281.

Cade, T. J. 1982. *The Falcons of the World*. Ithaca, NY: Cornell University Press. 188 pp.

Calder, W. A. 1973. Microhabitat selection during nesting of hummingbirds in the Rocky Mountains. *Ecology* 54: 127–134.

Canterbury, J. L., and P. A. Johnsgard. 2017. *Common Birds of the Brinton Museum and Bighorn Mountains Foothills*. Lincoln: University of Nebraska–Lincoln DigitalCommons and Zea Books. http://digitalcommons.unl.edu/zeabook/57/

Chalfaun, A., K. Gerow, J. Carlisle, and L. Sanders. 2013. *Analysis of Temporal and Spacial Patterns of Raptor Nest Occupancy in Areas of Coal-bed Methane Development in the Powder River Basin, Wyoming*. Buffalo, WY: Bureau of Land Management Office. 33 pp.

Chesser, R. T., et al. 2017. Fifty-eighth supplement to the American Ornithological Society's Check-list of North American Birds. *Auk* 134: 751–773.

Cochrane, J. F. 1983. Long-billed curlew habitat and land-use relationships in western Wyoming. MS thesis, University of Wyoming, Laramie. 136 pp.

Creighton, P. D. 1971. *Nesting of the Lark Bunting in North-central Colorado*. US International Biological Program, Grassland Biome Technical Report No. 68. Colorado State University, Fort Collins. 154 pp.

Dark-Smiley, D. N., and D. A. Keinath. 2004. *Species Assessment for Long-Billed Curlew* (Numenius americanus) *in Wyoming*. Cheyenne, WY: US Department of the Interior, Bureau of Land Management. 61 pp. http://www.uwyo.edu/wyndd/_files/docs/reports/speciesassessments/long-billedcurlew-jan2004.pdf

Davis, D. E. 1959. Observations on territorial behavior of least flycatchers. *Wilson Bulletin* 71: 73–85.

Davis, J., G. F. Fisher, and B. S. Davis. 1963. The breeding biology of the western flycatcher. *Condor* 65: 337–382.

Dechant, J. A., M. L. Sondreal, D. H. Johnson, L. D. Igl, C. M. Goldade, M. P. Nenneman, and B. R. Euliss. 1998 (revised 2002). *Effects of Management Practices on Grassland Birds: Mountain Plover*. US Geological Survey, Northern Prairie Wildlife Research Center, Jamestown, ND. 14 pp. http://digitalcommons.unl.edu/usgsnpwrc/139/ (This and the following related species accounts are available online at https://digitalcommons.unl.edu/empgb: American bittern, Baird's sparrow, bobolink, Brewer's sparrow, burrowing owl, chestnut-collared longspur, clay-colored sparrow, dickcissel, eastern meadowlark, ferruginous hawk, field sparrow, golden eagle, grasshopper sparrow, greater prairie-chicken, greater sage-grouse, Henslow's sparrow, horned lark, lark bunting, lark sparrow, Le Conte's sparrow, lesser prairie-chicken, loggerhead shrike, long-billed curlew, marbled godwit, McCown's longspur, merlin, Nelson's sharp-tailed sparrow, northern harrier, prairie falcon, Savannah sparrow, sedge wren, short-eared owl, Sprague's pipit, Swainson's hawk, upland sandpiper, vesper sparrow, western meadowlark, willet, and Wilson's phalarope.)

DeGraaf, R. M., tech. coordinator. 1978. *Proceedings of the Workshop on Nongame Bird Habitat Management in the Coniferous Forests of the Western United States*. Portland, OR: USDA Forest Service, General Technical Report PNW-65.

DeGraaf, R. M., and N. G. Tilghman, eds. 1980. *Workshop Proceedings: Management of Western Forests and Grasslands for Nongame Birds*. Ogden, UT: USDA Forest Service, General Technical Report INT–86. 546 pp. https://www.fs.fed.us/rm/pubs_int/int_gtr086.pdf

Desmond, M. J., and J. A. Savidge. 1995. Spatial patterns of burrowing owl (*Speotyto cunicularia*) nests within black-tailed prairie dog (*Cynomys ludovicianus*) towns. *Canadian Journal of Zoology* 73: 1375–1379.

Desmond, M. J., J. A. Savidge, and K, M. Eskridge. 2000. Correlations between burrowing owl and black-tailed prairie dog declines. *Journal of Wildlife Management* 64: 1067–1075.

Diem, K., and S. L. Zeveloff. 1980. Ponderosa pine bird communities. Pp. 170–197, in *Workshop Proceedings: Management of Western Forests and Grasslands for Nongame Birds* (R. M. DeGraaf and N. G. Tilghman, eds.). Ogden, UT: USDA Forest Service, General Technical Report INT–86.

Dinsmore, S. 2001. Population biology of mountain plover in southern Phillips County, Montana. PhD diss., Colorado State University, Ft. Collins.

Dobkin, D. S. 1994. *Conservation and Management of Neotropical Migrant Landbirds in the Northern Rockies and Great Plains*. Moscow: University of Idaho Press.

———. 1995. *Management and Conservation of Sage Grouse, Denominative Species for the Ecological Health of Shrubsteppe Ecosystems*. Portland, OR: US Department of the Interior, Bureau of Land Management.

Dorn, J. L. 1978. *Wyoming Ornithology: A History and Bibliography with Species and Wyoming Area Indexes*. Bureau of Land Management and Wyoming Game and Fish Department. 369 pp.

Dorn, J. L., and R. D. Dorn. 1990. *Wyoming Birds*. Cheyenne, WY. Mountain West. 138 pp.

Dow, D. D. 1965. The role of saliva in food storage of the gray jay. *Auk* 82: 139–154.

Dunham, D. W. 1964. Reproductive displays of the warbling vireo. *Wilson Bulletin* 76: 170–173.

Dunkle, F. W. 1977. Swainson's hawks on the Laramie Plains, Wyoming. *Auk* 94: 65–71.

Eckhardt, R. C. 1976. Polygyny in the western wood pewee. *Condor* 78: 561–562.

Edson, J. M. 1943. A study of the violet-green swallow. *Auk* 60: 396–403.

Ellison, A. E., and C. M. White. 2001. Breeding biology of mountain plovers (*Charadrius montanus*) in the Uinta Basin. *Western North American Naturalist* 61: 223–228.

Emlen, S. T., J. D. Rising, and W. L. Thompson. 1975. A behavioral and morphological study of sympatry in the indigo and lazuli buntings of the Great Plains. *Wilson Bulletin* 87: 145–179.

Ensign, J. T. 1983. Nest site selection, productivity, and food habits of ferruginous hawks in southeastern Montana. MS thesis, Montana State University, Bozeman.

Faulkner, D. C. 2010. *Birds of Wyoming*. Greenwood Village, CO: Ben Roberts. (Includes species accounts, range maps, and about 600 references on Wyoming birds published through 2009)

Felske, B. E. 1971. The population dynamics and productivity of McCown's longspur at Matador, Saskatchewan. MS thesis, University of Saskatchewan, Saskatoon.

Finch, D. M. 1991. *Threatened, Endangered, and Vulnerable Species of Terrestrial Vertebrates in the Rocky Mountain Region*. Fort Collins, CO: USDA Forest Service Rocky Mountain Forest and Range Experiment Station, General Technical Report RM-215. 38 pp.

Findholt, S. L. 1984. *Status and Distribution of Herons, Egrets, Ibis, and Related Species in Wyoming*. Nongame Special Report. Cheyenne: Wyoming Game and Fish Department. 27 pp.

Flack, J. A. D. 1976. *Bird Populations of Aspen Forests in Western North America*. Ornithological Monographs No. 19. American Ornithologists' Union. 97 pp.

Follet, D. 1986. *Birds of Yellowstone and Grand Teton National Parks*. Boulder, CO: Roberts Rinehart and Yellowstone Library and Museum Association.

Forsythe, D. M. 1972. Observations on the nesting biology of the long-billed curlew. Great Basin Naturalist 32: 88–90.

French, N. R. 1959. Distribution and migration of the black rosy finch. *Condor* 61: 18–29.

Frinzel, J. E. 1964. Avian populations of four herbaceous communities in southeastern Wyoming. *Condor* 66: 496–510.

Froiland, S. G. 1990. *Natural History of the Black Hills and Badlands.* Sioux Falls, SD: Augustana College, Center for Western Studies.

Gibbon, R. S. 1966. Observations on the behavior of nesting three-toed woodpeckers, *Picoides tridactylus*, in central New Brunswick. *Canadian Field-Naturalist* 80: 223–226.

Glinski, R. L., ed. 1998. *The Raptors of Arizona.* Tucson: University of Arizona Press.

Graul, W. D. 1973. Adaptive aspects of the mountain plover social system. *Living Bird* 12: 69–94.

———. 1975. Breeding biology of the mountain plover. *Wilson Bulletin* 87: 6–31.

Greer, R. D. 1988. Effects of habitat structure and productivity on grassland birds. PhD diss., University of Wyoming, Laramie. 137 pp.

Greer, R. D., and S. H. Anderson. 1989. Relationships between population demography of McCown's longspurs and habitat resources. *Condor* 91: 609–619.

Hahn, H. W. 1950. Nesting behavior of the American dipper in Colorado. *Condor* 52: 49–62.

Hancock, J., and H. Elliott. 1978. *The Herons of the World.* New York: Harper and Row. 304 pp.

Hanni, D. J., J. Birek, C. White, R. Sparks, and J. Blakesley. 2009. *Monitoring Wyoming's Birds: 2008 Field Report.* Brighton, CO: Rocky Mountain Bird Observatory. 91 pp.

Harjer, H. 1974. An analysis of some aspects of the ecology of dusky grouse. PhD diss., University of Wyoming, Laramie.

Hein, D. 1980. Management of lodgepole pine for birds. Pp. 238–246, in *Workshop Proceedings: Management of Western Forests and Grasslands for Nongame Birds* (R. M. DeGraaf and N. G. Tilghman, eds.). Ogden, UT: USDA Forest Service, General Technical Report INT–86.

Heinrich, B. 1979. *Ravens in Winter.* New York: Summit Books.

Hickey, J. J., ed. 1969. *Peregrine Falcon Populations: Their Biology and Decline.* Madison: University of Wisconsin Press. 446 pp.

Higgins, K. F., and L. M. Kirsch. 1975. Some aspects of the breeding biology of the upland sandpiper in North Dakota. *Wilson Bulletin* 87: 96–102.

Hubbard, J. P. 1969. The relationships and evolution of the *Dendroica coronata* complex. *Auk* 86: 393–432.

Hughes, A. J. 1993. Breeding density and habitat preference of the burrowing owl in northeastern Colorado. MS thesis, Colorado State University, Ft. Collins.

Hutto, R., and J. S. Young, 1999. *Habitat Relationships of Landbirds in the Northern Region.* Denver, CO: USDA Forest Service, General Technical Report RNRS-GTR-32. 72 pp.

Johnsgard, P. A. 1973. *Grouse and Quails of North America.* Lincoln: University of Nebraska Press. 553 pp. http://digitalcommons.unl.edu/bioscigrouse/1/ and http://digitalcommons.unl.edu/bioscigrouse/

———. 1979. *Birds of the Great Plains: Breeding Species and Their Distribution.* Lincoln: University of Nebraska Press. (A new expanded edition is available from the University of Nebraska–Lincoln DigitalCommons and Zea Books at http://digitalcommons.unl.edu/bioscibirdsgreatplains/1/)

———. 1981. *The Plovers, Sandpipers, and Snipes of the World.* Lincoln: University of Nebraska Press. 492 pp.

———. 1986. *Birds of the Rocky Mountains with Particular Reference to National Parks in the Northern Rocky Mountain Region.* Boulder: Colorado Associated University Press. 504 pp. (Rev. ed., 2009, at http://digitalcommons.unl.edu/bioscibirdsrockymtns/1/ and http://digitalcommons.unl.edu/bioscibirdsrockymtns/3)

———. 1988. *North American Owls: Biology and Natural History.* Washington, DC: Smithsonian Institution Press. 295 pp.

———. 1990. *Hawks, Eagles, and Falcons of North America: Biology and Natural History.* Washington, DC: Smithsonian Institution Press. 403 pp. http://digitalcommons.unl.edu/johnsgard/52/

———. 1994. *Arena Birds: Sexual Selection and Behavior.* Washington, DC: Smithsonian Institution Press. 330 pp.

———. 1997a. *The Avian Brood Parasites: Deception at the Nest.* New York: Oxford University Press. 409 pp.

———. 1997b. *The Hummingbirds of North America.* 2nd ed. Washington, DC: Smithsonian Institution Press. 277 pp.

———. 2001. *Prairie Birds: Fragile Splendor in the Great Plains.* Lawrence: University Press of Kansas.

———. 2002a. *Grassland Grouse and Their Conservation.* Washington, DC: Smithsonian Institution Press. 157 pp.

———. 2002b. *North American Owls: Biology and Natural History.* 2nd ed. Smithsonian Institution Press. 298 pp.

———. 2008. *The Platte: Channels in Time.* 2nd ed. Lincoln: University of Nebraska Press. 176 pp.

———. 2011a. *Rocky Mountain Birds: Birds and Birding in the Central and Northern Rockies.* Lincoln: University of Nebraska–Lincoln DigitalCommons and Zea Books. 274 pp. http://digitalcommons.unl.edu/zeabook/7/

———. 2011b. *Sandhill and Whooping Cranes: Ancient Voices over America's Wetlands.* Lincoln: University of Nebraska Press. 155 pp.

———. 2016a. *Swans: Their Biology and Natural History.* Lincoln: University of Nebraska–Lincoln DigitalCommons and Zea Books. 114 pp. http://digitalcommons.unl.edu/zeabook/38/

———. 2016b. *The North American Grouse: Their Biology and Behavior.* Lincoln: University of Nebraska–Lincoln DigitalCommons and Zea Books. 183 pp. http://digitalcommons.unl.edu/zeabook/41/

———. 2016c. Bittern surprise. *BirdWatching* 30(2): 36–39.

———. 2016d. *The North American Geese: Their Biology and Behavior.* Lincoln: University of Nebraska–Lincoln DigitalCommons and Zea Books. 159 pp. https://digitalcommons.unl.edu/zeabook/44/

———. 2016e. *The North American Sea Ducks: Their Biology and Behavior.* Lincoln: University of Nebraska–Lincoln DigitalCommons and Zea Books. 256 pp. https://digitalcommons.unl.edu/zeabook/50/

———. 2017a. *The North American Perching and Dabbling Ducks: Their Biology and Behavior.* Lincoln: University of Nebraska–Lincoln DigitalCommons and Zea Books. 228 pp. http://digitalcommons.unl.edu/zeabook/53/

———. 2017b. *North American Quails, Partridges, and Pheasants: Their Biology and Behavior.* Lincoln: University of Nebraska–Lincoln DigitalCommons and Zea Books. 133 pp. http://digitalcommons.unl.edu/zeabook/58/

———. 2017c. *The North American Whistling-Ducks, Pochards and Stifftails*. Lincoln: University of Nebraska–Lincoln DigitalCommons and Zea Books. 188 pp. http://digitalcommons.unl.edu/zeabook/54/

Johnson, R. R., L. T. Haight, M. F. Riffey, and J. M. Simpson. 1980. Sage-steppe bird populations. Pp. 98–112, in *Workshop Proceedings: Management of Western Forests and Grasslands for Nongame Birds* (R. M. DeGraaf and N. G. Tilghman, eds.). Ogden, UT: USDA Forest Service, General Technical Report INT-86.

Kangarise, C. M. 1979. Breeding biology of Wilson's phalarope in North Dakota. *Bird-Banding* 50: 12–22.

Kilham, L. 1968, 1972. Reproductive behavior in white-breasted nuthatches. *Auk* 85: 477–492; 89: 115–129.

———. 1973. Reproductive behavior in the red-breasted nuthatch. I. Courtship. *Auk* 90: 597–609.

Killpack, M. L. 1970. Notes on sage thrasher nestlings in Colorado. *Condor* 72: 486–488.

Kingery, H. E., ed. 1998. *Colorado Breeding Bird Atlas*. Denver: Colorado Bird Atlas Partnership and Colorado Division of Wildlife. 636 pp.

Knick, S. T., and J. W. Connelly. 2011. *Greater Sage-Grouse: Ecology and Conservation of a Landscape Species and Its Habitats*. Berkeley: University of California Press.

Knick, S. T., D. S. Dobkin, J. T. Rotenberry, M. A. Schroeder, W. M. Vander Haegen, and C. Van Riper. 2003. Teetering on the edge or too late? Conservation and research issues for avifauna of sagebrush habitats. *Condor* 105: 611–634.

Knopf, F., and J. R. Rupert. 1996. Reproduction and movements of mountain plovers breeding in Colorado. *Wilson Bulletin* 108: 504–506.

Knowles, C. J., and P. R. Knowles. 1984. Additional records of mountain plovers using prairie dog towns in Montana. *Prairie Naturalist* 16(4): 183–186.

Knowles, C. J., C. J. Stoner, and S. P. Gieb. 1982. Selective use of black-tailed prairie dog towns by mountain plovers. *Condor* 81: 71–74.

Korphany, N. M., L. W. Ayers, S. H. Anderson, and D. B. McDonald. 2001. A preliminary assessment of burrowing owl population status in Wyoming. *Journal of Raptor Research* 35: 337–343.

Kroodsma, R. L. 1970. North Dakota species pairs. I. Hybridization in buntings, grosbeaks, and orioles. II. Species' recognition behavior of territorial male rose-breasted and black-headed grosbeaks (*Pheucticus*). PhD diss., North Dakota State University, Fargo.

Laun, C. H. 1957. A life history study of the mountain plover, *Eupoda montana* Townsend, in the Laramie Plains, Albany County, Wyoming. MS thesis, University of Wyoming, Laramie.

Lederer, R. J. 1977. Winter feeding territories in the Townsend's solitaire. *Bird-Banding* 48: 11–18.

Lincer, J. L., and K. Steenhof, eds. 1997. *The Burrowing Owl, Its Biology and Management, including the Proceedings of the First International Burrowing Owl Symposium*. Raptor Research Reports No. 9. The Raptor Research Foundation.

Lokemoen, J. T., and H. F. Duebbert. 1976. Ferruginous hawk nesting ecology and raptor populations in northern South Dakota. *Condor* 78: 464–470.

Lumsden, H. G. 1965. *Displays of the Sharptail Grouse*. Maple: Ontario Department of Lands and Forests Technical Series Research Report no. 66.

Martin, D. J. 1973. Selective aspects of burrowing owl ecology and behavior. *Condor* 75: 446–456.

McLaren, P. A., S. H. Anderson, and D. E. Runde. 1988. Food habits and nest characteristics of breeding raptors in southwestern Wyoming. *Great Basin Naturalist* 48: 548–553.

Mengel, R. M., and J. S. Mengel. 1952. Sprague pipit and black rosy finch in north-central Wyoming in summer. *Condor* 54: 61–62.

Moriarty, L. J. 1965. A study of the breeding biology of the chestnut-collared longspur (*Calcarius ornatus*) in northeastern South Dakota. *South Dakota Bird Notes* 17: 76–79.

Nethersole-Thompson, D. 1975. *Pine Crossbills: A Scottish Contribution*. Berkhamstead, UK: T. and A. D. Poyser.

Newman, O. A. 1970. Cowbird parasitism and nesting success of lark sparrows in southern Oklahoma. *Wilson Bulletin* 82: 304–309.

Nicholoff, S. H., compiler. 2003. *Wyoming Bird Conservation Plan, Version 2.0*. Lander: Wyoming Partners in Flight and Wyoming Game and Fish Department.

Nye, D., M. Back, and H. Hinchman. 1979. *Birds of the Upper Wind River Valley*. USDA Forest Service, Shoshone National Forest. 34 pp.

Ohlendorf, R. R. 1975. *Golden Eagle Country*. New York: Knopf.

Orabona, A., and C. Rudd. 2016. *Wyoming Bird Checklist*. Wyoming Game and Fish Department Nongame Program. (444 species) https://wgfd.wyo.gov/WGFD/media/content/PDF/Education/Resources/Infographics%20and%20Activities/Wyoming-Bird-Checklist-(May-2016)-(Blue-Cover).pdf

Paige, C., and S. A. Ritter. 1999. *Birds in a Sagebrush Sea: Managing Sagebrush Habitats for Bird Communities*. Boise, ID: Partners in Flight Western Working Group. 47 pp.

Palmer, R. S., ed. 1962. *Handbook of North American Birds, Vol. I. Loons through Flamingos*. New Haven, CN: Yale University Press. 567 pp.

———. 1976. *Handbook of North American Birds, Vols. 2 and 3. Waterfowl*. New Haven, CN: Yale University Press. 521 pp. and 560 pp.

———. 1988. *Handbook of North American Birds, Vols. 4 and 5. Diurnal Raptors*. New Haven, CN: Yale University Press. 433 pp. and 465 pp.

Paothong, N. 2017. *Sage Grouse: Icon of the West*. Laguna Beach, CA: Laguna Wilderness Press.

Parrish, T. L., S. H. Anderson, and W. F. Oelklaus. 1993. Mountain plover habitat selection in the Powder River Basin, Wyoming. *Prairie Naturalist* 25: 219–226.

Patterson, R. 1952. *The Sage Grouse in Wyoming*. Denver, CO: Sage Books. 341 pp.

Paulin, K. M., J. J. Cook, and S. R. Dewey. 1999. Pinyon-juniper woodlands as sources of avian diversity. Pp. 240–243, in *Proceedings: Ecology and Management of Pinyon-Juniper Communities within the Interior West: Sustaining and Restoring a Diverse Ecosystem*. Ogden, UT: USDA Forest Rocky Mountain Research Station Proceedings RMRS-P-9. https://www.fs.fed.us/rm/pubs/rmrs_p009.pdf

Peterjohn, B. G., and J. R. Sauer. 1999. Population status of North American grassland birds from the North American Breeding Bird Survey, 1966–1996. Pp. 27–44, in *Ecology and Conservation of Grassland Birds of the Western Hemisphere* (P. D. Vickery and J. R. Herkert, eds.). Studies in Avian Biology No. 19. Camarillo, CA: Cooper Ornithological Society. https://sora.unm.edu/sites/default/files/journals/sab/sab_019.pdf

Peterson, R. A. 1995. *The South Dakota Breeding Bird Atlas*. Aberdeen: South Dakota Ornithologists' Union.

Phillips, R. L., A. H. Wheeler, J. M. Lockhart, T. P. McEneaney, and N. C. Forrester. 1990. *Nesting Ecology of Golden Eagles and Other Raptors in Southeastern Montana and Northern Wyoming.* Washington, DC: US Department of the Interior Fish and Wildlife Service. Fish and Wildlife Technical Report 26. 13 pp.

Poulin, R. G. 2003. Relationships between burrowing owls (*Athene cunicularia*), small mammals, and agriculture. PhD diss., University of Saskatchewan, Regina.

Power, H. W., III. 1966. Biology of the mountain bluebird in Montana. *Condor* 68: 351–371.

Rashid, S. 2010. *Small Mountain Owls.* Atglen, PA: Schiffler. 160 pp.

Raynes, B. 1984. *Birds of Grand Teton National Park and the Surrounding Area.* Moose, WY: Grand Teton Natural History Association. 90 pp. (Includes a regional checklist of 293 species)

Rehm-Lorber, J. A., J. Blakesley, D. C. Pavlacky, Jr., and D. J. Hanni. 2010. *Monitoring the Birds of Wyoming: 2009 Field Season Report (Revised).* Technical Report M-MWB-09-01. Brighton, CO: Rocky Mountain Bird Observatory. 64 pp.

Rich, R. 1980. Nest placement in sage thrasher, sage sparrow, and Brewer's sparrow in southeastern Idaho. *Wilson Bulletin* 92: 362–368.

Rich, T., M. J. Wisdom, and V. A. Saab. 2005. Conservation of priority birds in sagebrush ecosystems. Pp. 589–606, in *Bird Conservation Implementation and Integration in the Americas: Proceedings of the Third International Partners in Flight Conference* (C. J. Ralph and T. D. Rich, eds.). Albany, CA: USDA Forest Service Pacific Southwest Research Station, General Technical Report PSW-GTR-191.

Rich, T. C., C. J. Beardmore, H. Berlanga, P. J. Blancher, M. S. W. Bradstreet, G. S. Butcher, D. W. Demerest, E. H. Dunn, W. C. Hunter, E. E. Inig-Elias, J A. Kennedy, A. M. Martell, A. O. Punjabi, D. N. Pashley, K. V. Rosenburg, C. M. Rusta, J. S. Wendt, and T. C. Will. 2004. *North American Landbird Conservation Plan.* Ithaca, NY: Partners in Flight and Cornell University Laboratory of Ornithology. 84 pp.

Rising, J. D. 1983. The Great Plains hybrid zones. *Current Ornithology* 1: 131–157.

Rodewald, P., ed. 1992–2019. Birds of North America (online). Cornell Lab of Ornithology. https://birdsna.org/Species-Account/bna/home (Modern life histories of more than 600 species of North American birds, mostly updated from out-of-print versions from the1990s and early 2000s by the Academy of Natural Sciences, Philadelphia, and the American Ornithologists' Union, Washington, DC, originally edited by A. Poole.)

Rotenberry, J. T., and J. A. Wiens. 1980. Habitat structure, patchiness, and avian communities in North American steppe vegetation: A multivariate analysis. *Ecology* 61: 1228–1250.

———. 1989. Reproductive biology of shrubsteppe passerine birds: Geographical and temporal variation in clutch size, brood size, and fledging success. *Condor* 91: 1–14.

Salt, G. 1957. An analysis of avifauna in the Teton Mountains and Jackson Hole, Wyoming. *Condor* 59: 373–393.

Sanderson, G. C., ed. 1977. *Management of Migratory Shore and Upland Game Birds in North America.* Washington, DC: US Department of the Interior and International Association of Fish and Wildlife Agencies. 358 pp.

Sanderson, H. R., E. L. Bull, and P. J. Edgerton. 1980. Bird communities in mixed conifer forests of the interior Northwest. Pp. 224–237, in *Workshop Proceedings: Management of Western Forests and Grasslands for Nongame Birds* (R. M. DeGraaf and N. G. Tilghman, eds.). Ogden, UT: USDA Forest Service, General Technical Report INT-86.

Sauer, J. R., D. K. Niven, J. E. Hines, D. J. Ziolkowski, Jr., K. L. Pardieck, J. E. Fallon, and W. A. Link. 2017. *The North American Breeding Bird Survey, Results and Analysis 1966–2015.* Version 2.07. Laurel, MD: USGS Patuxent Wildlife Research Center.

Schaller, G. B. 1964. Breeding behavior of the white pelican at Yellowstone Lake, Wyoming. *Condor* 66: 3–23.

Scott, O. K. 1993. *A Birder's Guide to Wyoming.* Colorado Springs, CO: American Birding Association.

Scott, V. E., J. A. Whelan, and P. L. Svoboda. 1980. Cavity-nesting birds and forest management. Pp. 311–324, in *Workshop Proceedings: Management of Western Forests and Grasslands for Nongame Birds* (R. M. DeGraaf and N. G. Tilghman, eds.). Ogden, UT: USDA Forest Service, General Technical Report INT-86.

Sheffield, S. R. 1997. Current status, distribution, and conservation of the burrowing owl (*Speotyto cunicularia*) in midwestern and western North America. Pp. 399–407, in *Biology and Conservation of Owls of the Northern Hemisphere* (J. R. Duncan, D. H. Johnson, and T. H. Nicholls, eds.). St. Paul, MN: USDA Forest Service, General Technical Report NC–190.

Short, L. L., Jr. 1965. Hybridization in the flickers (*Colaptes*) of North America. *Bulletin of the American Museum of Natural History* 129: 309–428.

Sibley, C. G., and D. A. West. 1959. Hybridization in the rufous-sided towhees of the Great Plains. *Auk* 76: 326–338.

Sibley, C. G., and L. L. Short, Jr. 1959. Hybridization in the buntings (*Passerina*) of the Great Plains. *Auk* 76: 443–463.

———. 1964. Hybridization in the orioles of the Great Plains. *Condor* 66: 130–150.

Sidle, J. G., M. Ball, T. Byer, J. J. Chynoweth, G. Foli, R. Hodorff, G. Moravek, R. Peterson, and D. N. Svingen. 2001. Occurrence of burrowing owls in black-tailed prairie dog colonies on Great Plains national grasslands. *Journal of Raptor Research* 35: 316–321.

Smith, H., and D. A. Keinath. 2004a. *Species Assessment for Mountain Plover* (Charadrius montanus) *in Wyoming.* Cheyenne, WY: US Department of the Interior, Bureau of Land Management. 53 pp. https://www.uwyo.edu/wyndd/_files/docs/reports/speciesassessments/mountainplover-nov2004.pdf

Smith, H., and D. A. Keinath. 2004b. *Species Assessment for Northern Goshawk* (Accipiter gentilis) *in Wyoming.* Cheyenne, WY: US Department of the Interior, Bureau of Land Management. 48 pp.

Smith, J. W., and C. W. Benkman. 2007. A coevolutionary arms race causes ecological speciation in crossbills. *American Naturalist* 169: 455–465.

Smith, K. G. 1980. Nongame birds of the Rocky Mountain spruce-fir forests and their management. Pp. 258–278, in *Workshop Proceedings: Management of Western Forests and Grasslands for Nongame Birds* (R. M. DeGraaf and N. G. Tilghman, eds.). Ogden, UT: USDA Forest Service, General Technical Report INT–86.

Snow, C. 1973a. *Golden Eagle* (Aquila chrysaetos). Habitat Management Series for Unique or Endangered Species, Report No. 7. Denver, CO: US Department of the Interior, Bureau of Land Management Technical Note T-N-239.

———. 1973b. *Southern Bald Eagle* (Haliaeetus leucocephalus leucocephalus) *and Northern Bald Eagle* (Haliaeetus leucocephalus alascanus). Habitat Management Series for Endangered Species, Report No. 5. Denver, CO: US Department of the Interior, Bureau of Land Management Technical Note T-N-171.

———. 1974a. *Ferruginous Hawk* (Buteo regalis). Habitat Management Series for Unique or Endangered Species, Report No. 13. Denver, CO: US Department of the Interior, Bureau of Land Management Technical Note T-N-255.

———. 1974b. *Prairie Falcon* (Falco mexicanus). Habitat Management Series for Endangered Species, Report No. 8. Denver, CO: US Department of the Interior, Bureau of Land Management Technical Note T-N-240.

Squires, J. R., S. A. Anderson, and R. Oakleaf. 1989. Food habits of prairie falcons in Campbell County, Wyoming. *Journal of Raptor Research* 23: 157–161.

———. 1993. Home range size and habitat use of nesting prairie falcons near oil developments in northeastern Wyoming. *Journal of Field Ornithology* 64: 1–10.

Taylor, D. L., and W. J. Barymore, Jr. 1980. Postfire succession of avifauna in coniferous forests of Yellowstone and Grand Teton National Parks, Wyoming. Pp. 130–145, in *Workshop Proceedings: Management of Western Forests and Grasslands for Nongame Birds* (R. M. DeGraaf and N. G. Tilghman, eds.). Ogden, UT: USDA Forest Service, General Technical Report INT-86.

Taylor, S. V., and V. M. Ashe. 1976. The flight display and other behavior of male lark buntings (*Calamospiza melanochorys*). *Bulletin of the Psychonomic Society* 7: 527–529.

Thompson, C. D., and S. H. Anderson. 1988. Foraging behavior and food habits of burrowing owls in Wyoming. *Prairie Naturalist* 20: 23–28.

Tomback, D. F. 1983. Nutcrackers and pines: Coevolution or coadaptation? Pp. 179–223, in *Coevolution* (M. H. Nitecki, ed.). Chicago: University of Chicago Press.

Travsky, A., and G. P. Beauvais. 2004. *Species Assessment for the Trumpeter Swan* (Cygnus buccinator) *in Wyoming*. Cheyenne, WY: US Department of the Interior, Bureau of Land Management.

US Fish and Wildlife Service (USFWS). 2008. *Birds of Conservation Concern 2008*. Arlington, VA: USFWS, Division of Migratory Bird Management. 85 pp. https://www.fws.gov/migratorybirds/pdf/grants/birdsofconservationconcern2008.pdf

Vance, J., and N. Paothong. 2012. *Save the Last Dance: A Story of North American Grassland Grouse*. Columbia, MO: Noppadol Paothong Photography.

Vander Wall, S. B., and R. P. Balda. 1977. Coadaptations of the Clark's nutcracker and the piñon pine for efficient seed harvest and dispersal. *Ecological Monographs* 47: 89–111.

Welsch, B. L., F. J. Wagstaff, and J. A. Robertson. 1991. Preference of wintering sage grouse for big sagebrush. Journal of Range Management 44: 462–465.

West, D. A. 1962. Hybridization in grosbeaks (*Pheucticus*) of the Great Plains. *Auk* 79: 399–424.

Wiens, J. A., and J. T. Rotenberry. 1979. Diet niche relationships among North American grassland and shrubsteppe birds. *Oecologia* 42: 253–292.

———. 1981. Habitat associations and community structure of birds in shrubsteppe environments. *Ecological Monographs* 51: 21–41.

———. 1985. Response of breeding passerine birds to rangeland alteration in a North American shrubsteppe locality. *Journal of Applied Ecology* 22: 655–668.

Wyoming Game and Fish Department. 2003. *Wyoming Greater Sage-Grouse Conservation Plan*. Cheyenne: Wyoming Game and Fish Department.

———. 2008. *Wyoming Bird Checklist*. Lander: Wyoming Game and Fish Department. 16 pp.

Zardus, M. J. 1967. *Birds of Yellowstone and Grand Teton Park*. Salt Lake City, UT: Wheelwright Press.

Herpetiles and Fish

Ballinger, R. E., J. D. Lynch, and G. R. Smith. 2010. *Amphibians and Reptiles of Nebraska*. Oro Valley, AZ: Rusty Lizard Press, and Lincoln, NE: University of Nebraska Press. 400 pp. (13 amphibians and 48 reptiles, plus 151 distribution maps and diagrams)

Baxter, G. T., and M. D. Stone. 1985. *Amphibians and Reptiles of Wyoming*. 2nd ed. Cheyenne: Wyoming Game and Fish Department. 137 pp. (23 reptiles, 12 amphibians)

———. 1995. *Fishes of Wyoming*. Cheyenne: Wyoming Game and Fish Department. 290 pp.

Behler, J. L., and F. W. King. 1979. *The Audubon Society Field Guide to North American Reptiles and Amphibians*. New York: Knopf. 744 pp.

Collins, J. T. 1993. *Amphibians and Reptiles in Kansas*. Lawrence: University of Kansas, Museum of Natural History, Public Education Series No. 13. 397 pp.

Conant R., and J. Collins. 1998. *Reptiles and Amphibians of Eastern and Central North America*. Boston, MA: Houghton Mifflin. 640 pp.

Dodd, C. K. 2001. *North American Box Turtles: A Natural History*. Norman: University of Oklahoma Press. 256 pp.

Ernst, C. H., J. E. Lovich, and R. W. Barbour. 1994. *Turtles of the United States and Canada*. Washington DC: Smithsonian Institution Press. 578 pp.

Hammerson, G. A. 1982. *Amphibians and Reptiles in Colorado*. Denver: Colorado Division of Fish and Wildlife.

Iverson, J. B., and G. R. Smith. 1993. Reproductive ecology of the painted turtle (*Chrysemys picta*) in the Nebraska Sandhills and across its range. *Copeia* 1993: 1–21.

Joy, J. E., and D. Crews. 1985. Social dynamics of group courtship behavior in male red-sided garter snakes (*Thamnophis sirtalis parietalis*). *Journal of Comparative Psychology* 99(2): 145–149.

Kardong, K. V. 1980. Gopher snakes and rattlesnakes: Presumptive Batesian mimicry. *Northwest Science* 54: 1–4.

Keinath, D. and J. Bennett. 2000. *Distribution and Status of the Boreal Toad* (Bufo boreas boreas) *in Wyoming*. Cheyenne: US Fish and Wildlife Service, Wyoming Field Office.

Klauber, L. M. 1972. *Rattlesnakes: Their Habits, Life Histories, and Influence on Mankind*. 2nd ed. 2 vols. Berkeley: University of California Press.

Koch, E. D., and C. R. Peterson. 1995. *Amphibians and Reptiles of Yellowstone and Grand Teton National Parks*. Salt Lake City: University of Utah Press. (7 amphibians and 9 reptiles)

Krupa, J. 1994. Breeding biology of the Great Plains toad in Oklahoma. *Journal of Herpetology* 28: 217–224.

Kruse, K. C. 1981. Phonotactic responses of female leopard frogs (*Rana pipiens*) to *Rana blairi*, a presumed hybrid, and conspecific mating trills. *Journal of Herpetology* 15: 145–150.

Legler, J. M. 1960. *Natural history of the ornate box turtle*, Terrapene ornata ornata *Agassiz*. University of Kansas Publications, Museum of Natural History 11(10): 527–669.

Lewis, D. 2011. *A Field Guide to the Amphibians and Reptiles of Wyoming*. Douglas: The Wyoming Naturalist. http://wyomingnaturalist.com/html/herps/herp_book.html

Moriarty, J. J., ed. 2008. *Scientific and Standard English Names of Amphibians and Reptiles of North America North of Mexico, with Comments Regarding Confidence in Our Understanding*. 6th ed. Society for the Study of Amphibians and Reptiles Herpetological Circular 37. 84 pp.

Oliver, J. A. 1955. *The Natural History of North American Amphibians and Reptiles*. Princeton, NJ: Van Nostrand.

Parker, J., and S. Anderson. 2001. *Identification Guide to the Herptiles of Wyoming*. Laramie: Wyoming Cooperative Fish and Wildlife Research Unit. 40 pp.

Powell, R., R. Conant, and J. T. Collins. 2016. *Peterson Field Guide to the Reptiles and Amphibians of Eastern and Central North America*. Boston: Houghton Mifflin. 494 pp. (Includes more than 100 color photographs and 47 color plates illustrating more than 300 species)

Shaw, C. E., and S. Campbell. 1974. *Snakes of the American West*. New York: Knopf. 328 pp.

Smith, H., and R. D. Brodie, Jr. 2001. *Reptiles of North America: A Guide to Field Identification*. New York: St. Martin's Press. 240 pp.

Stebbens, R. C. 2018. *A Field Guide to Western Reptiles and Amphibians*. 4th ed. Boston, MA: Houghton Mifflin. 533 pp.

Stebbens, R. C., and N. W. Cohen. 1995. *A Natural History of Amphibians*. Princeton, NJ: Princeton University Press. 316 pp.

Wildlife-Watching, Ecology, and General Rocky Mountains

Alexander, R. R. 1985. *Major Habitat Types, Community Types and Plant Communities in the Rocky Mountains*. Fort Collins, CO: USDA Forest Service Rocky Mountain Forest and Range Experiment Station, General Technical Report RM-123. https://www.biodiversitylibrary.org/bibliography/99752

Canterbury, J. L., P. A. Johnsgard, and H. Downing. 2013. *Birds and Birding in Wyoming's Bighorn Mountains Region*. Lincoln: University of Nebraska–Lincoln DigitalCommons and Zea Books. 260 pp. http://digitalcommons.unl.edu/zeabook/18/

Elias, S. A. 1996. *The Ice-Age History of National Parks in the Rocky Mountains*. Washington, DC: Smithsonian Institution Press. 337 pp.

———. 2002. *Rocky Mountains*. Washington, DC: Smithsonian Institution Press. 167 pp.

Fenneman, N. M. 1931. *Physiography of Western United States*. New York: McGraw Hill.

Fischer, C., and H. Fischer. 1995. *Montana Wildlife Viewing Guide*. Rev. ed. Helena, MT: Falcon. 111 pp.

Hayward, C. L. 1945. Biotic communities of the southern Wasatch and Uinta Mountains, Utah. *Great Basin Naturalist* 6(1–4). 124 pp.

———. 1952. Alpine biotic communities of the Uinta Mountains, Utah. *Ecological Monographs* 22(2): 93–120.

Hutto, R. L., and J. S. Young. 1999. *Habitat Relationships of Landbirds in the Northern Region, USDA Forest Service*. USDA Forest Service, Rocky Mountain Research Station, General Technical Report RMRS-GTR-32. 72 pp. https://www.fs.fed.us/rm/pubs/rmrs_gtr032.pdf

Johnsgard, P. A. 2011a. *Rocky Mountain Birds: Birds and Birding in the Central and Northern Rockies*. Lincoln: University of Nebraska–Lincoln DigitalCommons and Zea Press. 274 pp. http://digitalcommons.unl.edu/zeabook/7/

Mathews, D. 2003. *Rocky Mountain Natural History: Grand Teton to Jasper*. Portland, OR: Raven Editions.

McEneaney, T. 1988. *Birds of Yellowstone*. Boulder, CO: Roberts Rinehart. 171 pp.

———. 1993. *The Birder's Guide to Montana*. Helena, MT: Falcon Press. 314 pp. (Describes 45 birding sites or areas, including northern Yellowstone Park)

———. 2007. *Yellowstone National Park Checklist of Birds*. Mammoth, WY: Yellowstone National Park and Yellowstone Association. (A 12-page foldout containing an annotated checklist of 323 species)

Pettingill, O. S., Jr., and N. R. Whitney, Jr. 1965. *Birds of the Black Hills*. Ithaca, NY: Cornell Laboratory of Ornithology, Special Publication No. 1.

Raynes, B. 2000. *A Pocket Guide to Birds of Jackson Hole: The Occurrence, Arrival and Departure Dates, and Preferred Habitat of Birds of the Jackson Hole, Wyoming, Area*. Moose, WY: Homestead Publishing. 8 pp. (Includes 301 species, a monthly abundance calendar, and habitat information)

Raynes, B., and D. Wile. 1994. *Finding the Birds of Jackson Hole*. Jackson, WY: D. Wile. 159 pp. (Includes Grand Teton National Park and a seasonal checklist)

Raynes, B., and G. Raynes. 2008. *Birds of Jackson Hole*. Moose, WY: Grand Teton Association, Wyoming Game and Fish Department, and US Fish and Wildlife Service. 16 pp. (A seasonal checklist of 340 species)

Scharf, R., ed. 1966. *Yellowstone and Grand Teton National Parks*. New York: David McKay.

Schmidt, J., and T. Schmidt. 1995. *The Northern Rockies: Idaho, Montana, Wyoming*. The Smithsonian Guides to Natural America. New York: Random House. 304 pp.

Scott, O. 1952. *A Birder's Guide to Wyoming*. Delaware City, DE: American Birding Association. 246 pp.

Scullery, P. 1997. *Searching for Yellowstone. Ecology and Wonder in the Last Wilderness*. Boston: Houghton Mifflin.

Tisdale, E. W., and M. Hironaka. 1981. *The Sagebrush-Grass Ecoregion: A Review of the Ecological Literature*. Forest, Wildlife and Range Experiment Station Contribution No. 209. University of Idaho, Moscow, ID. http://digital.lib.uidaho.edu/cdm/ref/collection/fwres/id/169

Trimble, S. 1989. *The Sagebrush Ocean*. Reno: University of Nevada Press.

West, N. E. 1988. Intermountain deserts, shrub steppes, and woodlands. Pp. 209–230, in *North American Terrestrial Vegetation* (M. G. Barbour and W. D. Billings, eds.). Cambridge, UK: Cambridge University Press.

White, M. 1999. *Guide to Birdwatching Sites: Western US*. Washington, DC: National Geographic Society. 240 pp. (22 locations mentioned)

Wilkinson, T. 2008. *Watching Yellowstone and Grand Teton Wildlife*. Helena, MT: Riverbend Publishing. 96 pp. (Locations are organized by species)

Wyoming Game and Fish Department. 1996. *Wyoming Wildlife Viewing Guide*. Helena, MT: Falcon Press. 216 pp. (Includes 55 locations)

Great Plains

Borroughs, R. D. 1961. *The Natural History of the Lewis and Clark Expedition*. East Lansing: Michigan State University Press. 340 pp.

Johnsgard, P. A. 1976. The grassy heartland. Pp. 234–264, in *Our Continent: A Natural History of North America*. Washington DC: National Geographic Society. 398 pp.

———. 2003. *Lewis and Clark on the Great Plains*. Lincoln: University of Nebraska Press. 143 pp.

———. 2009. *Birds of the Great Plains: Breeding Species and Their Distribution*. Rev. ed. Lincoln: University of Nebraska–Lincoln DigitalCommons and Zea Books. http://digitalcommons.unl.edu/bioscibirdsgreatplains/1/

———. 2012a. *Wetland Birds of the Central Plains: South Dakota, Nebraska, and Kansas*. Lincoln: University of Nebraska–Lincoln DigitalCommons and Zea Books. 275 pp. http://digitalcommons.unl.edu/zeabook/8/

———. 2012b. *Wings over the Great Plains: Bird Migrations in the Central Flyway*. Lincoln: University of Nebraska–Lincoln DigitalCommons and Zea Books. 249 pp. http://digitalcommons.unl.edu/zeabook/13/

———. 2018. *A Naturalist's Guide to the Great Plains*. Lincoln: University of Nebraska–Lincoln DigitalCommons and Zea Books. 161 pp. http://digitalcommons.unl.edu/zeabook/63/

Knopf, F. L., and F. B. Samson, eds. 1997. *Ecology and Conservation of Great Plains Vertebrates*. New York: Springer.

Lewis, A. 2004. Sagebrush steppe habitat and their associated bird species in South Dakota, North Dakota, and Wyoming: Life on the edge of the sagebrush ecosystem. PhD diss., South Dakota State University, Brookings.

Rohwer, S. A. 1971. Systematics and evolution of Great Plains meadowlarks, genus *Sturnella*. PhD diss., University of Kansas, Lawrence.

Sibley, C. G., and D. A. West. 1959. Hybridization in the rufous-sided towhees of the Great Plains. *Auk* 76: 326–338.

Sibley, C. G., and L. L. Short, Jr. 1959. Hybridization in the buntings (*Passerina*) of the Great Plains. *Auk* 76: 443–463.

———. 1964. Hybridization in orioles of the Great Plains. *Condor* 66: 130–150.

Western and National

Aldrich, J. 1966. *Life Areas of North America*. Poster 102. Washington, DC: US Department of the Interior, Bureau of Sport Fisheries and Wildlife.

Bailey, R. G. 1978. *Ecosystems of the United States*. Washington, DC: USDA Forest Service, RARE II Map B, 1:7,500.000.

———. 1995. *Description of the Ecoregions of the United States*. 2nd ed. Washington DC: USDA Forest Service Miscellaneous Publication No. 1391 (rev.). 198 pp. (Separate map)

Barbour, M. C., and W. D. Billings, eds. 1999. *North American Terrestrial Vegetation*. 2nd ed. Cambridge, UK: Cambridge University Press.

Bissell, J., and W. D. Graul. 1981. The latilong system of mapping wildlife distribution. *Wildlife Society Bulletin* 9: 185–189.

Boyle, W. J., and R. H. Bauer. 1994. *Birdfinding in Forty National Forests and Grasslands*. Delaware City, DE: American Birding Association. 186 pp.

Brower, K. 1997. *Our National Forests: American Legacy*. Washington, DC: National Geographic Society.

Crump, D. J., ed. 1984. *A Guide to our Federal Lands*. Washington, DC: National Geographic Society. 220 pp.

Daubenmire, R. F. 1968. *Plant Communities: A Textbook of Plant Synecology*. New York: Harper and Row. 300 pp.

Faber-Langendoen, D., T. Keeler-Wolf, D. Meidinger, C. Josse, A. Weakley, D. Tart, G. Navarro, B. Hoagland, S. Ponomarenko, J.-P. Saucier, G. Fults, and E. Helmer. 2012. *Classification and Description of World Formation Types*. Reston, VA: US Geological Survey, and Arlington, VA: NatureServe. 65 pp.

Graham, A. 1999. *Late Cretaceous and Cenozoic History of North American Vegetation North of Mexico*. New York: Oxford University Press.

Hitchcock, A. S. 1935. *Manual of the Grasses of the United States*. Washington, DC: US Department of Agriculture Publication No. 200. (Repr. 1971, New York: Dover)

Jones, J. O. 1990. *Where the Birds Are: A Guide to All 50 States and Canada*. New York: William Morrow. 400 pp.

Kendeigh, S. C. 1974. *Animal Ecology, with Special Reference to Animals and Man*. Engelwood Cliffs, NJ: Prentice Hall.

Küchler, A. W. 1964. *Potential Natural Vegetation of the Conterminous United States*. New York: American Geographical Society, Special Publication 36.

Leopold, E. B., and M. F. Denton. 1987. Comparative age of grassland and steppe east and west of the northern Rockies. *Annals of the Missouri Botanical Garden* 74: 841–867.

Martin, A. C., H. S. Zim, and A. L. Nelson. 1951. *American Wildlife and Plants*. New York: McGraw Hill. 512 pp.

Mengel, R. M. 1970. The North American Central Plains as an isolating agent in bird speciation. Pp. 280–340, in *Pleistocene and Recent Environments of the Central Great Plains* (W. Dort and J. K. Jones, eds.). Lawrence: University Press of Kansas. 433 pp.

Omernik, J. M. 1987. Ecoregions of the coterminous United States. *Annals of the Association of American Geographers* 77: 118–125.

Riccioti, E. 1995. *The Natural History of North America*. New York: Crescent Books. 224 pp.

Riley, L., and W. Riley. 1979. *Guide to the National Wildlife Refuges*. Garden City, NY: Anchor Press/Doubleday. 653 pp.

Scott, O. 1992. *A Birder's Guide to Wyoming*. ABA Lane Birdfinding Guides Series #478. Delaware City, DE: American Birding Association. 246 pp.

Shantz, H. L., and R. Zon. 1924. Natural vegetation. Pp. 1–28, in *Atlas of American Agriculture*. Washington, DC: US Department of Agriculture.

West, N. E. 1988. Intermountain deserts, shrub steppes, and woodlands. Pp. 209–30, in *North American Terrestrial Vegetation* (M. G. Barbour and W. D. Billings, eds.). Cambridge, UK: Cambridge University Press.

White, M. 1999. *Guide to Birdwatching Sites: Western US*. Washington, DC: National Geographic Society. 224 pp.

Whittaker, R. H. 1975. *Communities and Ecosystems*. New York: MacMillan.

Index to Species Profiles

Reptiles and Amphibians